STATISTICAL PROCESS MONITORING AND OPTIMIZATION

STATISTICS: Textbooks and Monographs

A Series Edited by

D. B. Owen, Founding Editor, 1972–1991

W. R. Schucany, Coordinating Editor
Department of Statistics
Southern Methodist University
Dallas, Texas

W. J. Kennedy, Associate Editor
for Statistical Computing
Iowa State University

A. M. Kshirsagar, Associate Editor
for Multivariate Analysis and for
Experimental Design
University of Michigan

E. G. Schilling, Associate Editor
for Statistical Quality Control
Rochester Institute of Technology

1. The Generalized Jackknife Statistic, *H. L. Gray and W. R. Schucany*
2. Multivariate Analysis, *Anant M. Kshirsagar*
3. Statistics and Society, *Walter T. Federer*
4. Multivariate Analysis: A Selected and Abstracted Bibliography, 1957–1972, *Kocherlakota Subrahmaniam and Kathleen Subrahmaniam*
5. Design of Experiments: A Realistic Approach, *Virgil L. Anderson and Robert A. McLean*
6. Statistical and Mathematical Aspects of Pollution Problems, *John W. Pratt*
7. Introduction to Probability and Statistics (in two parts), Part I: Probability; Part II: Statistics, *Narayan C. Giri*
8. Statistical Theory of the Analysis of Experimental Designs, *J. Ogawa*
9. Statistical Techniques in Simulation (in two parts), *Jack P. C. Kleijnen*
10. Data Quality Control and Editing, *Joseph I. Naus*
11. Cost of Living Index Numbers: Practice, Precision, and Theory, *Kali S. Banerjee*
12. Weighing Designs: For Chemistry, Medicine, Economics, Operations Research, Statistics, *Kali S. Banerjee*
13. The Search for Oil: Some Statistical Methods and Techniques, *edited by D. B. Owen*
14. Sample Size Choice: Charts for Experiments with Linear Models, *Robert E. Odeh and Martin Fox*
15. Statistical Methods for Engineers and Scientists, *Robert M. Bethea, Benjamin S. Duran, and Thomas L. Boullion*
16. Statistical Quality Control Methods, *Irving W. Burr*
17. On the History of Statistics and Probability, *edited by D. B. Owen*
18. Econometrics, *Peter Schmidt*
19. Sufficient Statistics: Selected Contributions, *Vasant S. Huzurbazar (edited by Anant M. Kshirsagar)*
20. Handbook of Statistical Distributions, *Jagdish K. Patel, C. H. Kapadia, and D. B. Owen*
21. Case Studies in Sample Design, *A. C. Rosander*

22. Pocket Book of Statistical Tables, *compiled by R. E. Odeh, D. B. Owen, Z. W. Birnbaum, and L. Fisher*
23. The Information in Contingency Tables, *D. V. Gokhale and Solomon Kullback*
24. Statistical Analysis of Reliability and Life-Testing Models: Theory and Methods, *Lee J. Bain*
25. Elementary Statistical Quality Control, *Irving W. Burr*
26. An Introduction to Probability and Statistics Using BASIC, *Richard A. Groeneveld*
27. Basic Applied Statistics, *B. L. Raktoe and J. J. Hubert*
28. A Primer in Probability, *Kathleen Subrahmaniam*
29. Random Processes: A First Look, *R. Syski*
30. Regression Methods: A Tool for Data Analysis, *Rudolf J. Freund and Paul D. Minton*
31. Randomization Tests, *Eugene S. Edgington*
32. Tables for Normal Tolerance Limits, Sampling Plans and Screening, *Robert E. Odeh and D. B. Owen*
33. Statistical Computing, *William J. Kennedy, Jr., and James E. Gentle*
34. Regression Analysis and Its Application: A Data-Oriented Approach, *Richard F. Gunst and Robert L. Mason*
35. Scientific Strategies to Save Your Life, *I. D. J. Bross*
36. Statistics in the Pharmaceutical Industry, *edited by C. Ralph Buncher and Jia-Yeong Tsay*
37. Sampling from a Finite Population, *J. Hajek*
38. Statistical Modeling Techniques, *S. S. Shapiro and A. J. Gross*
39. Statistical Theory and Inference in Research, *T. A. Bancroft and C.-P. Han*
40. Handbook of the Normal Distribution, *Jagdish K. Patel and Campbell B. Read*
41. Recent Advances in Regression Methods, *Hrishikesh D. Vinod and Aman Ullah*
42. Acceptance Sampling in Quality Control, *Edward G. Schilling*
43. The Randomized Clinical Trial and Therapeutic Decisions, *edited by Niels Tygstrup, John M Lachin, and Erik Juhl*
44. Regression Analysis of Survival Data in Cancer Chemotherapy, *Walter H. Carter, Jr., Galen L. Wampler, and Donald M. Stablein*
45. A Course in Linear Models, *Anant M. Kshirsagar*
46. Clinical Trials: Issues and Approaches, *edited by Stanley H. Shapiro and Thomas H. Louis*
47. Statistical Analysis of DNA Sequence Data, *edited by B. S. Weir*
48. Nonlinear Regression Modeling: A Unified Practical Approach, *David A. Ratkowsky*
49. Attribute Sampling Plans, Tables of Tests and Confidence Limits for Proportions, *Robert E. Odeh and D. B. Owen*
50. Experimental Design, Statistical Models, and Genetic Statistics, *edited by Klaus Hinkelmann*
51. Statistical Methods for Cancer Studies, *edited by Richard G. Cornell*
52. Practical Statistical Sampling for Auditors, *Arthur J. Wilburn*
53. Statistical Methods for Cancer Studies, *edited by Edward J. Wegman and James G. Smith*
54. Self-Organizing Methods in Modeling: GMDH Type Algorithms, *edited by Stanley J. Farlow*
55. Applied Factorial and Fractional Designs, *Robert A. McLean and Virgil L. Anderson*
56. Design of Experiments: Ranking and Selection, *edited by Thomas J. Santner and Ajit C. Tamhane*
57. Statistical Methods for Engineers and Scientists: Second Edition, Revised and Expanded, *Robert M. Bethea, Benjamin S. Duran, and Thomas L. Boullion*
58. Ensemble Modeling: Inference from Small-Scale Properties to Large-Scale Systems, *Alan E. Gelfand and Crayton C. Walker*
59. Computer Modeling for Business and Industry, *Bruce L. Bowerman and Richard T. O'Connell*
60. Bayesian Analysis of Linear Models, *Lyle D. Broemeling*
61. Methodological Issues for Health Care Surveys, *Brenda Cox and Steven Cohen*

62. Applied Regression Analysis and Experimental Design, *Richard J. Brook and Gregory C. Arnold*
63. Statpal: A Statistical Package for Microcomputers—PC-DOS Version for the IBM PC and Compatibles, *Bruce J. Chalmer and David G. Whitmore*
64. Statpal: A Statistical Package for Microcomputers—Apple Version for the II, II+, and IIe, *David G. Whitmore and Bruce J. Chalmer*
65. Nonparametric Statistical Inference: Second Edition, Revised and Expanded, *Jean Dickinson Gibbons*
66. Design and Analysis of Experiments, *Roger G. Petersen*
67. Statistical Methods for Pharmaceutical Research Planning, *Sten W. Bergman and John C. Gittins*
68. Goodness-of-Fit Techniques, *edited by Ralph B. D'Agostino and Michael A. Stephens*
69. Statistical Methods in Discrimination Litigation, *edited by D. H. Kaye and Mikel Aickin*
70. Truncated and Censored Samples from Normal Populations, *Helmut Schneider*
71. Robust Inference, *M. L. Tiku, W. Y. Tan, and N. Balakrishnan*
72. Statistical Image Processing and Graphics, *edited by Edward J. Wegman and Douglas J. DePriest*
73. Assignment Methods in Combinatorial Data Analysis, *Lawrence J. Hubert*
74. Econometrics and Structural Change, *Lyle D. Broemeling and Hiroki Tsurumi*
75. Multivariate Interpretation of Clinical Laboratory Data, *Adelin Albert and Eugene K. Harris*
76. Statistical Tools for Simulation Practitioners, *Jack P. C. Kleijnen*
77. Randomization Tests: Second Edition, *Eugene S. Edgington*
78. A Folio of Distributions: A Collection of Theoretical Quantile-Quantile Plots, *Edward B. Fowlkes*
79. Applied Categorical Data Analysis, *Daniel H. Freeman, Jr.*
80. Seemingly Unrelated Regression Equations Models: Estimation and Inference, *Virendra K. Srivastava and David E. A. Giles*
81. Response Surfaces: Designs and Analyses, *Andre I. Khuri and John A. Cornell*
82. Nonlinear Parameter Estimation: An Integrated System in BASIC, *John C. Nash and Mary Walker-Smith*
83. Cancer Modeling, *edited by James R. Thompson and Barry W. Brown*
84. Mixture Models: Inference and Applications to Clustering, *Geoffrey J. McLachlan and Kaye E. Basford*
85. Randomized Response: Theory and Techniques, *Arijit Chaudhuri and Rahul Mukerjee*
86. Biopharmaceutical Statistics for Drug Development, *edited by Karl E. Peace*
87. Parts per Million Values for Estimating Quality Levels, *Robert E. Odeh and D. B. Owen*
88. Lognormal Distributions: Theory and Applications, *edited by Edwin L. Crow and Kunio Shimizu*
89. Properties of Estimators for the Gamma Distribution, *K. O. Bowman and L. R. Shenton*
90. Spline Smoothing and Nonparametric Regression, *Randall L. Eubank*
91. Linear Least Squares Computations, *R. W. Farebrother*
92. Exploring Statistics, *Damaraju Raghavarao*
93. Applied Time Series Analysis for Business and Economic Forecasting, *Sufi M. Nazem*
94. Bayesian Analysis of Time Series and Dynamic Models, *edited by James C. Spall*
95. The Inverse Gaussian Distribution: Theory, Methodology, and Applications, *Raj S. Chhikara and J. Leroy Folks*
96. Parameter Estimation in Reliability and Life Span Models, *A. Clifford Cohen and Betty Jones Whitten*
97. Pooled Cross-Sectional and Time Series Data Analysis, *Terry E. Dielman*
98. Random Processes: A First Look, Second Edition, Revised and Expanded, *R. Syski*
99. Generalized Poisson Distributions: Properties and Applications, *P. C. Consul*
100. Nonlinear L_p-Norm Estimation, *Rene Gonin and Arthur H. Money*
101. Model Discrimination for Nonlinear Regression Models, *Dale S. Borowiak*
102. Applied Regression Analysis in Econometrics, *Howard E. Doran*
103. Continued Fractions in Statistical Applications, *K. O. Bowman and L. R. Shenton*

104. Statistical Methodology in the Pharmaceutical Sciences, *Donald A. Berry*
105. Experimental Design in Biotechnology, *Perry D. Haaland*
106. Statistical Issues in Drug Research and Development, *edited by Karl E. Peace*
107. Handbook of Nonlinear Regression Models, *David A. Ratkowsky*
108. Robust Regression: Analysis and Applications, *edited by Kenneth D. Lawrence and Jeffrey L. Arthur*
109. Statistical Design and Analysis of Industrial Experiments, *edited by Subir Ghosh*
110. *U*-Statistics: Theory and Practice, *A. J. Lee*
111. A Primer in Probability: Second Edition, Revised and Expanded, *Kathleen Subrahmaniam*
112. Data Quality Control: Theory and Pragmatics, *edited by Gunar E. Liepins and V. R. R. Uppuluri*
113. Engineering Quality by Design: Interpreting the Taguchi Approach, *Thomas B. Barker*
114. Survivorship Analysis for Clinical Studies, *Eugene K. Harris and Adelin Albert*
115. Statistical Analysis of Reliability and Life-Testing Models: Second Edition, *Lee J. Bain and Max Engelhardt*
116. Stochastic Models of Carcinogenesis, *Wai-Yuan Tan*
117. Statistics and Society: Data Collection and Interpretation, Second Edition, Revised and Expanded, *Walter T. Federer*
118. Handbook of Sequential Analysis, *B. K. Ghosh and P. K. Sen*
119. Truncated and Censored Samples: Theory and Applications, *A. Clifford Cohen*
120. Survey Sampling Principles, *E. K. Foreman*
121. Applied Engineering Statistics, *Robert M. Bethea and R. Russell Rhinehart*
122. Sample Size Choice: Charts for Experiments with Linear Models: Second Edition, *Robert E. Odeh and Martin Fox*
123. Handbook of the Logistic Distribution, *edited by N. Balakrishnan*
124. Fundamentals of Biostatistical Inference, *Chap T. Le*
125. Correspondence Analysis Handbook, *J.-P. Benzécri*
126. Quadratic Forms in Random Variables: Theory and Applications, *A. M. Mathai and Serge B. Provost*
127. Confidence Intervals on Variance Components, *Richard K. Burdick and Franklin A. Graybill*
128. Biopharmaceutical Sequential Statistical Applications, *edited by Karl E. Peace*
129. Item Response Theory: Parameter Estimation Techniques, *Frank B. Baker*
130. Survey Sampling: Theory and Methods, *Arijit Chaudhuri and Horst Stenger*
131. Nonparametric Statistical Inference: Third Edition, Revised and Expanded, *Jean Dickinson Gibbons and Subhabrata Chakraborti*
132. Bivariate Discrete Distribution, *Subrahmaniam Kocherlakota and Kathleen Kocherlakota*
133. Design and Analysis of Bioavailability and Bioequivalence Studies, *Shein-Chung Chow and Jen-pei Liu*
134. Multiple Comparisons, Selection, and Applications in Biometry, *edited by Fred M. Hoppe*
135. Cross-Over Experiments: Design, Analysis, and Application, *David A. Ratkowsky, Marc A. Evans, and J. Richard Alldredge*
136. Introduction to Probability and Statistics: Second Edition, Revised and Expanded, *Narayan C. Giri*
137. Applied Analysis of Variance in Behavioral Science, *edited by Lynne K. Edwards*
138. Drug Safety Assessment in Clinical Trials, *edited by Gene S. Gilbert*
139. Design of Experiments: A No-Name Approach, *Thomas J. Lorenzen and Virgil L. Anderson*
140. Statistics in the Pharmaceutical Industry: Second Edition, Revised and Expanded, *edited by C. Ralph Buncher and Jia-Yeong Tsay*
141. Advanced Linear Models: Theory and Applications, *Song-Gui Wang and Shein-Chung Chow*

142. Multistage Selection and Ranking Procedures: Second-Order Asymptotics, *Nitis Mukhopadhyay and Tumulesh K. S. Solanky*
143. Statistical Design and Analysis in Pharmaceutical Science: Validation, Process Controls, and Stability, *Shein-Chung Chow and Jen-pei Liu*
144. Statistical Methods for Engineers and Scientists: Third Edition, Revised and Expanded, *Robert M. Bethea, Benjamin S. Duran, and Thomas L. Boullion*
145. Growth Curves, *Anant M. Kshirsagar and William Boyce Smith*
146. Statistical Bases of Reference Values in Laboratory Medicine, *Eugene K. Harris and James C. Boyd*
147. Randomization Tests: Third Edition, Revised and Expanded, *Eugene S. Edgington*
148. Practical Sampling Techniques: Second Edition, Revised and Expanded, *Ranjan K. Som*
149. Multivariate Statistical Analysis, *Narayan C. Giri*
150. Handbook of the Normal Distribution: Second Edition, Revised and Expanded, *Jagdish K. Patel and Campbell B. Read*
151. Bayesian Biostatistics, *edited by Donald A. Berry and Dalene K. Stangl*
152. Response Surfaces: Designs and Analyses, Second Edition, Revised and Expanded, *André I. Khuri and John A. Cornell*
153. Statistics of Quality, *edited by Subir Ghosh, William R. Schucany, and William B. Smith*
154. Linear and Nonlinear Models for the Analysis of Repeated Measurements, *Edward F. Vonesh and Vernon M. Chinchilli*
155. Handbook of Applied Economic Statistics, *Aman Ullah and David E. A. Giles*
156. Improving Efficiency by Shrinkage: The James-Stein and Ridge Regression Estimators, *Marvin H. J. Gruber*
157. Nonparametric Regression and Spline Smoothing: Second Edition, *Randall L. Eubank*
158. Asymptotics, Nonparametrics, and Time Series, *edited by Subir Ghosh*
159. Multivariate Analysis, Design of Experiments, and Survey Sampling, *edited by Subir Ghosh*
160. Statistical Process Monitoring and Control, *edited by Sung H. Park and G. Geoffrey Vining*
161. Statistics for the 21st Century: Methodologies for Applications of the Future, *C. R. Rao and Gabor J. Szekely*

Additional Volumes in Preparation

STATISTICAL PROCESS MONITORING AND OPTIMIZATION

edited by

Sung H. Park
Seoul National University
Seoul, Korea

G. Geoffrey Vining
Virginia Polytechnic Institute
and State University
Blacksburg, Virginia

MARCEL DEKKER, INC. NEW YORK • BASEL

Library of Congress Cataloging-in-Publication Data

Statistical process monitoring and optimization / edited by Sung H. Park and G. Geoffrey Vining
p. cm. — (Statistics, textbooks and monographs; v. 160)
Includes bibliographical references.
ISBN: 0-8247-6007-7 (alk. paper)
1. Process control—Statistical methods. 2. Quality control—Statistical methods. 3. Optimal designs (Statistics) I. Park, Sung H. II. Vining, G. Geoffrey. III. Series.
TS156.8.S7537 1999
658.5′62—dc21 99-38518
CIP

This book is printed on acid-free paper.

Headquarters
Marcel Dekker, Inc.
270 Madison Avenue, New York, NY 10016
tel: 212-696-9000; fax: 212-685-4540

Eastern Hemisphere Distribution
Marcel Dekker AG
Hutgasse 4, Postfach 812, CH-4001 Basel, Switzerland
tel: 41-61-261-8482; fax: 41-61-261-8896

World Wide Web
http://www.dekker.com

The publisher offers discounts on this book when ordered in bulk quantities. For more information, write to Special Sales/Professional Marketing at the headquarters address above.

Current printing (last digit):
10 9 8 7 6 5 4 3 2 1

PRINTED IN THE UNITED STATES OF AMERICA

Preface

This book has been written primarily for engineers and researchers who want to use some advanced statistical methods for process monitoring and optimization in order to improve quality and productivity in industry, and also for statisticians who want to learn more about recent topics in this general area. The book covers recent advanced topics in statistical reasoning in quality management, control charts, multivariate process monitoring, process capability index, design of experiments (DOE) and analysis for process control, and empirical model building for process optimization. It will also be of interest to managers, quality improvement specialists, graduate students, and other professionals with an interest in statistical process control (SPC) and its related areas.

In August 1995, the International Conference on Statistical Methods and Statistical Computing for Quality and Productivity Improvement (ICSQP'95) was held in Seoul, Korea, and many of the authors of this book participated. A year later after the conference, the editors agreed to edit this book and invited some key conference participants and some other major contributors in the field who did not attend the conference. Authors from 15 nations have joined in this project, making this truly a multinational book. The authors are all well-known scholars in SPC and DOE areas. The book provides useful information for those who are eager to learn about recent developments in statistical process monitoring and optimization. It also provides an opportunity for joint discussion all over the world in the general areas of SPC and DOE.

We would like to thank Elizabeth Curione, production editor of Marcel Dekker, Inc., for her kind help and guidance, and Maria Allegra, acquisitions editor and manager at Marcel Dekker, Inc., for making the publication of this book possible. We very much appreciate the valuable contributions and cooperation of the authors which made the book a reality. We sincerely hope that it is a useful source of valuable information for statistical process monitoring and optimization.

We want to dedicate this book to God for giving us the necessary energy, health, and inspiration to write our chapters and to edit this book successfully.

Sung H. Park
G. Geoffrey Vining

Contents

Contributors

Bo Bergman, Ph.D. Linköping University, Linköping, and Department of TQM, School of Technology Management, Chalmers University of Technology, Gothenburg, Sweden

Søren Bisgaard, Ph.D. Institute for Technology Management, University of St. Gallen, St. Gallen, Switzerland

Min-Te Chao, Ph.D. Institute of Statistical Science, Academia Sinica, Taipei, Taiwan, Republic of China

Je H. Choi, Ph.D. Statistical Analysis Group, Samsung Display Devices Co., Ltd., Suwon, Korea

Jens J. Dahlgaard, Dr. Merc. Department of Information Science, The Aarhus School of Business, Aarhus, Denmark

Su Mi Park Dahlgaard, M.Sc., Lic. Orcon. Department of Information Science, The Aarhus School of Business, Aarhus, Denmark

Enrique Del Castillo, Ph.D. Department of Industrial and Manufacturing Engineering, The Pennsylvania State University, University Park, Pennsylvania

Pasquale Erto, M.D. Department of Aeronautical Design, University of Naples Federico II, Naples, Italy

Norbert Gaffke, Ph.D. Department of Mathematics, Universität Magdeburg, Magdeburg, Germany

Fah Fatt Gan, Ph.D. Department of Statistics and Applied Probability, National University of Singapore, Singapore, Republic of Singapore

R. Gnanadesikan, Ph.D. Department of Statistics, Rutgers University, New Brunswick, New Jersey

Rainer Göb, Dr. Institute of Applied Mathematics and Statistics, University of Wuerzburg, Wuerzburg, Germany

A. J. Hayter, Ph.D. Department of Industrial and Systems Engineering, Georgia Institute of Technology, Atlanta, Georgia

Berthold Heiligers, Ph.D. Department of Mathematics, Universität Magdeburg, Magdeburg, Germany

Chihiro Hirotsu, Ph.D. Department of Mathematical Engineering and Information Physics, University of Tokyo, Tokyo, Japan

Anders Hynén, Ph.D. Department of Systems Engineering, ABB Corporate Research, Västerås, Sweden

G. K. Kanji, B.Sc., M.Sc., Ph.D. Department of Statistics, Sheffield Business School, Sheffield Hallam University, Sheffield, England

J. R. Kettenring, Ph.D. Department of Mathematical Sciences Research Center, Telcordia Technologies, Morristown, New Jersey

André I. Khuri, Ph.D. Department of Statistics, University of Florida, Gainesville, Florida

Kai Kristensen, Dr. Merc. Department of Information Science, The Aarhus School of Business, Aarhus, Denmark

Youngjo Lee, Ph.D. Department of Statistics, Seoul National University, Seoul, Korea

Dennis K. J. Lin, Ph.D. Department of Management and Information Systems, The Pennsylvania State University, University Park, Pennsylvania

Robert L. Mason, Ph.D. Statistical Analysis Section, Southwest Research Institute, San Antonio, Texas

Christina M. Mastrangelo, Ph.D. Department of Systems Engineering, University of Virginia, Charlottesville, Virginia

Carl Modigh Arkwright Enterprises Ltd., Paris, France

Douglas C. Montgomery, Ph.D. Department of Industrial Engineering, Arizona State University, Tempe, Arizona

Raymond H. Myers, Ph.D. Department of Statistics, Virginia Polytechnic Institute and State University, Blacksburg, Virginia

Angela R. Neff, Ph.D. Department of Corporate Research and Development, General Electric, Schenectady, New York

John A. Nelder, D.Sc., F.R.S. Department of Mathematics, Imperial College, London, England

Anders Nørgaard, M.Sc. Department of Information Science, The Aarhus School of Business, Aarhus, Denmark, and Bülon Management, Viby, Denmark

Sung H. Park, Ph.D. Department of Statistics, Seoul National University, Seoul, Korea

M. F. Ramalhoto, Ph.D. Department of Mathematics, Technical University of Lisbon, "Instituto Superior Téchnico," Lisbon, Portugal

Marion R. Reynolds, Jr., Ph.D. Departments of Statistics and Forestry, Virginia Polytechnic Institute and State University, Blacksburg, Virginia

Diane A. Schaub, Ph.D. Department of Industrial and Systems Engineering, University of Florida, Gainesville, Florida

Fred A. Spiring, Ph.D. Department of Statistics, The University of Manitoba, and Department of Quality, Pollard Banknote Limited, Winnipeg, Manitoba, Canada

David M. Steinberg, Ph.D. Department of Statistics and Operations Research, Tel Aviv University, Tel Aviv, Israel

Zachary G. Stoumbos, Ph.D. Department of Management Science and Information Systems, and Rutgers Center for Operations Research (RUTCOR), Rutgers University, Newark, New Jersey

Genichi Taguchi, D.Sc. Ohken Associate, Tokyo, Japan

Elsie S. Valeroso, Ph.D. Department of Mathematics and Statistics, Montana State University, Bozeman, Montana

Alan Veevers, B.Sc., Ph.D. Department of Mathematical and Information Sciences, Commonwealth Scientific and Industrial Research Organization, Clayton, Victoria, Australia

G. Geoffrey Vining, Ph.D. Department of Statistics, Virginia Polytechnic Institute and State University, Blacksburg, Virginia

Elart von Collani, Dr. rer. nat., Dr. rer. nat. habil., School of Economics, University of Wuerzburg, Wuerzburg, Germany

John C. Young, Ph.D. Department of Mathematics and Computer Science, McNeese State University, Lake Charles, Louisiana

Gongxu Zhang Research Institute of Management Science, Beijing University of Science and Technology, Beijing, People's Republic of China

1
On-Line Quality Control System Designs

Genichi Taguchi
Ohken Associate, Tokyo, Japan

1. INTRODUCTION

It is the responsibility of a production department to produce a product that meets a designed quality level at the lowest cost. However, it is important to not merely have the product quality meet specifications but to also endeavor to bring quality as close as possible to the ideal value.

2. JAPANESE PRODUCTS AND AMERICAN PRODUCTS

Many Japanese read an article on April 17, 1979, on the front page of *Asahi Shinbun*, one of the most widely circulated newspapers in Japan, regarding a comparison of the quality of color television sets produced by the Sony factory in Japan with that of TVs produced by the Sony factory in San Diego, California. The comparison was made on the basis of the color distribution, which is related to the color balance. Although both factories used the same design, the TVs from the San Diego factory had a bad reputation, and Americans preferred the products from Japan. Based on this fact, Mr. Yamada, the vice president of Sony United States at that time, described the difference in the article.

The difference in the quality characteristic distributions is shown in Figure 1. It is seen from the figure that the color quality of Japanese-made

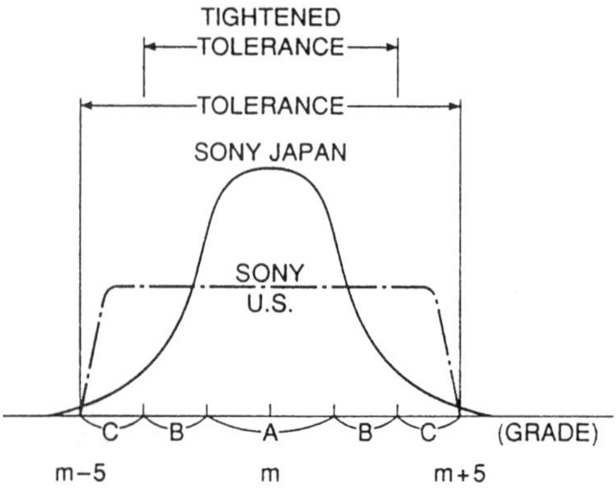

Figure 1 Distribution of color quality in television sets of Sony U.S. and Sony Japan.

TVs shown by the solid curve have approximately a normal distribution with the target value at the center; its standard deviation is about one-sixth of the tolerance or 10 in certain units.

In quality control, the index of tolerance divided by six standard deviations is called the process capability index, denoted by C_p:

$$C_p = \frac{\text{tolerance}}{6 \times \text{standard deviation}} \tag{1}$$

The process capability of the Japanese-made TVs is therefore 1, and the average quality level coincides with the target value.

The quality distribution of the sets produced in San Diego, shown by the dash-dot curve, on the other hand, has less out-of-specification product than that of the Japanese-made sets and is quite similar to the uniform distribution for those products that are within the tolerance. Since the standard deviation of the uniform distribution is given by $1/\sqrt{12}$ of the tolerance, its process capability index is given by

$$C_p = \frac{\text{tolerance}}{(\text{tolerance}/\sqrt{12}) \times 6} = 0.577 \tag{2}$$

which shows that its process capability index is worse than that of the Japanese-made product.

A product with out-of-tolerance quality is a bad product. It is an unpassed product, so it should not be shipped out. From the opposite point of view, a product within tolerance should be considered good and should be shipped. In a school examination, a score above 60 with 100 as the full mark is considered to be a passing grade. A product quality that coincides with its target value should have a full mark. Quality gradually becomes worse when it deviates from the target value, and fails when it exceeds the specification limits, or ±5 in this example.

In a school examination, a score of 59 or below 59 is failing, 60 or above 60 is passing. The scores are normally classified into the following grades:

60– 69	D
70– 79	C
80– 89	B
90–100	A

I put grades A, B, and C in Figure 1. It is seen that the Japanese-made TVs have more A's and fewer B's and C's.

To reduce the Japan–United States difference, Mr. Yamada dictated a narrower tolerance for the San Diego factory, specifying B as the lowest allowable quality limit. This is wrong, since specifying a more severe tolerance because of inferior process capability is similar to raising the passing score from 60 to 70 because of the incapability of students. In schools, teachers do not raise the limit for such students. Instead, teachers used to lower the passing limit.

As stated above, loss is caused when the quality characteristic (denoted by y) deviates from the target value (denoted by m) regardless of how small the deviation is. Let the loss be denoted by $L(y)$. $L(y)$ is the minimum when y coincides with the target value m, and we may put the loss to be 0.

$$L(m) = 0 \tag{3}$$

When $y = m$, $L(y)$ is zero or minimum, and its differential coefficient is, accordingly, zero.

$$L'(m) = 0 \tag{4}$$

Using the Taylor expansion, the loss function $L(y)$ is expanded as

$$\begin{aligned} L(y) &= L(m) + \frac{L'(m)}{1!}(y-m) + \frac{L''(m)}{2!}(y-m)^2 + \cdots \\ &= \frac{L''(m)}{2!}(y-m)^2 \end{aligned} \tag{5}$$

The constant and linear terms (differential terms) become zero from Eqs. (3) and (4). If the third-order term and the following terms can be omitted, the loss function is then

$$L = k(y - m)^2 \tag{6}$$

Let the allowance or the deviation of y from the middle value by Δ. The more y deviates from m, the middle value, the more loss is caused. A product whose deviation is less than its allowance Δ should pass inspection; otherwise the company will lose more. When the deviation exceeds the allowance, the product should not be passed. therefore, when the deviation is equal to the allowance, its loss is equal to the loss due to the disposal of the failed product.

Let A (yen) signify the loss caused by disposing of a failed product. Putting A and allowance Δ in Eq. (6), k is obtained as

$$k = \frac{\text{loss of disposing of a failed product}}{(\text{allowance})^2} = \frac{A}{\Delta^2} \tag{7}$$

Assume that the cost of repairing a failed color TV set is 600 yen. k is then calculated as

$$k = \frac{600}{5^2} = 24.0\,(\text{yen}) \tag{8}$$

The loss function is therefore

$$L = 24.0(y - m)^2 \tag{9}$$

This equation applies to the case when a single product is manufactured. An electrical manufacturing company in India (BHEL) said to me "Our company manufactures only one product, a certain type of nuclear power station. There is no second machinery of the same type producted. Since the variation of a single product is zero, standard deviation in statistics is not applicable in our case."

Variation is measured by the deviation from a target value or an ideal value. Therefore, it can be obtained from Eq. (6) even when only one product is produced. When there is more than one product, the average of Eq. (6) is calculated. Variance (σ^2), the average of the square of differences between y and the target value, is used for this purpose. σ^2 is correctly called the average error squared, but we will call it variance for simplicity.

$$\sigma^2 = \text{average of } (y - m)^2 \tag{10}$$

The loss function is given by

$$L = k\sigma^2 \tag{11}$$

The quality level difference between Sony U.S. and Sony Japan is calculated from Eq. (11) as shown in Table 1.

Table 1 shows that although the fraction defective of the Japanese Sony factory is larger, its loss is one-third that of the U.S. Sony factory. In other words, the Japanese quality level is three times higher. If Vice President Yamada specified a narrower tolerance such as $10 \times 2/3$, the quality level would be improved (assuming a uniform distribution within the tolerance limits):

$$L = 24.0 \times \left[\frac{1}{\sqrt{12}} \times 10 \times \frac{2}{3}\right]^2 = 88.9\,(\text{yen}) \tag{12}$$

This shows that there is a 111 (= 200.0 – 88.9) yen improvement, but that the Sony U.S. quality level is still 22.1 (= 88.9 – 66.7) yen worse than that of Sony Japan.

If such an improvement were attained by repairing or adjusting failed products whose quality level exceeds $m \pm 10/3$ but lies within $m \pm 5$, holding 33.3% of the total production as seen from Figure 1, at a cost of 600 yen per unit, then the cost of repair per unit would be

$$600 \times 0.333 = 200\,(\text{yen}) \tag{13}$$

An 111.1 yen quality improvement at a cost of 200 yen is not profitable. The correct solution to this problem is to apply both on-line and off-line quality control techniques.

I visited the Louisville factory of the General Electric Company in September 1989. On the production line, workers were instructed to use

Table 1 Quality Comparison Between Sony Japan and Sony U.S.

Country	Average	Standard deviation	Variance	Loss L (yen)	Fraction defective (%)
Japan	m	10/6	$(10/6)^2$	66.7	0.27
United States	m	$10/\sqrt{12}$	100/12	200.0	0.00

go–no go gauges, which determine only pass or fail; there was a lack of consciousness of the importance of the quality distribution within tolerance. It was proposed that Shewhart's control charts be used to control quality by the distribution of quality characteristics on production lines as a substitute for a method using specification and inspection. Inspectors tend to consider production quality as perfect if the fraction defective is zero. In Japan, none of the companies that product JIS (Japanese Industrial Standards) products are satisfied producing products whose quality level marginally passes the JIS specifications. Instead, the companies always attempt to reduce the quality distribution within the tolerance range. Nippon Denso Company, for example, demands that its production lines and vendors improve their process capability indexes above 1.33.

To determine the process capability index, data $y_1, y_2, \ldots, y_n$ are collected once or a few times a day for 3 months. The standard deviation is obtained from the following equation, where m is the target value.

$$\sigma = \left\{\frac{1}{n}\left[(y_1 - m)^2 + (y_2 - m)^2 + \cdots + (y_n - m)^2\right]\right\}^{1/2}$$

The process capability index C_p is calculated as

$$C_p = \frac{\text{tolerance}}{6\sigma} \tag{15}$$

The loss function $L(y)$ is then determined as

$$L = k\sigma^2 \tag{16}$$

3. WHAT IS ON-LINE QUALITY CONTROL?

Manufacturers contribute to society and grow through a series of activities including product planning, product development, design, production, and sales. Within these steps, routine quality control activity on production lines is called on-line quality control. It includes the following three activities:

1. Diagnosis and adjustment of processes. This is called process control. A manufacturing process is diagnosed at constant intervals. When the process is judged to be normal, production is continued; otherwise the cause of abnormality is investigated, the abnormal condition is corrected, and production is restarted.

Preventive activities such as adjusting a manufacturing process when it appears to become abnormal are also included in this case.

2. Prediction and modification. In order to control a variable quality characteristic in a production line, measurements are made at constant intervals. From the measurement results, the average quality of the products to be produced is predicted. If the predicted value deviates from the target value, corrective action is taken by moving the level of a variable (called a signal factor in on-line quality control) to bring the deviation back to the target value. This is called feedback control.
3. Measurement and disposition. This is also called inspection. Every product from a production line is measured, and its disposition, such as scrapping or repair, is decided on when the result shows the product to be out of specification.

Case 3 is different from cases 1 and 2 in that a manufacturing process is the major object of treatment for cases 1 and 2, while products are the sole object of disposition in case 3.

The above cases are explained by an example of controlling the sensors or measuring systems used in robots or in automatic control. Measurement and disposition, case 3, concern products, classifying them into pass and fail categories and disposing of them. In a measuring system, it is important to inspect the measuring equipment and to determine whether the system should be passed or failed. This is different from the calibration of equipment. Calibration is meant to correct the deviation of parameters of a piece of measuring equipment after a certain period of use that corresponds to the concept implied in case 2, prediction and modification.

When measuring equipment falls out of calibration, either gradually or suddenly, it is replaced or repaired, which corresponds to the concept in case 1, diagnosis and adjustment. It is difficult in many cases to decide if the equipment should be repaired or scrapped. Generally, the decision to repair or replace is made when the error of the measuring equipment exceeds the allowance of the product quality characteristic.

When measuring equipment cannot be adjusted by calibration and has to be repaired or scrapped (called adjustment in on-line quality control), and when there is a judging procedure (called diagnosis in on-line quality control) for these actions, it is more important to design a diagnosis and adjustment system than to design a calibrating system.

Radical countermeasures such as determining the cause of the variation, followed by taking action to prevent a relapse (which are described in control chart methods and called off line quality control countermeasures) are not discussed in this chapter. I am confident that a thorough on-line quality control system design is the way to keep production lines from falling out of control. It is the objective of this chapter to briefly describe on-line quality control methods and give their theoretical background.

4. EQUATION AND AN EXAMPLE FOR DIAGNOSIS AND ADJUSTMENT

In I Motor Company in the 1970s, there are 28 steps in the truck engine cylinder block production line. Quality control activity is necessary to ensure normal production at each step. One of the steps, called boring by reamers, is explained as an example, which is also described in detail in Ref. 1.

Approximately 10 holes are bored at a time in each cylinder block by reamers. A cylinder block is scrapped as defective if there is any hole bore that is misaligned by more than 10 μm, causing an 8000 yen loss, which is denoted by A. The diagnosis cost to know whether holes are being bored straight, designated by B, is 400 yen, and the diagnosing interval, denoted by n, is 30 units. In the past half-year, 18,000 units were produced, and there were seven quality control problems.

The average problem occurrence interval, denoted by $\bar{u}$, is then

$$\bar{u} = \frac{18{,}000}{7} \fallingdotseq 2570 \text{ (units)} \tag{17}$$

When there is a problem such as a crooked hole, production is stopped, reamers are replaced, the first hole bored after the replacement is checked, and if it is normal, the production is continued. The total cost, including the cost of stopping the production line, tool replacement, and labor, is called the adjustment cost; it is denoted by C and is equal to 20,000 yen in this example.

In such a process adjustment for on-line quality control, the parameters characterizing the three system element—the process, diagnosing method, and adjusting method—include $A, B, C, \bar{u}$, and ℓ (the time lag caused by diagnosis). The quality control cost when the diagnosis interval is n is given by the theory described in Sections 5 and 6 as follows:

$$L = \frac{B}{n} + \frac{n+1}{2}\left(\frac{A}{\bar{u}}\right) + \frac{C}{\bar{u}} + \frac{\ell A}{\bar{u}} \tag{18}$$

Putting $n = 30$, $A = 8000$ yen, $B = 400$ yen, $C = 20{,}000$ yen, $\bar{u} = 2570$, and $\ell = 1$ unit in the above equation, the quality control cost per unit product of this example would be

$$L = \frac{400}{30} + \frac{30+1}{2}\left(\frac{8000}{2570}\right) + \frac{20{,}000}{2570} + \frac{1 \times 8000}{2570}$$
$$= 13.3 + 48.2 + 7.8 + 3.1$$
$$= 74.2 \text{ (yen)} \tag{19}$$

With annual production of 36,000 units, the total cost would be $72.4\times 36{,}000 = 2{,}610{,}000$ yen. The improvement in quality control is needed to reduce the quality control cost given by Eq. (19). For this purpose, there are two methods: one from the pertinent techniques and one from managerial techniques. The former countermeasures include simplification of the diagnosis method or reduction of adjusting the cost, which must be specifically researched case by case. For this, see Chapters 4–8 of Ref. 1.

There are methods to reduce quality control cost while keeping current process, current diagnosis, and adjustment methods unchanged. These managerial techniques are soft techniques applicable to all kinds of production processes. Two of these techniques are introduced in this chapter. One is the determination of the diagnosis interval, and the other is the introduction of preventive maintenance such a periodic replacement.

The optimum diagnosis interval is given by

$$n = \left[\frac{2(\bar{u}+\ell)B}{A - C/\bar{u}}\right]^{1/2} \tag{20}$$

In the example of the boring process,

$$n = \left[\frac{2(2570+1)\times 400}{8000 - 20{,}000/2570}\right]^{1/2} \fallingdotseq 16 \text{ (units)} \tag{21}$$

The quality control cost from Eq. (19) when the diagnosis interval is 16 is

$$L = \frac{400}{16} + \frac{16+1}{2}\left(\frac{8000}{2570}\right) + \frac{20{,}000}{2570} + \frac{1 \times 8000}{2570}$$
$$= 25.0 + 26.5 + 7.8 + 3.1 = 62.4 \text{ (yen)} \tag{22}$$

There is a savings of $72.4 - 62.4 = 10.0$ yen per unit product, or 360,000 yen per year. The value of L does not change significantly even when n varies by 20%. When $n = 20$, for example,

$$L = \frac{400}{20} + \frac{21}{2}\left(\frac{8000}{2570}\right) + \frac{20{,}000}{2570} + \frac{8000}{2570}$$
$$= 63.6 \text{ (yen)} \tag{23}$$

The difference from Eq. (22) is only 1.2 yen. It is permissible to allow about 20% error for the values of system parameters $A, B, C, \bar{u}$, and ℓ, or it is permissible to adjust n within the range of 20% after the optimum diagnosis interval is determined.

Next, the introduction of a preventive maintenance system is explained. In preventive maintenance activities, there are periodic checks and periodic replacement. In periodic replacement, a component part (which could be the cause of the trouble) is replaced with a new one at a certain interval. For example, a tool with an average life of 3000 units of product is replaced after producing 2000 units without checking.

Periodic checking is done to inspect products at a certain interval and replace tools if product quality is within specification at the time inspected but there is the possibility that it might become out-of-specification before the next inspection. In this chapter, periodic replacement is described.

In the case of reamer boring, a majority of the problems are caused by tools. The average problem-causing interval is $\bar{u} = 2570$ units, and periodic replacement is made at an interval of $\bar{u}' = 1500$, which is much shorter than the average life. Therefore, the probability of the process causing trouble becomes very small. Assume that the replacement cost, denoted by C', is approximately the same as the adjustment cost C, or 18,000 yen. Assume that the probability of the process causing trouble is 0.02. This probability includes the instance of a reamer being bent by the pinholes existing in the cylinder block, or some other cause. Then the true average problem-causing interval will be improved from the current 2570 units to

$$\bar{u} = \frac{1500}{0.02} = 75{,}000 \tag{24}$$

The optimum diagnosis interval n would be

$$n = \left[\frac{2 \times (75{,}000 + 1) \times 400}{8000 - 20{,}000/75{,}000}\right]^{1/2} \fallingdotseq 87 \fallingdotseq 100 \text{ (units)} \tag{25}$$

The quality control cost is then

$$
\begin{aligned}
L &= (\text{preventive maintenance cost}) + (\text{diagnosis and adjustment cost}) \\
&= \frac{c'}{u'} + \left[\frac{B}{n} + \frac{n+1}{2}\left(\frac{A}{\bar{u}}\right) + \frac{C}{\bar{u}} + \frac{\ell A}{\bar{u}}\right] \\
&= \frac{18{,}000}{1500} + \left[\frac{400}{100} + \frac{101}{2}\left(\frac{8000}{75{,}000}\right) + \frac{20{,}000}{75{,}000} + \frac{1 \times 8000}{75{,}000}\right] \\
&= 12.0 + (4.0 + 5.4 + 0.3 + 0.1) \\
&= 12.0 + 9.8 = 21.8 \text{ (yen)}
\end{aligned} \tag{26}
$$

This is an improvement of 63.6 – 21.8 = 41.8 yen per unit compared to the case without preventive maintenance, which is equivalent to 1,500,000 yen per annum. If there were similar improvements in each of the 27 cylinder block production steps, it would be an improvement of 42 million yen per annum.

Such a quality control improvement is equivalent to the savings that might be obtained from extending the average interval been problems 6.3 times without increasing any cost. In other words, this preventive maintenance method has a merit parallel to that of an engineering technology that is so fantastic that it could extend the problem-causing interval by 6.3 times without increasing any cost. For details, see Chapters 4–6 of Ref. 1.

Equations (18) and (20) may be approximately applied with satisfaction regardless of the distribution of the production quantity before the problem and despite variations in the fraction defective during the problem period. These statements are proved in Sections 5 and 6.

5. PROOF OF EQUATIONS FOR NONSPECIFIC DISTRIBUTION

Parameters A, B, C, $\bar{u}$, ℓ, and n in the previous section are used similarly in this section. Let P_i $(i = 1, 2, \ldots)$ be the probability of causing trouble for the first time after the production was started at the ith unit. The probability of causing trouble for the first time at the kth diagnosis is

$$
P_{n(k-1)+1} + P_{n(k-1)+2} + \cdots + P_{n(k-1)+n} \tag{27}
$$

When a problem is caused at the kth diagnosis, the number of defectives varies from the maximum of n units to 1; its average number of defective units is given by

$$\frac{n \times P_{n(k-1)+1} + (n-1) \times P_{n(k-1)+2} + \cdots + 1 \times P_{n(k-1)+n}}{P_{n(k-1)+1} + P_{n(k-1)+2} + \cdots + P_{n(k-1)+n}} \tag{28}$$

Assuming that

$$P_{n(k-1)+1} \fallingdotseq P_{n(k-1)+2} \fallingdotseq \cdots \fallingdotseq P_{n(k-1)+n} \tag{29}$$

the average number of defective units will be $(n+1)/2$.

In the loss function L, the average number of defectives in the second term is

$$\sum_{k=1}^{\infty} \frac{n+1}{2} \left(P_{n(k-1)+1} + P_{n(k-1)+2} + \cdots + P_{n(k-1)+n}\right) = \frac{n+1}{2} \tag{30}$$

Since the first, third, and fourth terms of Eq. (18) are self-explanatory, the loss function is given by

$$L = \frac{B}{n} + \frac{n+1}{2}\left(\frac{A}{\bar{u}}\right) + \frac{C}{\bar{u}} + \frac{\ell A}{\bar{u}} \tag{31}$$

Next, the equation for the optimum diagnosis interval is derived. The average problem-causing interval is $\bar{u}$. Since the diagnosis is made at n-unit intervals, it is more correctly calculated to consider the losses from actual recovery actions or by the time lag caused once every $\bar{u} + n/2$ units. Therefore, $\bar{u} + n/2$ is substituted for $\bar{u}$ in Eq. (31).

$$L = \frac{B}{n} + \frac{n+1}{2}\left(\frac{A}{\bar{u} + n/2}\right) + \frac{C}{\bar{u} + n/2} + \frac{\ell A}{\bar{u} + n/2} \tag{32}$$

It is easily understood from the previous example that $\bar{u}$ is much larger than $n/2$. Also, since there is n in the second term of the equation, $\bar{u} + n/2$ in the denominator may be approximated to be $\bar{u} + n/2 \fallingdotseq \bar{u}$. If the approximation

$$\frac{1}{\bar{u} + n/2} \fallingdotseq \frac{1}{\bar{u}}\left(1 - \frac{n}{2\bar{u}}\right)$$

is made, the equation of L is then approximated as

$$L \fallingdotseq \frac{B}{n} + \frac{n+1}{2}\left(\frac{A}{\bar{u}}\right) + \frac{C}{\bar{u}}\left(1 - \frac{n}{\bar{u}}\right) + \frac{\ell A}{u}\left(1 - \frac{n}{\bar{u}}\right) \tag{33}$$

the above equation and then putting it to zero gives

$$-\frac{B}{n^2}+\frac{1}{2}\left(\frac{A}{\bar{u}}\right)-\frac{C}{\bar{u}}\left(\frac{1}{2\bar{u}}\right)-\frac{\ell A}{\bar{u}}\left(\frac{1}{2\bar{u}}\right)=0$$

Solving this equation, n is obtained as

$$n=\left[\frac{2\bar{u}B}{A-C/\bar{u}-\ell A/\bar{u}}\right]^{1/2} \tag{34}$$

Since

$$A \gg \frac{C}{\bar{u}} \qquad \text{and} \qquad A \gg \frac{\ell A}{\bar{u}}$$

the following approximation is made:

$$\begin{aligned}\frac{1}{A-\dfrac{C}{\bar{u}}-\dfrac{\ell A}{\bar{u}}} &= \frac{1}{\left(A-\dfrac{C}{\bar{u}}\right)\left(\dfrac{1-\ell A/\bar{u}}{A-C/\bar{u}}\right)}\\ &\fallingdotseq \frac{1}{A-C/\bar{u}}\left(\frac{1+\ell A/\bar{u}}{A-C/\bar{u}}\right)\\ &= \frac{1}{A-C/\bar{u}}\left(1+\frac{\ell}{\bar{u}}\right)\end{aligned} \tag{35}$$

Putting (35) to (34), n is

$$n=\left(\frac{2(\bar{u}+\ell)B}{A-C/\bar{u}}\right)^{1/2} \tag{36}$$

6. PROOF OF EQUATIONS FOR A LARGE FRACTION OF DEFECTIVES DURING TROUBLE PERIOD

In this section, it is to be proved that Eqs. (18) and (20) can be used approximately even if the fraction of defectives during the trouble period is larger than 0.

When a process is under normal conditions, it may be deemed that there are no defectives. Assume that the fraction defective under abnormal conditions is p and the loss when a defective units is not disposed of but is sent on to the following steps is D yen. After the process causes trouble, the probability of detecting the trouble at the diagnosis is p and the probability of failing to detect the trouble is $1 - p$. Accordingly, the average number of problems at the time the trouble is detected is $(n + 1)p/2$. The probability of detecting a problem at the second diagnosis after missing the detection at the first diagnosis is $(1 - p)p$; then the average number of defectives the inspection fails to detect is $(n + 1)/2$, and the number detected is np units. Thus we obtain Table 2.

From Table 2, the average loss by defectives when a process is in trouble is

$$\left\{\frac{(n+1)p}{2} \times p + np[(1-p)p + (1-p)^2 p + \cdots]\right\} A$$
$$+ \left\{\frac{(n+1)p}{2} + [(1-p)p + (1-p)^2 p + \cdots]\right.$$
$$\left. + np[(1-p)^2 p + 2(1-p)^3 p + \cdots + (i-2)(1-p)^{i-1} p]\right\} D$$
$$= \left[\frac{n+1}{2} p^2 + np(1-p)\right] A + \left[\frac{n+1}{2} p(1-p) + n(1-p)^2\right] D \qquad (37)$$

D is normally much larger than A. The amount of loss in Eq. (37) is minimum when $p = 1$ and becomes larger when p is close to zero. Putting $p = 0$ in Eq. (37) gives nD, showing that the equations for L and n should be changed from Eqs. (18) and (20) to

$$L = \frac{B}{n} + \frac{n+1}{2}\left(\frac{2D}{\bar{u}}\right) + \frac{C}{\bar{u}} + \frac{\ell A}{\bar{u}} \qquad (38)$$

Table 2 Diagnosis and Probability of Problem Detection

Diagnosis	Probability of detecting troubles	Number of defectives found	Number of defectives missed
1st	p	$(n+1)p/2$	0
2nd	$(1-p)p$	np	$(n+1)p/2$
3rd	$(1-p)^2 p$	np	$(n+1)p/2 + np$
4th	$(1-p)^3 p$	np	$(n+1)p/2 + np$
⋮	⋮	⋮	⋮
ith	$(1-p)^{i-1} p$	np	$(n+1)p/2 + (i-2)np$

and

$$n = \left[\frac{2(\bar{u} + \ell)B}{2D - C/\bar{u}}\right]^{1/2} \tag{39}$$

where $(n + 1) \doteqdot n$ is approximated.

When the fraction defective during the trouble period is not 100%, it is normal to trace back and find defectives when a trouble is found. In this case, there are no undetected defectives, so $D = A$. Equation (37) is therefore

$$\left[\frac{n+1}{2} p^2 + np(1-p) + \frac{n+1}{2} p(1-p) + n(1-p)^2\right]A \tag{40}$$

Putting $n + 1 \doteqdot n$, Eq. (40) becomes

$$\left[\frac{n}{2} p^2 + \frac{3}{2} np(1-p) + n(1-p)^2\right]A = n\left(1 - \frac{p}{2}\right)A \tag{41}$$

Therefore, the loss after tracing back to find defectives is nA at maximum and $nA/2$ at minimum. If the equations for L and n were determined as

$$L = \frac{B}{n} + \frac{n+1}{2}\left(\frac{2A}{\bar{u}}\right) + \frac{C}{\bar{u}} + \frac{\ell A}{\bar{u}} \tag{42}$$

and

$$n = \left[\frac{2(u + \ell)B}{2A - C/\bar{u}}\right]^{1/2} \tag{43}$$

it would become overdiagnosis, which is too costly. Although the fraction defective can have any value, it would be good enough to consider about 0.5 for p. In that case, $1.5A$ is used instead of A. As described before, L and n are not significantly affected by the error in A up to 50%, so Eqs. (18) and (20) can be satisfactorily applied.

7. PREDICTION AND MODIFICATION

In the control of a variable quality characteristic, a signal factor is used for correcting the deviation of the characteristic from a target value. For example, pressure of a press is a signal factor to control the thickness of steel sheets, and flow of fuel is a signal factor to control temperature. For such a control, the following three steps must be taken:

1. Determine the optimum measuring interval.
2. Forecast the average quality of products produced before the next measurement.
3. Determine the optimum modifying quantity against the deviation of the forecasted value from the target value.

After the above parameters are determined,

4. Modify the quality characteristic made by varying the level of the signal factor.

To determine the optimum modifying quantity, an analysis of variance method, called cyclic analysis, and the following loss equation (caused by variation) are useful.

$$L = k\sigma^2 \tag{44}$$

where

$$k = \frac{\text{loss caused by out-of-specification}}{(\text{allowance})^2} \tag{45}$$

$$\sigma^2 = \text{average of the error from target value squared} \tag{46}$$

For Eq. (44), see Ref 1, Chapters 1 and 2. The simplest prediction method is to consider the measured value itself as the average quality of all products to be produced before the next measurement. There are many methods for this purpose. However, it is important to determine σ_p^2, the error variance of such a prediction.

The optimum modifying quantity in step 3 is determined by forecasting the average in step 2, which is denoted by y, and calculate the following quantity:

$$\text{Optimum modifying quantity} = -\beta(y - y_0) \tag{47}$$

where y_0 is the target value and

$$\beta = \begin{cases} 0 & \text{when } F_0 = \dfrac{(y - y_0)^2}{\sigma_p^2} \leq 1 \\ 1 - \dfrac{1}{F_0} & \text{when } F_0 = \dfrac{(y - y_0)^2}{\sigma_p^2} > 1 \end{cases} \tag{48}$$

Recently, more and more production systems using automatic machinery and robots that handle the four steps listed above have been developed. For such systems, the center of quality control is the calibration of sensors (measuring devices) employed by automatic machinery or robots and the diagnosis of hunting phenomena. Steps 1–4 are therefore required.

A simple example is illustrated in the following. The specification of the thickness of a metal sheet is $m \pm 5\,\mu\text{m}$. The loss caused by defects is 300 yen per meter. The daily production is 20,000 m, and the production line operates 5 days a week or 40 hr a week. Currently, measurement is made once every 2 hr, costing 2000 yen for the measurement and adjustment (correction or calibration). There is a tendency for the average and variation of thickeness to increase during the course of production. The average thickness increases $3\,\mu\text{m}$ every 2 hr, and the error variance increases $8\,\mu\text{m}^2$ in 2 hr.

Since the production is 5000 m in 2 hr, the average variance σ^2 of the products during 2 hr assuming that adjustment is correctly made at the time of measurement is

$$\sigma^2 = \frac{1}{5000}\int_0^{5000}\left[\left(\frac{3}{5000}\right)^2 t^2 + \frac{8}{5000}\,t\right]dt = 7.0 \tag{49}$$

The daily loss in the loss function L, including the correcting cost, is

$$L = \frac{300}{5^2} \times 7.0 \times 20{,}000 + 4 \times 2000 = 1{,}688{,}000 \text{ (yen)} \tag{50}$$

Letting the optimum measuring and adjusting interval be n,

$$L = \frac{300}{5^2} \times 20{,}000 \times \left[\frac{1}{n}\int_0^n \left\{\left(\frac{3}{5000}\right)^2 t^2 + \frac{8}{5000}\,t\right\}dt\right] + 2000 \times \frac{20{,}000}{n} = 0.0288n^2 + 192n + \frac{40{,}000{,}000}{n} \tag{51}$$

The optimum n that minimizes Eq. (51) is about 430. Then the loss due to prediction and correction is

$$\begin{aligned} L &= 0.0288 \times 430^2 + 192 \times 430 + \frac{40{,}000{,}000}{430} \\ &= 181{,}000 \text{ (yen)} \end{aligned} \tag{52}$$

There is an improvement of 1.507 million yen per day. There is additional improvement due to the reduction of the prediction error. For this, see Chapter 9 of Ref. 1.

REFERENCE

1. Taguchi, G. Taguchi methods. On-line Production. ASI, 1993.
2. Taguchi, G. Taguchi on Robust Technology Development, ASME Press 1993.

2

Statistical Monitoring and Optimization in Total Quality Management

Kai Kristensen
The Aarhus School of Business, Aarhus, Denmark

1. MEASUREMENT WITHIN TOTAL QUALITY MANAGEMENT

Modern measurement of quality should, of course, be closely related to the definition of quality. The ultimate judge of quality is the customer, which means that a system of quality measurement should focus on the entire process that leads to customer satisfaction in the company, from the supplier to the end user.

Total quality management (TQM) argues that a basic factor in the creation of customer satisfaction is leadership, and it is generally accepted that a basic aspect of leadership is the ability to deal with the future. This has been demonstrated very nicely by, among others, Mr. Jan Leschly, president of Smith Kline, who in a recent speech in Denmark compared his actual way of leading with the ideal as he saw it. His points are demonstrated in Figure 1. It appears that Mr. Leschly argues that today he spends approximately 60% of his time on "firefighting," 25% on control, and 15% on the future. In his own view a much more appropriate way of leading would be to turn the figure upside down, so to speak, and spend 60% of his time on the future, 25% on control, and only 15% on firefighting.

The situation described by Mr. Leschly holds true of many leaders in the Western world. There is a clear tendency for leaders in general to be much more focused on short-term profits than on the process that creates profit. This again may lead to firefighting and to the possible disturbance of processes that may be in statistical control. The result of this may very well

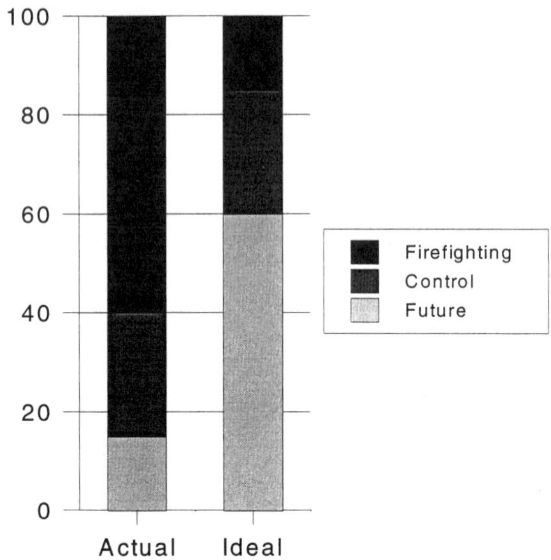

Figure 1 Leadership today and tomorrow. (Courtesy of Jan Leschly, Smith Kline.)

be an increase in the variability of the company's performance and hence and increase in quality costs. In this way "the short-term leader" who demonstrates leadership by fighting fires all over the company may very well be achieving quite the opposite of what he wants to achieve.

To be more specific, "short-term leadership" may be synonymous with low quality leadership, and in the future it will be necessary to adopt a different leadership style in order to survive, a leadership style that in its nature is long-term and that focuses on the processes that lead to the results rather than the results themselves. This does not, of course, mean that the results are uninteresting per se, but rather that when the results are there you can do nothing about them. They are the results of actions taken a long time ago.

All this is much easier said than done. In the modern business environment leaders may not be able to do anything but act on the short-term basis because they do not have the necessary information to do otherwise. To act on a long-term basis requires that you have an information system that provides early warning and that makes it possible for you to make the necessary adjustments to the processes and gives you time to make them before they turn into unwanted business results. This is what modern measurement of total quality is all about.

This idea is in very good accordance with the official thoughts in Europe. In a recent working document from the European Commission, DGIII, the following is said about quality and quality management (European Commission, 1995):

> The use of the new methodologies of total quality management is for the leaders of the European companies a leading means to help them in the current economic scenario, which involves not only dealing with changes, but especially anticipating them.

Thus, to the European Commission, quality is primarily a question of changes and early warning.

To create an interrelated system of quality measurement it has been decided to define the measurement system according to Table 1, where measurements are classified according to two criteria: the interested party (the stakeholder) and whether we are talking about processes or results. Other types of measurement systems are given in Kaplan and Norton (1996).

As Table 1 illustrates, we distinguish between measurements related to the process and measurements related to the results. The reason for this is obvious in the light of what has been said above and in the light of the definition of TQM. Furthermore we distinguish between three "interested parties:" the company itself, the customer, and the society. The first two should obviously be part of a measurement system according to the definition of TQM, and the third has been included because there is no doubt that

Table 1 Measurement of Quality – The Extended Concept

	The company	The customer	The society
The process	Employee satisfaction (ESI) Checkpoints concerning the internal structure	Control- and check-points concerning the internal definition of product and service quality	Control and checkpoints concerning e.g. environment, life cycles etc.
The result	Business results Financial ratios	Customer satisfaction (CSI) Checkpoints describing the customer satisfaction	'Ethical accounts' Environmental accounts

the focus on companies in relation to their effect on society will be increased in the future and it is expected that very soon we are going to see a lot of new legislation within this area.

Traditional measurements have focused on the lower left-hand corner of this table, i.e., the business results, and we have built up extremely detailed reporting systems that can provide information about all possible ways of breaking down the business results. However, as mentioned above, this type of information is pointing backwards in time, and at this stage it is too late to do anything about the results. What we need is something that can tell us about what is going to happen with business results in the future. This type of information we find in the rest of the table, and we especially believe and also have documentation to illustrate that the top set of entries in the table are related in a closed loop that may be called the improvement circle. This loop is demonstrated in Figure 2.

The improvement is particularly due to an increase in customer loyalty stemming from an increase in customer satisfaction. The relationship between customer satisfaction and customer loyalty has been documented empirically several times. One example is Rank Xerox, Denmark, who in their application for the Danish Quality Award reported that when they analyzed customer satisfaction on a five-point scale where 1 is very dissatisfied and 5 is very satisfied they observed that on average 93% of those customers who were very satisfied (a score of 5) came back as customers, while only 60% of those who gave a 4 came back.

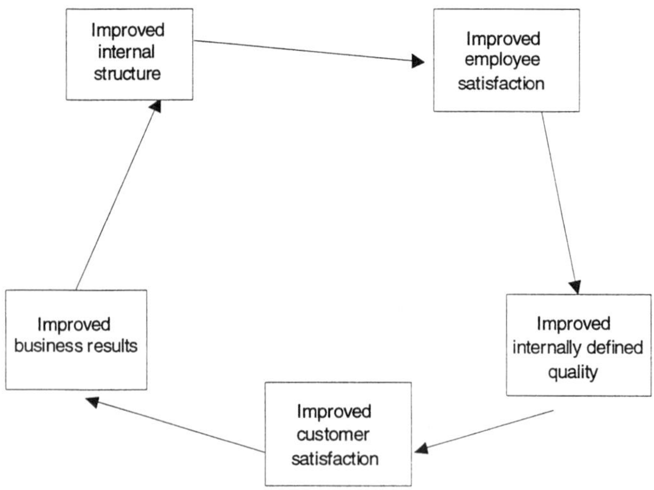

Figure 2 The improvement circle.

Another example is a large Danish real estate company who in a customer satisfaction survey asked approximately 2500 customers to evaluate the company on 20 different parameters. From this evaluation an average of customer satisfaction (customer satisfaction index) was calculated. The entire evaluation took place on a five-point scale with 5 as the best score, which means that the customer satisfaction index will have values in the interval from 1 to 5. In addition to the questions on parameters, a series of questions concerning loyalty were asked, and from this a loyalty index was computed and related to the customer satisfaction index. This analysis revealed some very interesting results, which are summarized in Figure 3, in which the customer satisfaction index is related to the probability of using the real estate agent once again (probability of being loyal). It appears that there is a very close relationship between customer satisfaction and customer loyalty. The relationship is beautifully described by a logistic model.

Furthermore, it appears from Figure 3 that in this case the loyalty is around 35% when the customer satisfaction index is 3, i.e., neither good nor bad. When the customer satisfaction increases to 4, a dramatic increase in loyalty is observed. In this case the loyalty is more than 90%. Thus the area between 3 and 4 is very important, and it appears that even very small changes in customer satisfaction in this area may lead to large changes in the probability of loyalty.

The observed relationship between business results and customer loyalty on the one hand and customer satisfaction on the other is very impor-

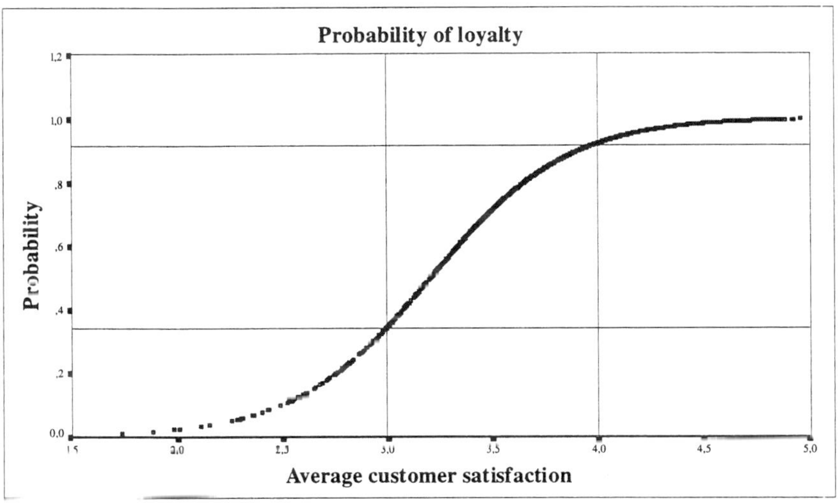

Figure 3 Probability of loyalty as a function of customer satisfaction.

tant information for modern management. This information provides an early warning about future business results and thus provides management with an instrument to correct failures before they affect business results.

The next logical step will be to take the analysis one step further back and find internal indicators of quality that are closely related to customer satisfaction. In this case the warning system will be even better. These indicators, which in Table 1 are named control points and checkpoints, will, of course, be company-specific even if some generic measures are defined.

Moving even further back, we come to employee satisfaction and other measures of the process in the company. We expect these to be closely related to the internally defined quality. This is actually one of the basic assumptions of TQM. The more satisfied and more motivated your employees, the higher the quality in the company [see Kristensen (1996)]. An indicator of this has been established in the world's largest service company, the International Service System (ISS), where employee satisfaction and customer satisfaction have been measured on a regular basis for some years now [see Kristensen and Dahlgaard (1997)]. In order to verify the hypothesis of the improvement circle in Figure 2, employee satisfaction and customer satisfaction were measured for 19 different districts in the cleaning division of the company in 1993. The results were measured on a traditional five-point scale, and the employee satisfaction and customer satisfaction indices were both computed as weighted averages of the individual parameters. The results are shown in Figure 4.

These interesting figures show a clear linear relationship between employee satisfaction and customer satisfaction. The higher the employee satisfaction, the higher the customer satisfaction. The equation of the relationship is as follows:

$$\mathrm{CSI} = 0.75 + 0.89\,\mathrm{ESI}, \qquad R^2 = 0.85$$

The coefficients of the equation are highly significant. Thus the standard deviation of the constant term is 0.33, and that of the slope is 0.09. Furthermore, we cannot reject a hypothesis that the slope is equal to 1.

It appears from this that a unit change in employee satisfaction gives more or less the same change in customer satisfaction. We cannot, from these figures alone, claim that this is a causal relationship, but we believe that combined with other information this is strong evidence for the existence of an improvement circle like the one described in Figure 2. To us, therefore, the creation of a measurement system along the lines given in Table 1 is necessary. Only in this way will management be able to lead

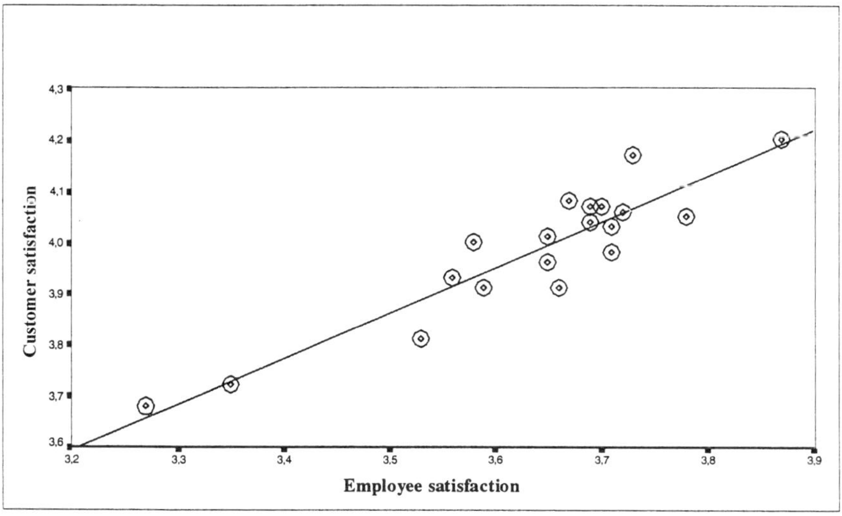

Figure 4 Relationship between ESI and CSI, 19 districts.

the company upstream and thus prevent the disasters that inevitably follow the firefighting of short-term management.

An example of an actual TQM measurement system is given in Figure 5 for a Danish medical company. It will be seen that the system follows the methodology given in the Process section of Table 1.

2. MEASURING AND MONITORING EMPLOYEE AND CUSTOMER SATISFACTION

Since optimization and monitoring of the internal quality are dealt with elsewhere in this book we are going to concentrate on the optimization and monitoring of customers whether they are internal (employees) or external. First a theoretical, microeconomic model of satisfaction and loyalty is constructed and then we establish a "control chart" for the managerial control of satisfaction.

2.1. A Model of Satisfaction and Loyalty

Since exactly the same model applies to both customer satisfaction and employee satisfaction we can without loss of generality base the entire discussion on a model for customer satisfaction.

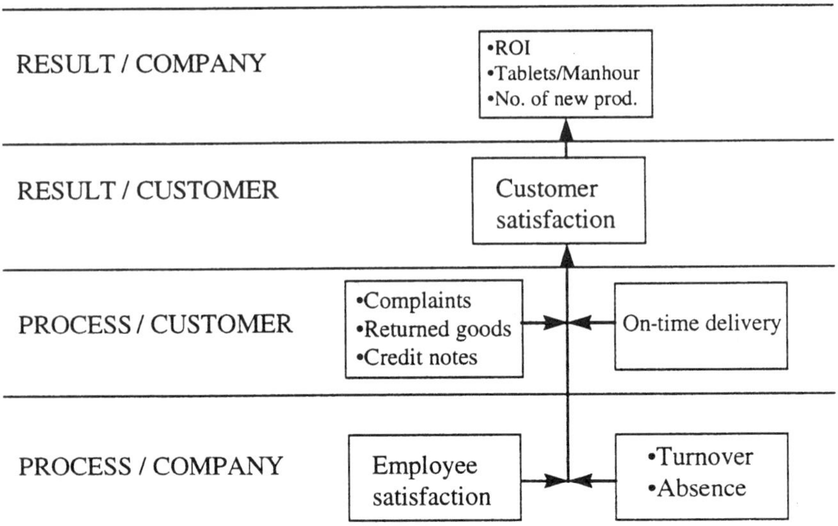

Figure 5 An example from a Danish medical company.

In Kristensen et al. (1992) a model linking customer satisfaction to company profit was established. In this model, customer satisfaction was defined as

$$\mathrm{CSI} = \sum_{i=1}^{n} w_i c_i \tag{1}$$

where n is the number of quality parameters, w_i is the importance of a given parameter, and c_i is the evaluation. It was assumed that the profit of the company could be described as

$$\Pi = \varphi\left(\sum_{i=1}^{n} w_i c_i\right) - \sum_{i=1}^{n} k_i c_i^2 \tag{2}$$

where φ is an increasing function linking customer satisfaction to company earnings and the second factor on the right-hand side is a quadratic cost function with k_i as a cost parameter.

By maximizing (2) with respect to the individual satisfactions (c_i's) it can be shown that for identical cost parameters, i.e.,

$$k_i = k_j \quad \forall_{ij} \tag{3}$$

the optimum allocation of resources will occur when

$$\frac{c_i}{w_i} = \frac{c_j}{w_j} \quad \forall_{ij} \tag{4}$$

i.e., when the degree of fulfilment of customer expectations is identical for all areas. This is based on the fact that the first-order condition for maximization of Eq. (2) is equal to

$$\frac{c_i}{w_i} = \frac{\varphi'_{\mathrm{CSI}}}{2k_i} \tag{5}$$

From this it will be seen that if the right-hand side of Eq. (5) is equal to 1 then a very simple rule for optimum customer satisfaction will emerge:

$$c_i = w_i \tag{6}$$

This result, even if it is based on rather strong assumptions, has become very popular among business people, and a graphical representation known as the quality map where each parameter is plotted with w on the x axis and c on the y axis has become a more-or-less standard tool for monitoring customer and employee satisfaction. This is the reason we later on elaborate a little on the quality map

But even if this is the case we intend to take model (2) a step further in order to incorporate customer loyalty. The reason for this is that customer loyalty has gained a lot of interest among quality management researchers recently because it seems so obvious that loyalty and quality are related, but we still need a sensible model for relating customer loyalty to profit [see Kristensen and Martensen (1996)].

We start by assuming that profit can be described as follows:

$$\Pi = \frac{\text{likelihood of}}{\text{buying}} \times \frac{\text{quantity}}{\text{bought}} - \text{costs} \tag{7}$$

where quantity bought is measured in sales prices.

The likelihood of buying is, of course, the loyalty function. We assume that this function can be described as follows:

$$L = L(\zeta_1, \zeta_2, \ldots, \zeta_n) \tag{8}$$

where

$$\zeta_i = w_i(c_i - c_i^*) \tag{9}$$

where c_i^* is the satisfaction of parameter i for the main competitor. Thus the elements of the loyalty function are related to the competitive position of a given parameter combined with the importance of the parameter. We assume that the quantity bought given loyalty is a function of the customer satisfaction index. This means that we will model the income or revenue of the company as

$$L(\zeta_1, \ldots, \zeta_n)\varphi\left(\sum_{i=1}^{n} w_i c_i\right) \tag{10}$$

This tells us that you may be very satisfied and still not buy very much, because competition is very tough and hence loyalty is low. On the other hand, when competition is very low, you may be dissatisfied and still buy from the company even though you try to limit your buying as much as possible.

Combining (10) with the original model in (2), we come to the following model for the company profit:

$$\Pi = L(\zeta_1, \ldots, \zeta_n)\varphi\left(\sum_{i=1}^{n} w_i c_i\right) - \sum_{i=1}^{n} k_i c_i^2 \tag{11}$$

Hence the optimum allocation of resources will be found by maximizing this function with respect to c_i, which is the only parameter that the company can affect in the short run. Long-run optimization will, of course, be different, but this is not part of the situation we consider here.

The first-order condition for the optimization of Eq. (11) is

$$\frac{\delta \Pi}{\delta c_i} = L\varphi' w_i + \varphi L_i' w_i - 2k_i c_i \tag{12}$$

By equating this to zero we get the following characterization result:

$$\frac{c_i}{w_i} = \frac{\varphi}{2k_i} L_i' + L \frac{\varphi'}{2k_i} \tag{13}$$

To make practical use of this result we assume that

$$k_i = k_j \quad \forall_{ij} \tag{14}$$

which means that we may write the characterization result as

$$\frac{c_i}{w_i} = \alpha + \beta L_i' \tag{15}$$

To put it differently, we have shown that if company resources have been allocated optimally, then the degree to which you live up to customer expectations should be a linear function of the contribution to loyalty. This seems to be a very logical conclusion that will improve the interpretation of the results of customer satisfaction studies.

Practical use of results (4), (6), and (15) will be easy, because in their present form you only need market information to use them. Once you collect information about c_i, c_i^*, w_i, and the customers' buying intentions, the models can be estimated. In the case of a loyalty model you will most likely use a logit specification for L and then L_i' will be easy to calculate.

2.2. Statistical Monitoring of the Satisfaction Process

Let

$$x = \left(\frac{c}{w}\right) \tag{16}$$

where c is an $n \times 1$ vector of evaluations and w is an $n \times 1$ vector of importances. Assume that x is multivariate normal with covariance matrix

$$\sum = \left(\begin{array}{c|c} \sum_c & \sum_{cw} \\ \hline \sum_{wc} & \sum_w \end{array}\right) \tag{17}$$

and expectation

$$\mu = \left(\frac{\mu_c}{\mu_w}\right) \tag{18}$$

According to the theoretical development we want to test the hypothesis

$$H_0{:} \quad \mu_c = \mu_w \tag{19}$$

Assume a sample of N units, and let the estimates of (17) and (18) be

$$\bar{x} = \begin{pmatrix} \bar{c} \\ \hline \bar{w} \end{pmatrix} \tag{20}$$

and

$$S = \left(\begin{array}{c|c} S_c & S_{cw} \\ \hline S'_{cw} & S_w \end{array}\right) \tag{21}$$

Let I be the identity matrix of order n. Then our hypothesis may be written

$$H_0\colon \quad (I| -I)\begin{pmatrix} \mu_c \\ \hline \mu_w \end{pmatrix} = 0 \tag{22}$$

From this it is seen that the T^2 statistic is equal to

$$T^2 = N(\bar{c} - \bar{w})'[S_c + S_w - S_{cw} - S'_{cw}]^{-1}(\bar{c} - \bar{w}) \tag{23}$$

If the hypothesis is true, then

$$F = \frac{N - n}{(N - 1)n}\, T^2 \tag{24}$$

has an F-distribution with n and $N - n$ degrees of freedom. The hypothesis is rejected if the computed F-statistic exceeds the critical value $F_{\alpha;n,N-n}$. Let

$$S_d = S_c + S_w - S_{cw} - S'_{cw} \tag{25}$$

Then simultaneous confidence intervals for the differences between μ_c and μ_w may be written as follows for any vector $l' = (l_1, l_2, \ldots, l_n)$:

$$\begin{aligned} l'(\bar{c} - \bar{w}) - \left[\frac{1}{N}\, l'S_d l\, \frac{(N-1)n}{N-n}\, F_{\alpha;n,N-n}\right]^{1/2} &\le l'(\mu_c - \mu_w) \\ \le l'(\bar{c} - \bar{w}) + \left[\frac{1}{N}\, l'S_d l\, \frac{(N-1)n}{N-n}\, F_{\alpha;n,N-n}\right]^{1/2} & \end{aligned} \tag{26}$$

Now assume that the hypothesis is true, and let

$$l' = (0, \ldots, 0, 1, 0, \ldots, 0) \tag{27}$$

Then we may write

$$c_i - \left[\frac{1}{N} l' S_d l \frac{(N-1)n}{N-n} F_\alpha\right]^{1/2} \leq w_i \leq c_i + \left[\frac{1}{N} l' S_d l \frac{(N-1)n}{N-n} F_\alpha\right]^{1/2} \tag{28}$$

or

$$c_i - \left(\frac{s_{d_i}^2}{N}\right)^{1/2} \left[\frac{(N-1)n}{N-n} F_\alpha\right]^{1/2} \leq w_i \leq c_i + \left(\frac{s_{d_i}^2}{N}\right)^{1/2} \left[\frac{(N-1)n}{N-n} F_\alpha\right]^{1/2} \tag{29}$$

To simplify, let us assume that all differences have the same theoretical variance. Then we may substitute the average $\bar{s}_d^2$ for $s_{d_i}^2$, which means that the interval for monitoring satisfaction will be constant. In that case we may set up the "control" chart shown in Figure 6 for monitoring satisfaction, where the limits are given by

$$\pm\left(\frac{\bar{s}_d^2}{N}\right)^{1/2} \left[\frac{(N-1)n}{N-n} F_{\alpha;n,N-n}\right]^{1/2} \tag{30}$$

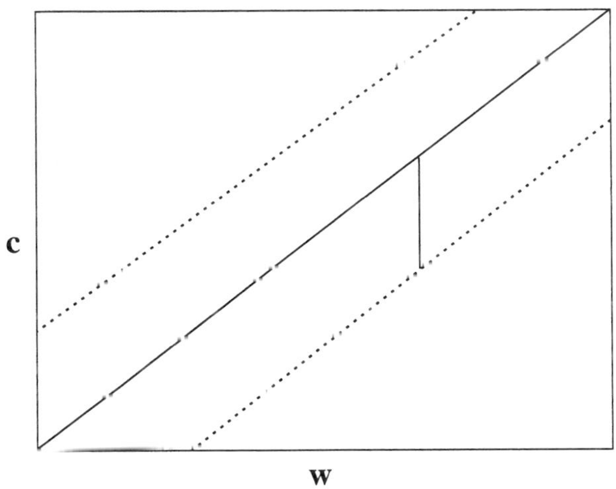

Figure 6 Quality map.

If a parameter falls between the dotted lines we cannot reject the hypothesis that we have optimal allocation of resources. If, on the other hand, a parameter falls outside the limits, the process needs adjustment.

We should remember that the limits are simultaneous. If we want individual control limits, which, of course, will be much narrower, we may substitute $t_{\alpha,N-1}$ for

$$\left[\frac{(N-1)n}{N-n} F_{\alpha;n,N-n}\right]^{1/2} \tag{31}$$

2.3. An Example

An actual data set from a Danish company is presented in Table 2. Seven parameters were measured on a seven-point rating scale.

Now we are ready to set up the control chart for customer satisfaction. We use formula (30) to get the limits,

$$\begin{aligned} L &= \pm\sqrt{\frac{2.18}{64}}\sqrt{\frac{(64-1)7}{64-7} F_{0.05;7,57}} \\ &= \pm(0.18)\sqrt{7.74 \times 2.18} = \pm 0.74 \end{aligned} \tag{32}$$

From the control chart (Figure 7) we can see that most of the parameters are in control but one parameter needs attention. The importance of the environmental parameter is significantly greater than that of the evaluation of

Table 2 Data Set (Customer Satisfaction for a Printer)

Parameter	Importance W_i	Satisfaction C_i	Sample size	Variance of difference
Operation	6.68	6.06	64	1.66
User friendliness	5.85	5.67	↓	1.82
Print quality	5.99	5.48		1.80
Service	5.32	5.38		2.56
Speed	3.91	4.94		2.62
Price	4.64	5.02		2.69
Environmentally friendly	5.17	4.18		2.16
Average				2.18

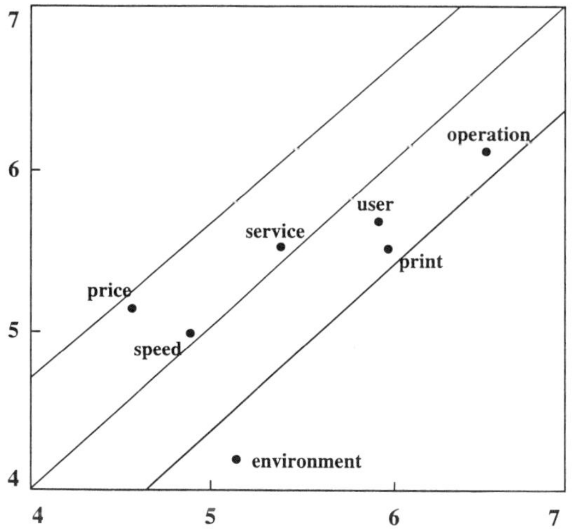

Figure 7 Control chart for customer satisfaction.

company performance. Hence the quality of this parameter must be improved.

3. CONCLUSION

The use of the concept of total quality management expands the need for measurement in the company. The measurement of quality will no longer be limited to the production process. Now we need to monitor "processes" such as customer satisfaction and employee satisfaction. In this chapter I have given a managerial model for the control of these processes, and we have considered a practical "control" chart that will help management choose the right parameters for improvement.

REFERENCES

European Commission, DG III. A European Quality Promotion Policy. Bruxelles, Feb. 17, 1995.

Kaplan RS, Norton DP. The Balanced Scorecard. Harvard Business School Press, Boston, MA, 1996.

Kristensen K. Relating employee performance to customer satisfaction. World Class Europe—Striving for Excellence. EFQM, Edinburgh, 1996, pp. 53–57.

Kristensen K, Dahlgaard JJ. ISS International Service System A/S—Denmark. The European Way to Excellence. Case Study Series. Directorate General III, European Commission, Bruxelles, 1997.

Kristensen K, Martensen A. Linking customer satisfaction to loyalty and performance. ESOMAR Pub. Ser. 204: 159–169, 1996.

Kristensen K, Dahlgaard JJ, Kanji GK. On measurement of customer satisfaction. Total Quality Management 3(2): 123–128, 1992.

3

Quality Improvement Methods and Statistical Reasoning*

G.K. Kanji
Sheffield Hallam University, Sheffield, England

1. PRINCIPLES OF TOTAL QUALITY MANAGEMENT

Total quality management (TQM) is about continuous performance improvement of individuals, groups, and organizations. What differentiates total quality management from other management processes is the emphasis on continuous improvement. Total quality is not a quick fix; it is about changing the way things are done—forever.

Seen in this way, total quality management is about continuous performance improvement. To improve performance, people need to know what to do and how to do it, have the right tools to do it, be able to measure performance, and receive feedback on current levels of achievement.

Total quality management (Kanji and Asher, 1993) provides this by adhering to a set of general governing principles. They are:

1. Delight the customer
2. Management by fact
3. People-based management
4. Continuous improvement

*For an extended version of this paper, see Kanji GK. Total Quality Management 5: 105, 1994.

Each of these principles can be used to drive the improvement process. To achieve this, each principle is translated into practice by using two core concepts, which show how to make the principle happen.

These concepts are:

Customer satisfaction
Internal customers are real
All work is a process
Measurement
Teamwork
People make quality
Continuous improvement cycle
Prevention

Further details of the four principles with the core concepts follow. The pyramid principles of TQM are shown in Figure 1.

1.1. Delight the Customer

The first principle focuses on the external customers and asks "what would delight them?" This implies understanding needs—both of product and service, tangible and intangible—and agreeing with requirements and meeting them. Delighting the customer means being best at what matters most to customers, and this changes over time. Being in touch with these changes and delighting the customer now and in the future form an integral part of total quality management.

The core concepts of total quality that relate to the principle of delighting the customer are "customer satisfaction" and "internal customers are real."

1.2. Management by Fact

Knowing the current performance levels of our products or services in our customers' hands and of all our employees is the first stage in being able to improve. If we know where we are starting from, we can measure our improvement.

Having the facts necessary to manage the business at all levels is the second principle of total quality. Giving that information to people so that decisions are based upon fact rather than "gut feel" is essential for continuous improvement.

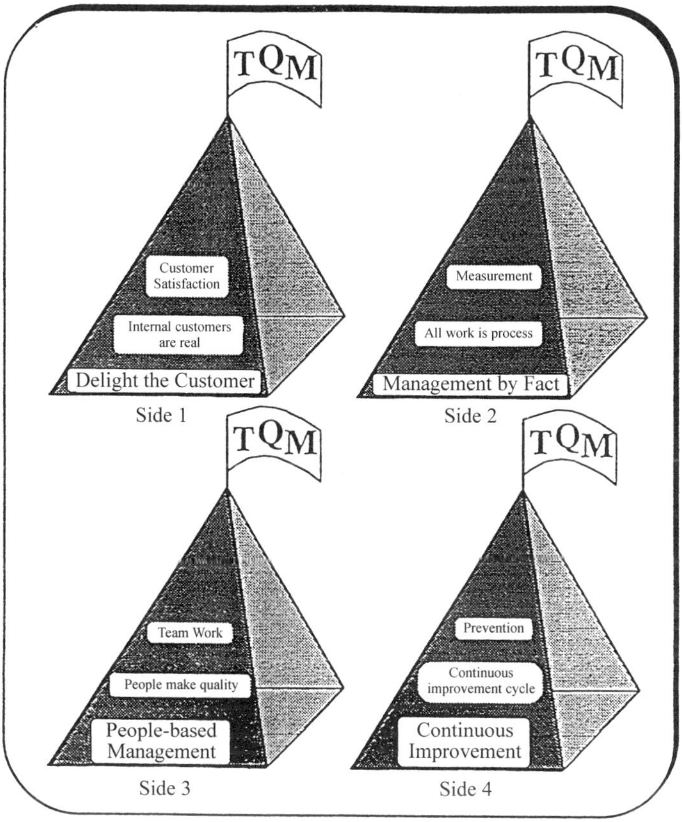

Figure 1 The pyramid principles of TQM. (From Kanji and Asher, 1993.)

The core concepts that relate to management by fact are "all work is a process" and "measurement."

1.3. People-based Management

Knowing what to do and how to do it and getting feedback on performance form one part of encouraging people to take responsibility for the quality of their own work. Involvement and commitment to customer satisfaction are ways to generate this. The third principle of total quality management recognizes that systems, standards, and technology in themselves do not mean quality. The role of people is vital.

The core concepts that relate to people-based management are "teamwork" and "people make quality."

1.4. Continuous Improvement

Total quality cannot be a quick fix or a short-term goal that will be reached when a target has been met. Total quality is not a program or a project. It is a management process that recognizes that however much we may improve, our competitors will continue to improve and our customers will expect more from us. The link between customer and supplier with process improvement can be seen in Kanji (1990).

Here, continuous improvement—incremental change, not major breakthroughs—must be the aim of all who wish to move toward—total quality.

The core concepts that relate to the company's continuous improvement are "the continuous improvement cycle" and "prevention."

Each concept is now discussed, together with an example of how that concept was used by a company to bring about improvement.

2. CORE CONCEPTS OF TQM

2.1. Internal Customers Are Real

The definition of quality [see Kanji (1990)], "satisfying agreed customer requirements," relates equally to internal and external customers. Many writers refer to the customer–supplier chain and the need to get the internal relationships working in order to satisfy the external customer.

Whether you are supplying information, products, or a service, the people you supply internally depend on their internal suppliers for quality work. Their requirements are as real as those of external customers; they may be speed, accuracy, or measurement.

Internal customers constitute one of the "big ideas" of total quality management. Making the most of this idea can be very time-consuming, and many structured approaches take a long time and can be complicated. However, one successful approach is to take the "cost of quality" and obtain information about the organization's performance and analyze it. Dahlgaard et al. (1993) used statistical methods to discuss the relationship between the total quality cost and the number of employees in an organization.

2.2. All Work Is a Process

The previous section looked at internal customers and how to use the idea that they are real as a focus for improvement.

Another possible focus is that of business processes. By "process" we mean any relationship such as billing customers or issuing credit notes—anything that has an input, steps to follow, and an output. A process is a combination of methods, materials, manpower, machinery, etc., which taken together produce a product or service.

All processes contain inherent variability, and one approach to quality improvement is progressively to reduce variation, first by removing variation due to special causes and second by driving down common cause variation, thus bringing the process into control and then improving its capability.

Various statistical methods, e.g., histograms, Pareto analysis, control charts, and scatter diagrams, are widely used by quality managers and others for process improvement.

2.3. Measurement

The third core concept of total quality management is measurement. Having a measure of how we are doing is the first stage in being able to improve. Measures can focus internally, i.e., on internal customer satisfaction (Kristensen et al., 1993), or externally, i.e., on meeting external customer requirements.

Examples of internal quality measurements are

Production
Breach of promise
Reject level
Accidents
Process in control
Yield/scrap (and plus value)

Kristensen et al. (1993), when discussing a measurement of customer satisfaction, used the usual guidelines for questionnaire design and surveys and statistical analysis to obtain the customer satisfaction index.

2.4. Prevention

The core concept of prevention is central to total quality management and is one way to move toward continuous improvement.

Prevention means not letting problems happen. The continual process of driving possible failure out of the system can, over time, breed a culture of continuous improvement.

There are two distinct ways to approach this. The first is to concentrate on the design of the product itself (whether a hard product or a service); the second is to work on the production process. However, the most important aspect of prevention is quality by design using statistical reasoning.

There are several frequently used tools, and failure mode and effect analysis (FMEA) is one of the better known ones. It is associated with both design (design FMEA) and process (process FMEA).

Other frequently used methods are failure prevention analysis, which was pioneered by Kepner Tregoe, and foolproofing (or Pokaoki). The advantage of all of these methods is that they provide a structure or thought process for carrying the work through.

2.5. Customer Satisfaction

Many companies, when they begin quality improvement processes, become very introspective and concentrate on their own internal problems almost at the expense of their external customers.

Other companies, particularly in the service sector, have deliberately gone out to their customers, first to survey what is important to the customer and then to measure their own performance against customer targets (Kristensen et al., 1993). The idea of asking one's customers to set customer satisfaction goals is a clear sign of an outward-looking company.

One example is Federal Express, who surveyed their customer base to identify the top 10 causes of aggravation. The points were weighted according to customer views of how important they were. A complete check was made of all occurrences, and a weekly satisfaction index was compiled. This allowed the company to keep a weekly monitor of customer satisfaction as measured by the customer. An understanding of survey and statistical methods is therefore needed for the measurement of customer satisfaction.

2.6. Teamwork

Teamwork can provide an opportunity for people to work together in their pursuit of total quality in ways in which they have not worked together before.

People who work on their own or in small, discrete work groups often have a picture of their organization and the work that it does that is very compartmentalized. They are often unaware of the work that is done even by people who work very close to them. Under these circumstances they are usually unaware of the consequences of poor quality in the work they themselves do.

By bringing people together in terms with a common goal, quality improvement becomes easier to communicate over departmental or functional walls. In this way the slow breaking down of barriers acts as a platform for change.

We defined culture as "the way we do things here," and cultural change as "changing the way we do things here." This change implies significant personal change in the way people react and in their attitudes. A benchmarking approach can also help to change the way they do things.

Teamwork can be improved by benchmarking, a method that is similar to the statistical understanding of outliers.

2.7. People Make Quality

Deming has stated that the majority of quality-related problems within an organization are not within the control of the individual employee. As many as 80% of these problems are caused by the way the company is organized and managed.

Examples where the system gets in the way of people trying to do a good job are easy to find, and in all cases simply telling employees to do better will not solve the problem.

It is important that the organization develop its quality management system, and it should customize the system to suit its own requirements. Each element will likely encompass several programs. As a matter of fact, this is where the role of statistics is most evident.

2.8. The Continuous Improvement Cycle

The continuous cycle of establishing customer requirements, meeting those requirements, measuring success, and continuing to improve can be used both externally and internally to fuel the engine of continuous improvement.

By continually checking with customer requirements, a company can keep finding areas in which improvements can be made. This continual supply of opportunity can be used to keep quality improvement plans up-to-date and to reinforce the idea that the total quality journey is never-ending.

In order to practice a continuous improvement cycle it is necessary to obtain continuous information about customer requirements, i.e., do market research. However, we know that market research requires a deep statistical understanding for the proper analysis of the market situation.

3. STATISTICAL UNDERSTANDING

The role of statistical concepts in the development of total quality management it nothing new. For example, Florence Nightingale, the 19th century statistician and famous nurse, was known as the mother of continuous health care quality improvement. In 1854 she demonstrated that a statistical approach by graphical methods could be persuasive in reducing the cost of poor quality care by 90% within a short period of time. Later, in 1930, Walter Shewhart, another prominent statistician, also suggested that the same kind of result could be achieved by using statistical quality control methods.

The fundamental aspect of statistical understanding is the variation that exists in every process, and the decisions are made on that basis. If the variation in a process is not known, then the required output of that process will be difficult to manage.

It is also very important to understand that every process has an inherent capability and that the process will be doing well if it operates within that capability. However, sometimes one can observe that resources are being wasted in solving a problem, and simply not realize that the process is working at its maximum capability.

In order to understand variability and the control of variation, it is necessary to understand basic statistical concepts. These concepts are simple to understand and learn and provide powerful management tools for higher productivity and excellent service.

In this complex business world, managers normally operate in an uncertain environment, and therefore their major emphasis is on the immediate problems. In their everyday life they deal with problems where the application of statistics occurs in pursuit of organizational objectives.

However, as we know, the business world is changing, and managers along with other workers are adopting this change and also learning how to manage it. For many people, the best way of adopting this change is to focus on statistical understanding because it permeates all aspects of total quality management.

We have already learned that "all work is a process" and therefore identification and reduction of a variation of processes provides opportunity for improvement. Here, the improvement process, which recognizes that variation is everywhere, gets help from the statistical world for this quality journey.

In general, managers can take many actions to reduce variation to improve quality. Snee (1990) pointed out that managers can reduce variation by maintaining the constant purpose of their employees to pursue a common quality goal.

4. CONCLUSIONS

In recent years, particularly in Japan and the United States, there has been a strong movement for greater emphasis on total quality management in which statistical understanding has been seen to be a major contributor for management development.

It is clear that statistical understanding plays a major role in product and service quality, care of customers through statistical process control, customer surveys, process capability, cost of quality, etc. The value of statistical design of experiments, which distinguishes between special cause and common cause variation, is also well established in the area of quality improvement.

If we also accept that "all work is process," that all processes are variable, and that there is a relationship between management action and quality, then statistical understanding is an essential aspect of the quality improvement process.

Further, in the areas of leadership, quality culture, teamwork, etc., development can be seen in various ways by the use of statistical understanding.

In conclusion, I believe that total quality management and statistical understanding go hand in hand. People embarking on the quality journey must therefore venture onto the road of total statistical understanding and follow the lead of total quality statisticians.

REFERENCES

Dahlgaard JJ, Kristensen K, Kanji GK. Quality cost and total quality management. Total Quality Management 3(3): 211–222, 1993.

Kanji GK. Total quality management: The second industrial revolution. Total Quality Management 1(1): 3–12, 1990.

Kanji GK, Asher M. Total Quality Management: A Systemic Approach. Carfax Publishing Company, Oxfordshire, U.K., 1993.

Kristensen K, Kanji GK, Dahlgaard JJ. On measurement of customer satisfaction. Total Quality Management 3(2): 123–128, 1993.

Snee RD. Statistical thinking and its contribution to total quality. Am Stat 44(2): 116–121, 1990.

4

Leadership Profiles and the Implementation of Total Quality Management for Business Excellence

Jens J. Dahlgaard, Su Mi Park Dahlgaard, and Anders Nørgaard
The Aarhus School of Business, Aarhus, Denmark

1. INTRODUCTION

Total quality management (TQM) is defined by Kanji and Asher, 1993 as

> A company culture which is characterized by everybody's participation in continuous improvements of customer satisfaction.

To build the TQM culture it is important that every staff member—top managers, middle managers, and other employees understand and apply the five basic principles of TQM. These can be visualized in terms of the TQM pyramid (Dahlgaard and Kristensen, 1992, 1994) presented in Figure 1.

As can be seen from Figure 1, the foundation of the TQM pyramid is leadership. All staff members need leaders who can explain the importance of TQM principles and who can show how those principles can be continuously practiced so that the organization gradually achieves business excellence.

Each staff member and each group must continuously focus on the customer (external as well as internal customers). They must continuously try to understand the customers' needs, expectations, and experiences so that they can delight the customer. To be able to delight the customer, continuous improvement is necessary. World class companies are continuously trying to improve existing products or develop new ones. They are

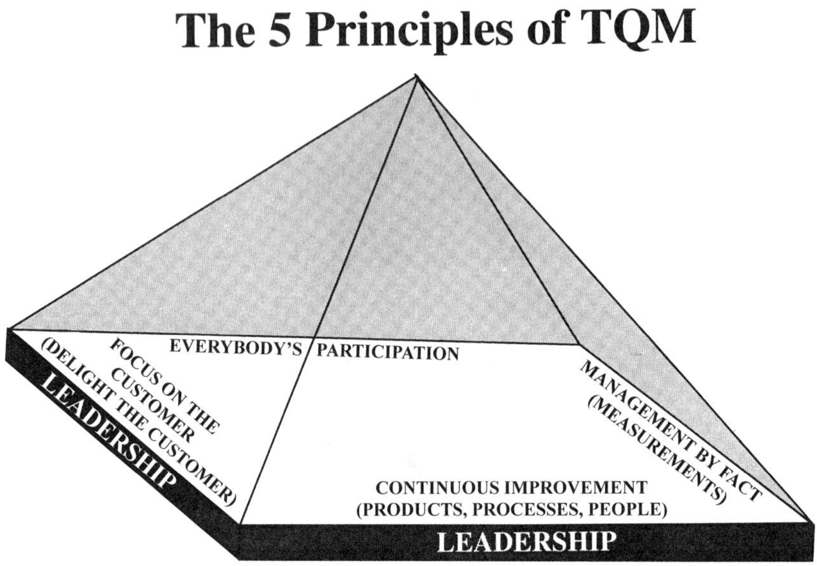

Figure 1 The TQM pyramid.

continuously trying to work smarter, not harder, by improving their processes, and they understand that the most important asset to improve is their people. To support everybody in continuous improvements, measurements are of vital importance. To improve products we need feedback from customers (measurements of customer satisfaction and other customer facts). To improve processes we need feedback from the various processes (process measurements of defects, wastage, quality costs, etc.). To improve people we need feedback from employees (measurements of employee satisfaction and other facts related to improvement of people). Statistical methods can be used in many of these measurements. The application of statistical methods is often the best way to ensure high reliability of the measurements, and for complex measurements such as measurements of people's mind-sets it may be the only way to generate reliable facts (see Section 2).

Continuously applying the five principles of TQM will gradually result in business excellence. But what is business excellence? Business excellence has many definitions. One example is (Raisbeck, 1997)

> The overall way of working that results in balanced stakeholder (customers, employees, society, shareholders) satisfaction so increasing the probability of long term success as a Business.

In 1992 the European Foundation for Quality Management (EFQM) launched the European Quality Award and a model to be used for assessment of the applicants for the award. The model, which is seen in Figure 2, has gradually been accepted as an efficient self-assessment tool that companies can use to improve the strategic planning process in order to achieve business excellence. Since 1996 the model has been called the *European model for TQM and business excellence*.

It is not the aim of this chapter to explain the detailed logic behind the model in Figure 2; the model closely resembles the Malcolm Baldridge Quality Award model that was launched in 1988. The model signals very clearly to its user that if you want good business results you have to understand their relationships to other results—people satisfaction, customer satisfaction, impact on society—and, of course, to the enablers. The model gives a good overview of How (enablers) you may get desired results (=What). How to use the model in a strategic planning process, monitored by Shewhart's and Deming's PDCA cycle, is explained in Section 3.

Comparing this model with the TQM pyramid of Figure 1, we recognize that both models have leadership as an important element. There are good reasons for that. Good leadership and strong management commitment have long been recognized as the most essential preconditions for any organization aspiring to be world class. As a result, much effort has been devoted to the pursuit of a "business excellence" approach to leading and managing an organization in order to achieve world class performance.

Combining the principles of the TQM pyramid with the principles (values) behind the European model for TQM business and excellence, we

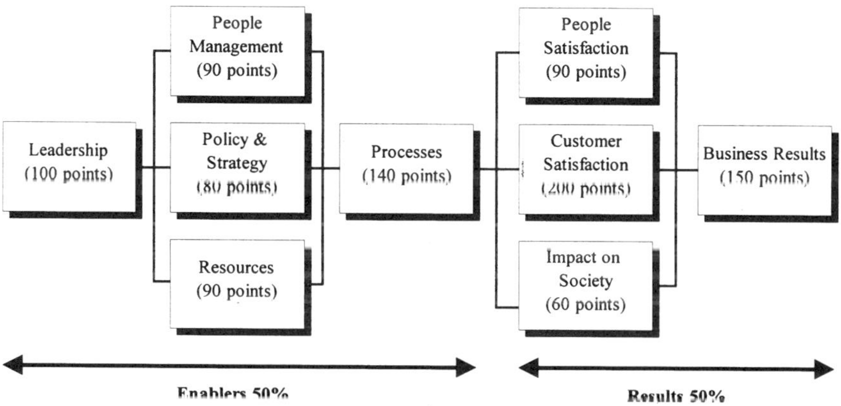

Figure 2 The European model for TQM and business excellence.

propose in this chapter that the fundamental principles of business excellence be taken to be the basic principles of total quality management supplemented by the principles of the learning organization and the creative organization.* The results are the following six principles for business excellence:

1. A focus on customers and their needs
2. Continuous improvement
3. The empowerment and participation of all staff members
4. A focus on facts
5. A commitment to creativity
6 A focus on continuous learning

A lot has been written about leadership and management's responsibilities for the implementation of these principles and the related concepts, but there has not been much concern about identifying the different leadership profiles in today's business world and their relations to the above principles and the success criteria for business excellence. If a manager's leadership profile does not correlate positively with the six principles listed above, then the manager may be a barrier to the implementation of TQM. In this case you obviously have only three options:

1. Fire the manager.
2. Forget TQM.
3. Educate the manager.

It is our belief that education of the manager is a feasible solution in most cases. For that purpose we have developed an integrated approach for management development that is based on quality function deployment (QFD; see Section 2). By applying the QFD technique to this area it is possible to gain information about the effect of different leadership profiles on the success criteria for business excellence. Without a profound understanding of this relationship, we cannot achieve business excellence.

The aims of this chapter are

1. To show an example of how statistical methods can be used to control and develop the softer parts of total quality management—the leadership styles (Section 2).

*Success criteria taken from the EQA business excellence model have been supplemented with success criteria from the creative and learning organizations because although creativity and learning are implicitly included in total quality management, theory on total quality management has to a certain degree neglected these two important disciplines. The aspect that unites all of the chosen success criteria is that they all demand a strong commitment from the senior management of an organization.

2. To provide an overview of the role and application of statistical methods in monitoring the implementation of total quality management to achieve business excellence (Section 3).

2. THE EUROPEAN EMPIRICAL STUDY ON LEADERSHIP STYLES

To achieve our first aim an empirical study was carried out that involved more than 200 leaders and managers of European companies and some 1200 of their employees. The format of the study was as follows.

1. Four hundred chief executive officers from France, Germany, Holland, Belgium, the United Kingdom, and Denmark were randomly selected from various European databases. The selection criteria were that they had to be from private companies (100% state-owned companies were excluded) with more than 50 employees.
2. The selected leaders were asked to complete an 86-point questionnaire* composed of two sections:
   a. 49 questions asking leaders to rate the importance of a number of aspects of modern business management†
   b. 37 questions asking leaders to rate the importance of a number of statements or success criteria on business excellence
3. By analyzing the material supplied by the leaders in response to the first 49 questions, it was possible to plot the "leadership profile" of each individual respondent. These leadership profiles are expressed in eight different leadership "styles".
4. The success criteria, which form the focus of the second section (37 questions), indicate the key leadership qualities required to achieve business excellence. The higher the leaders scored on these questions, the more they could be said to possess these qualities.

---

*The complete Leadership Profile questionnaire in fact consisted of 106 questions. The additional 20 questions covered cultural issues that do not form part of this chapter. The questions were developed by Geert Hofstede in 1994.

†The aspects of management were identified by a Danish focus group participating in a pilot version of this survey in 1995, developed by Anders Nørgaard and Heine Zahll Larsen. The focus group consisted of nine directors representing various areas of business, who were asked to identify the key attributes of a good business leader. The attributes so identified were classified on the basis of an affinity analysis, and as a result 49 variables were established. These variables could then be used to plot any individual leadership profile.

5. For each leader, 10 employees were also selected to participate in the survey. These employees were asked to rate the importance of the 49 management aspects, in order to give a picture of what the employees considered desirable for ideal leaders.

2.1. Description of the Leadership Model

The leadership model that was developed as the basis for this analysis is designed to shed light on the relationship between the business leadership styles of today's leaders and the requirements to achieve business excellence. By plotting the leadership profile of any individual leader, the model provides a tool to assess the extent to which he or she is working toward the successful achievement of business excellence.

Success Criteria

As described in Section 1, the success criteria for business excellence used in this research comprise three main elements—total quality management, creativity, and learning. However, since the interaction between an organization's leadership and its employees has a major impact on whether these criteria are achieved or not, this interaction becomes, in effect, a fourth success criterion.

As Figure 3 shows, the achievement of these success factors is affected by the leadership profiles of those in charge of the organization. Although

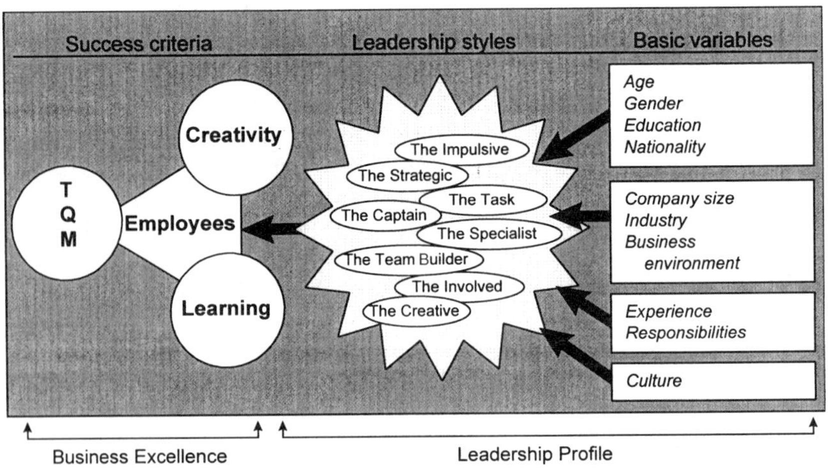

Figure 3 The leadership model.

not included within the scope of this chapter, it is reasonable to assume that these leadership profiles are in turn influenced by a number of "basic variables" such as leader's age, education, and experience and the size of the company or the sector in which it operates.

The First Success Criterion: Total Quality Management. Total quality management is regarded as the main criterion for business excellence. Focusing on achieving continuous improvements in an effort to enhance the company's strengths and eliminate its weaknesses, TQM covers all areas of the business, including its policies and strategies, its management of people, and its work processes. The core values of the total quality–oriented organization are a focus on the customer, the empowerment of its people, a focus on fact-based management, and a commitment to continuous improvement.

Since the European Quality Award (EQA) model is the most authoritative and most widely used method of assessing TQM in Europe, core aspects of this model have been used to determine the performance of the surveyed leaders with regard to the first success criterion. The higher the score the leaders achieved in this part of the questionnaire, the more positively they can be said to be working with total quality management.

The Second Success Criterion: Creativity. To achieve business excellence, organizations must also focus strongly on developing creativity. Urban* (1995, p. 56) has stated, "If all companies are high-quality and low-cost, creativity will be the differentiating factor."

Creativity is an important criterion for business excellence because it is a vital stimulus for improvement and innovation. It is a prerequisite for business excellence that an organization and its leaders be both committed to, and capable of, putting in place an organizational structure that fosters a creative environment. At the same time, they must be able to control and make use of that creativity. Since creative ideas do not just surface spontaneously, it is essential to implement a creative planning process. The creative organization aims to establish an effective basis for innovation and continuous improvement by adopting a systematic approach to the various aspects of creativity, such as the evaluation of ideas and procedures for communication.

For the purposes of this study, European leaders' performance with regard to this success criterion—the extent to which they are proactively working to generate and retain creativity—was assessed according to the theory of managing ideas set out by Simon Majaro.

*Glen L. Urban, Dean of the Sloan School of Management, Massachusetts Institute of Technology.

The Third Success Criterion: Learning. To quote Peter Senge (Senge, 1990, p. 4): "The organizations that excel will be those that discover how to tap their people's commitment and capacity to learn at all levels in an organisation."

The successful organization of the future will be a learning organization—one that has the ability to take on new ideas and adapt faster than its competitors. The model of the learning organization used for this study follows the five learning disciplines set out by Senge. These disciplines have therefore served as the basis for evaluating the European leaders' performance with respect to this third success criterion.

The Fourth Success Criterion: Leader–Employee Interaction. The three success criteria above all depend critically on the interaction between the leaders and their employees. For successful work with total quality management, learning, and creativity, it is important for leaders to get their subordinates "on board" and to harness their energies in the pursuit of these success criteria. A comparison of the views of the employees (through the profile they provided of their "ideal leader") with the actual performance of the leaders themselves was therefore used as a measurement of this interaction.

2.2. Leadership Styles

As described earlier, the answers the leaders provided to the questionnaire formed the basis of an assessment of them in terms of eight different leadership "styles." The eight leadership styles were identified by a factor analysis. The 49 questions regarding leadership capabilities were reduced to eight latent factors. It is essential to bear in mind that a leader is not defined simply as belonging to one or another of these styles but in terms of an overall profile that contains varying degrees of all eight. In other words, it is the relative predominance of some styles over others that determines the overall leadership profile of any given individual. The eight leadership styles are described in the following paragraphs.

The Captain

Key attributes: Commands respect and trust; leads from the front; is professionally competent, communicative, reliable, and fair.

The Captain is in many ways a "natural" leader. He commands the respect and trust of his employees and leads from the front. He has a confidence based on his own professional competence, and when a decision is made it is always carried out. He has an open relationship with his employees. He treats them all equally, is usually prepared to listen to their opinions, and usually ensures that information they require is communicated to them.

The Creative Leader

Key attributes: Is innovative, visionary, courageous, inspiring; has a strong sense of ego.

The Creative leader is full of ideas and is an active problem solver and a tireless seeker after continuous improvement. He has a clear image of the direction the company should pursue in the future. He is courageous and is willing to initiate new projects despite the risk of failure. He is a source of inspiration to his employees. He has a tendency to act on inspiration rather than on rational analysis and is driven by a strong sense of ego.

The Involved Leader

Key attributes: Shows empathy, practices a "hands-on" approach, does not delegate, focuses on procedures.

The Involved leader possesses good people skills, is well attuned to the mood of his staff, and takes time to listen to their problems and ideas. His close involvement with his employees gives him a good overview of the tasks they are working on. This level of involvement, however, makes it hard for him to delegate tasks rather than participate personally. He is focused on procedures and routines in teamwork and is consequently less well suited to take an overall leadership role.

The Task Leader

Key attributes: Is analytical, "bottom line"–driven, result-oriented, impersonal, persevering, intolerant of mistakes.

The Task leader believes success is measured by bottom-line financial results. Day-to-day business in the organization is carried out on the basis of impersonal, rational analysis. The Task leader is result-oriented and tends to be extremely persevering and determined once a course of action has been decided. The reliance on a rational attitude toward work and procedures means that this leader has difficulty accepting mistakes made by employees, with employee morale and performance consequently tending to suffer when they fail to meet the leader's expectations. The Task leader lacks personal skills when it comes to dealing with the problems or opinions of employees.

The Strategic Leader

Key attributes: Focuses on strategic goals, takes a holistic view of the organization, is a good planner, avoids day-to-day details, is process-oriented, trustworthy.

The Strategic leader has an overall view of the organization, focusing on longer term goals rather than day-to-day issues. This leader is process-oriented, believing that consistent work processes are essential for positive results. He is very efficient, setting clear objectives for what needs to be achieved. His comprehensive overview of the organization and his personal efficiency make him a highly trustworthy leader of his employees.

The Impulsive Leader

Key attributes: Obsessed with new ideas, unfocused, curious, energetic, participative.

The Impulsive leader's most salient characteristic is an obsession with new ideas combined with an unfocused energy. He is constantly "on fire" and lets nothing get in the way of his enthusiasm. As a result, he tends to take an interest in a wide range of issues and opportunities without necessarily having the capability to pursue the possibilities this process generates. In his fanaticism to push through his latest ideas, he tends to appear autocratic and domineering to his employees.

The Specialist Leader

Key attributes: Is expert, solitary, lacks inspirational ability, is resistant to change, calm.

The Specialist leader is an expert in his field who prefers to work alone. His leadership is expressed through the quality of his expertise rather than through any "people" skills. He is not good at teamwork, lacking the ability to inspire others and having a tendency to be pedantic and uncompromising. He appears calm, assured, and in control.

The Team Builder Leader

Key attributes: Is tolerant, gives feedback, acts as a coach, motivates, inspires, is supportive.

The Team Builder leader perceives himself primarily as a coach aiming to maximize the advantages of teamwork. He gives constructive feedback concerning his employees' work and behavior. He is also very tolerant and understands the need to support and inspire employees in critical situations.

2.3. The Relationship Between Success Criteria and Leadership Styles

The three success criteria and the eight leadership styles are estimated in this study to determine the precise demands that European leaders face when they seek business excellence. By estimating the relationships among the three success criteria and the eight leadership styles it is possible to isolate the leadership styles resulting in the greatest impact on the success criteria.

With the data of 202 European leaders we have been able to empirically prove that the Team Builder, the Captain, the Strategic, the Creative, and the Impulsive leadership styles all have a positive impact on one, two, or all three success criteria. The leadership styles are ranked according to their degree of influence on the success criteria. The more success criteria the leadership styles influence, the more important they are to achieving business excellence, i.e., the Team Builder is the most important one (impacts on three success criteria; see Fig. 4), whereas the Impulsive leader is the least important (impacts on one success criterion, Quality). The remaining leadership styles—the Involved, the Task, and the Specialist leaders—have no influence on achieving business excellence.

However, it is not enough to have knowledge of the correlation between the success criteria and the leadership styles. European leaders

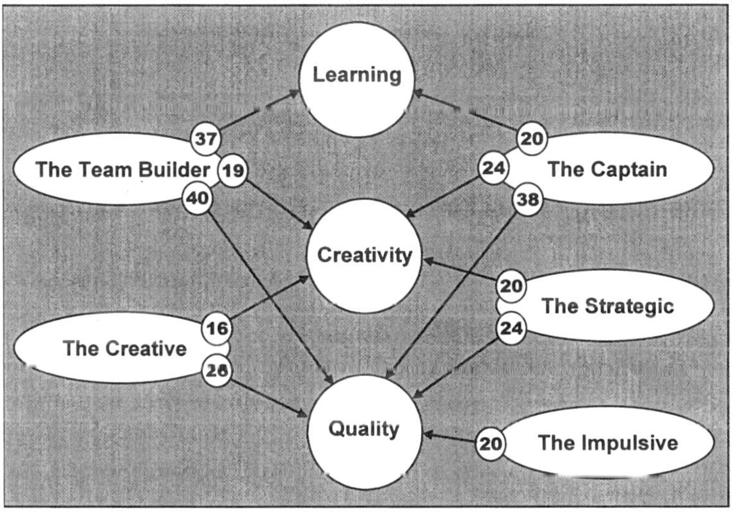

Figure 4 The correlation between success criteria and leadership styles. The numbers indicate the strength of the relationships.

must also take into consideration the Ideal Leadership profile outlined by the employees. By using quality function deployment (see Section 2.4) it is possible for managers to work with the demands of the employees.

2.4. Model for Measuring Excellent Leadership

An Excellent Leadership model should integrate the demands that the successful leader must consider when trying to achieve business excellence. The model should clarify what the leader should do to improve his performance as a leader in relation to the success criteria for achieving business excellence.

A *product improvement technique* called quality function deployment (QFD) is used as a tool for measurement of Excellent Leadership. The essence of this technique consists of combining a set of subjective variables, normally set out by the demands of customers, with a set of objective variables provided by the manufacturers' product developers. As a result of this interactive process a number of focus areas for developing high quality products become apparent, enabling manufacturers to adapt their products more precisely to customer demands.

Treating the leaders as "products" and the employees as "customers," QFD is used as a technique for determining Excellent Leadership. This is the reason for making the parallel between leaders and products. In QFD, the voice of the customer is used to develop the product. A leader has many "customers" such as employees and stakeholders. In this project, the employees are selected as our link to the customer part in QFD. This means that the voice of the employees will serve as an important guideline for leaders today in developing the right leadership qualities.

The information required for the QFD construction consists of

Employee demands of an ideal leader. The employees' Ideal Leader profile represents the customers' demands of the "product" in QFD.

The relationship between success criteria for achieving business excellence.

The relationship between success criteria and different leadership styles.

The individual leader's score on the success criteria and leadership styles.

Information about the "best in class" leaders within the areas of performance (quality, learning, and creativity).

The QFD technique provides the possibility to work with the following aspects:

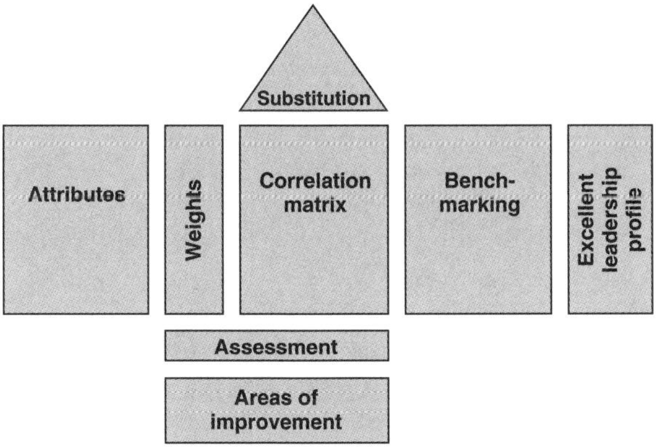

Figure 5 The Excellent Leadership model.

1. Assessment of the leader's performance on the success criteria.
2. Benchmarking—a comparison with "best in class"—leaders that have the highest score on the success criteria.
3. Estimation of an Excellent Leadership profile (ELP). The ELP is used to evaluate whether or not any leader matches the requirements for achieving business excellence.

The integrated QFD model is described below and is referred to hereafter as the **Excellent Leadership model** (Fig. 5). The description provides an explanation of the model but does not explain its full potential. Only the relevant parts of the model's matrix are explained in order to clarify how QFD can be used in this specific managerial perspective.

The QFD technique consists of a number of different matrices (collections of large numbers of quantifiable data), which makes the technique systematic and rational. Using each matrix as a foundation for analyzing the empirical data on European leaders makes it possible to work with the data in an easy and understandable way. Each of the matrices in Figure 5 is discussed in the following subsections.

Attributes—Leadership Styles

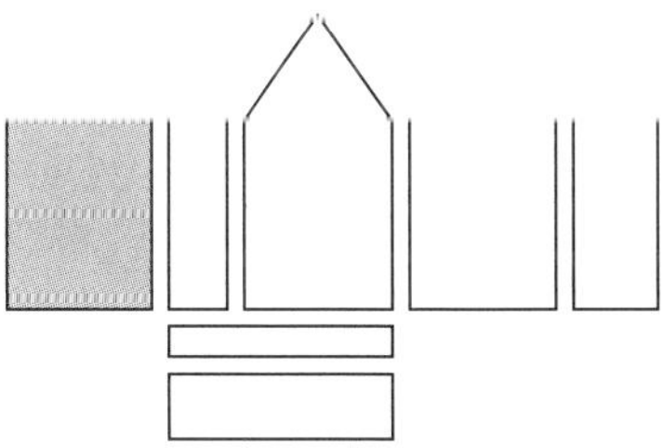

The attributes matrix (far left in Fig. 5) includes the different attributes of leadership. Eight leadership styles have been identified in relation to this study. As explained earlier, the eight leadership styles were created on the basis of rating the importance of 49 aspects of modern business management. However, in keeping a general view it is evident that the focus is on the eight latent leadership factors. Furthermore, the development of future leadership is based on different leadership styles, so it is not essential to have a high degree of detail until a later stage.

Weights—A Rating of the Eight Leadership Styles

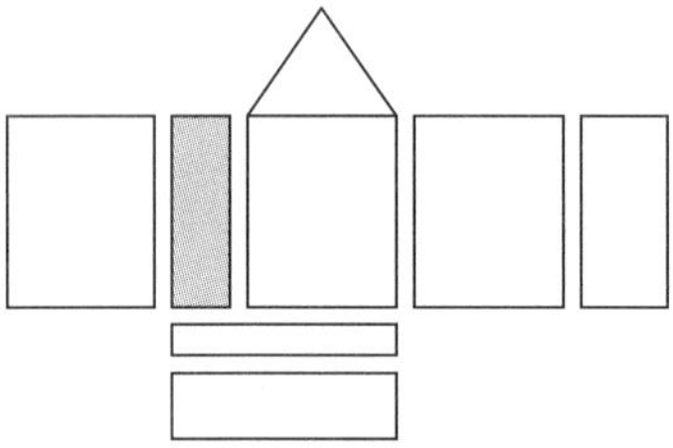

The 1150 employees who participated in the survey also evaluated the importance of the 49 aspects of modern business management under consideration to their concept of an ideal leader. This employees' Ideal Leader profile provides a rating or a weight of importance for each of the eight leadership styles. With this information the leader can identify possible areas of improvement in meeting employee demands for an ideal leader.

Correlation Matrix

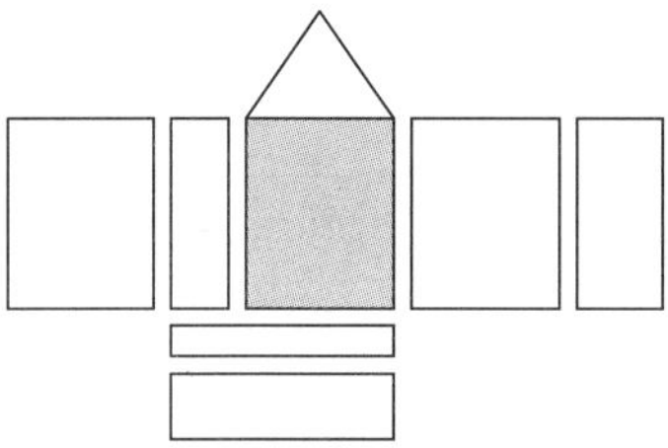

The correlation matrix is the heart of the Excellent Leadership model. In this part of the model the correlation between the individual leader's profile and the employees' Ideal Leader profile is estimated. Correlating the three success criteria with the eight leadership styles yields a picture illustrating the effects that each of the individual leadership styles has on the success criteria for achieving business excellence.

Substitution

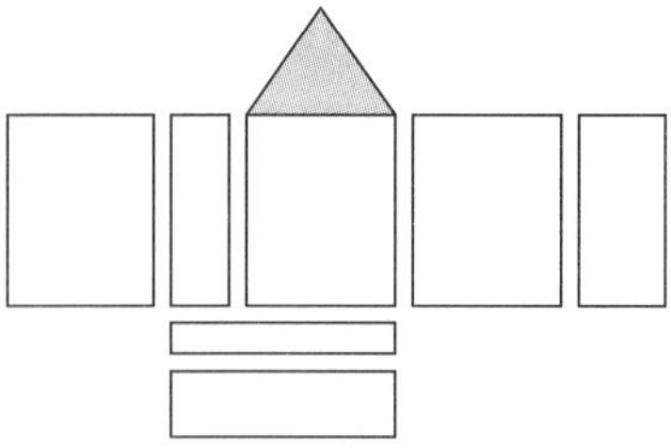

The roof of the QFD house, (Fig. 5) consists of a correlation matrix that illustrates the correlation between the three success criteria. This part of the model is relevant in determining potential substitution opportunities between the criteria. Only three criteria are included in this project, which gives only limited information on substitution. Using the 37 elements of the success criteria might make it possible to come up with a more differentiated view of substitution between the elements.

Assessment

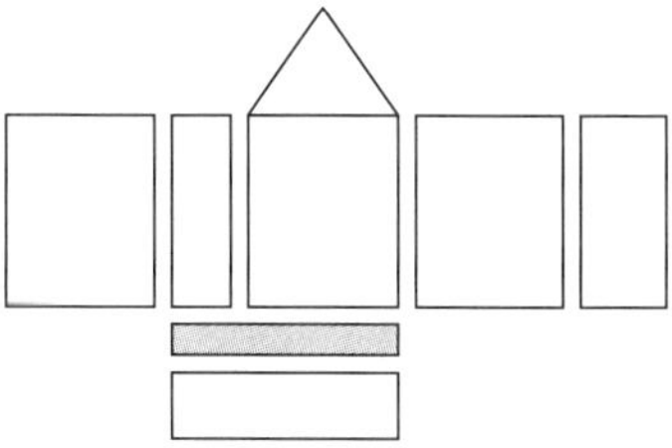

The leader's performance is measured on the basis of the three success criteria. This assessment is carried out by means of a self-evaluation, during which the leader answers 37 questions. The answers to these questions indicate the leader's and/or organization's level of activity on the success criteria (quality, learning, and creativity), for achieving business excellence, illustrated by an individual score. This assessment provides the leaders with a score of their current performance and critical areas in which further allocation of resources is required for the development of business excellence. It is important to have knowledge of one's current level if one is to set relevant objectives for the future. The three successive critieria should be individually evaluated. A global approach is required, as they are strongly correlated

Benchmarking

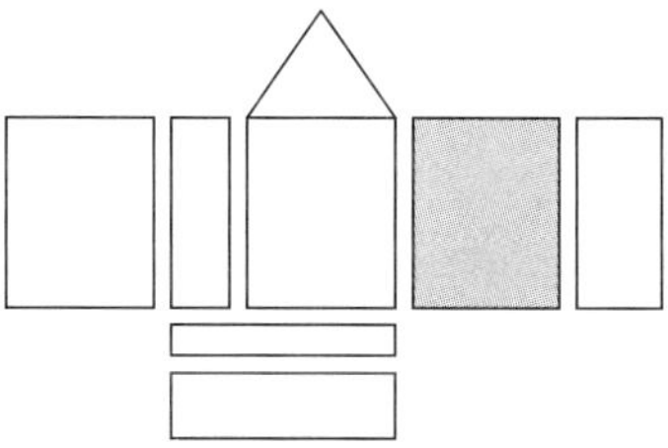

The right-hand side of the model illustrates the profiles for "best in class" within the three success criteria. These profiles can be used as a benchmark against "best in industry," which can generate new ideas for improvement. These profiles serve as a foundation for the Excellent Leadership profile, which takes into account the three success criteria and employees' demands of an ideal leader.

Areas of Improvement

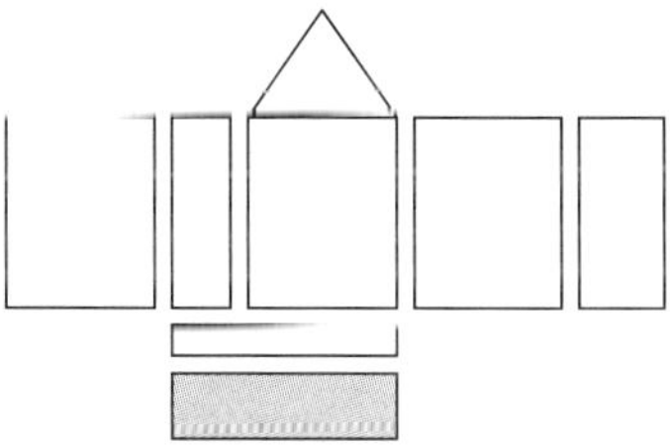

The bottom matrix in Figure 5 illustrates the "result" of the process. Multiplying the weights of the employees with the relationships between the leadership styles and the three success criteria creates this end product. Taking the view of the employees, the areas of improve-ment for the leader can be identified. In other words, the leader is provided with concrete ideas of ways in which the respective areas of improvement are weighted according to employee demands.

Excellent Leadership Profile

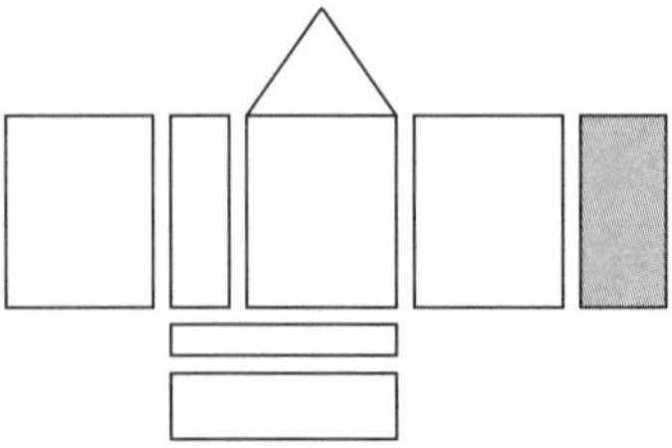

The Excellent Leadership profile (also known as the Success Profile), serves as a benchmark for the leaders. It is the ideal leadership profile if the leader wants to succeed in managing quality, learning, and creativity. In this project the overall objective was to create one profile of an excellent leader working actively with the management disciplines included in the success criteria. From this perspective this matrix at the far right in Figure 5, is considered the most important one in our use of QFD.

The QFD technique has served as the basis for our research and resulted in the identification of the Excellent Leadership profile. The five crucial *drivers* (leadership styles) for achieving excellent leadership were identified by a factor analysis. By correlating leadership styles with success criteria for business excellence it was possible to identify the styles most positively correlated to business excellence. Expanding the theoretical foundation, as seen in this chapter to treat the empirical data on European leaders with QFD and thereby take into consideration "employees' ideality" has resulted in a more accurate picture of the true *drivers* in the achievement of business excellence.

The Excellent Leadership profile shown in the rightmost matrix in the QFD-model can be benchmarked against any segment or group of leaders, i.e., leaders from different countries or sectors, of different ages, and so on. Two segments have been selected for further analysis:

1. European leaders' leadership profile versus the Excellent Leadership profile.
2. Country-by-country comparison of European leaders' leadership profile.

2.5. The Excellent Leadership Profile

In order to evaluate whether or not a leader is equipped to lead an organization to business excellence, a benchmark Excellent Leadership profile (ELP) must be developed. This illustrates the leadership profile that is best oriented toward the achievement of all three of the main business excellence success criteria.

The leadership profile benchmark is based on three groups of leaders, the 20 leaders who scored highest on creativity, quality, and learning. It is then used to develop the Excellent Leadership profile.

A Note on the Leadership Profile Graphs Used in this Study

1. The eight leadership styles that make up the leadership profiles are measured on a scale of 0 to 100 (vertical axis of Fig. 6).
2. Scores above or below 50 points represent deviations from the average of each leadership style.
3. The closer a leader gets to 100, the more strongly his or her leadership profile is characterized by the elements identified in the description of that particular leadership style.
4. Conversely, the further a score falls below 50, the less applicable those elements are as a description of the leader's profile.

As Figure 6 illustrates, two leadership styles have the predominant influence within the Excellence Leadership profile—the Strategic and the Task.

The Strategic is clearly the most important leadership style when it comes to identifying the characteristics required of a leader seeking

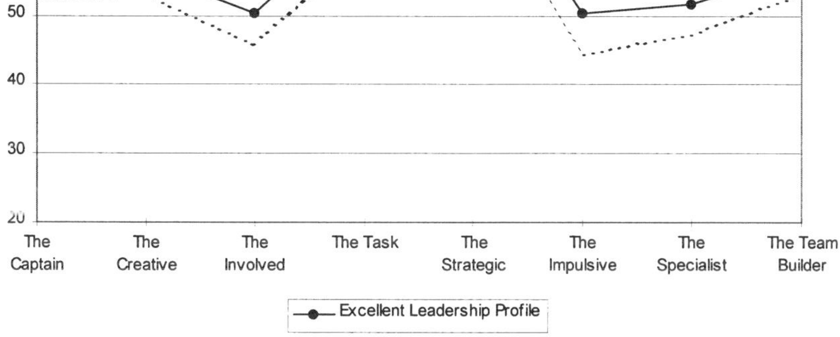

Figure 6 The Excellent Leadership profile. Dotted lines represent the band of deviation from the excellent leadership profile.

business excellence. The competencies of the Strategic leader are therefore ones that any leader hoping to achieve business excellence must continuously develop. This means that an overall view of the business is essential, with practical details of daily work not being allowed to prevent a focus on strategic goals or get in the way of setting clear organizational objectives.

The strong presence of the Task leadership style within the ELP underlines the fact that a highly developed analytical capability and an extremely result-oriented approach are both necessary for the achievement of business excellence. The Captain, the Creative, and the Team Builder styles also play an important part in achieving business excellence. The ELP confidence interval is above 50, and these styles are therefore important to the Excellent Leadership profile.

Compared to the results in Figure 4 it may seem surprising that the Task leadership style has such a strong weight in the Excellent Leadership profile. The explanation for that is that our benchmarks consisted of the 20 leaders who had the highest scores on quality, creativity, and learning. A characteristic of those leaders was that they also showed a relatively high score on the questions that correlated positively with the Task leadership style.

The remaining three styles—the Involved, the Impulsive, and the Specialist—are not regarded as important in the context of the Excellent Leadership profile. As can be seen from Figure 6, they are all broadly "neutral," reaching a score around average. This does not mean that they can be safely disregarded, however, since a score below average (i.e., below 50) would certainly represent a deviation from the ELP. In other words, while leaders need to strive actively to achieve the Strategic and Task leadership competencies and also the softer leadership attributes of the Creative, the Team Builder, and the Captain, they should not ignore the other leadership styles or seek to eliminate them from their profile altogether.

2.6. European Leaders Versus the Excellent Leadership Profile

In Figure 7 two profiles are illustrated: the Excellent Leadership profile interval (dotted lines) and the European leaders' profile (bold line), the latter being the average profile of all 202 European leaders participating in the study. The graphs show that there are large deviations between the European leaders' profile and the ELP on two leadership styles: the Captain and the Strategic.

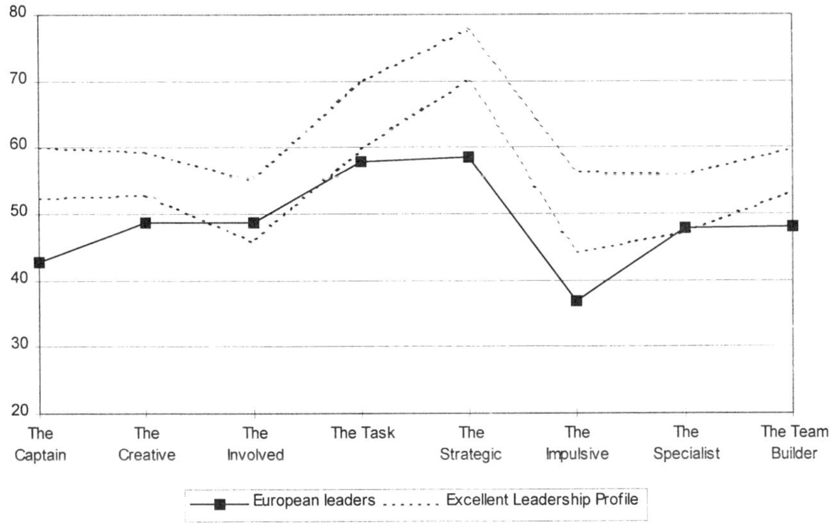

Figure 7 European leaders versus Excellent Leadership profile.

The Strategic:

1. A score of almost 60 indicates that European leaders do place importance on the skills of the Strategic leader and put them into practice, by taking a long-term view of the company and its direction, setting clear objectives, and being focused on maintaining consistent work processes.
2. They need to develop these competencies even further, however, if they wish to match the ELP.
3. The significant deviation between the leaders' actual performance and the requirements of the ELP is of considerable importance, given that the Strategic leadership style is the most crucial element of the ELP.

The Captain:

1. The European leaders' low score on the Captain style category indicates that they are not "natural" leaders. At best, they learn leadership skills as they grow into their assignment.
2. The below 50 score indicates that these leaders are not strongly characterized by the competencies of this particular leadership style—providing leadership from the front, encouraging open communication, and commanding the respect and trust of employees.

3. Although the Captain is not as crucial to the overall ELP as, for example, the Strategic leadership style, the deviation here is still an important one in terms of providing the balance of leadership styles that is needed to achieve business excellence.

2.7. European Leaders Versus Employees' Ideality

The employees' Ideal Leadership profile embodies the preferences expressed by the 1150 employees who participated in the survey. Direct subordinates to chief executives and managing directors were asked to use their answers to the first 49 questions of the survey to describe their "ideal" leader—someone for whom they would be willing to make an extra effort in their work. Comparing the leaders' profile with the employees' Ideal Leadership profile shows whether the employees are in harmony with the leader for achieving business excellence and where they are in conflict.

Figure 8 highlights four main areas of leadership where European employees' expectations differ significantly from the actual performances of the leaders: the Captain, the Creative leader, the Involved leader, and the Specialist leader. (A difference of 10 points or more is significant). The two styles positively correlated to achieving business excellence are included in the analysis.

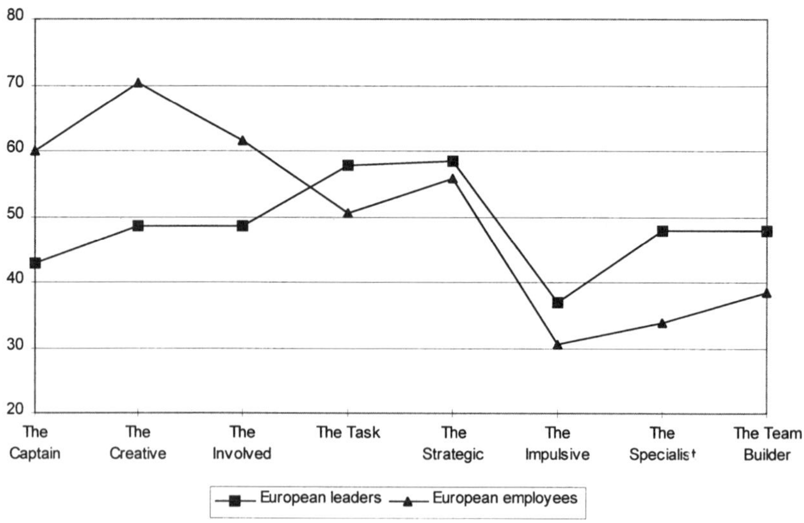

Figure 8 European leaders versus employees' ideality.

The Captain

Figure 8 shows a difference of approximately 18 points between employees' expectations and actual performance in the Captain category.

The European leaders' low score in the Captain style category indicates that they are not "natural" leaders. At best, they learn leadership skills as they grow into their assignment.

The below 50 score indicates that the leaders are not strongly characterized by the competencies of this particular leadership style—providing leadership from the front, encouraging open communication, and commanding the respect and trust of employees.

Employees place a much greater value on the leadership characteristics of the Captain than their leaders do.

The employees' score of 60 indicates that they react positively to a strong "natural" leader who can guide them and to whom they can look with respect, and that they appreciate the continuous flow of information provided by the Captain.

The Creative

Figure 8 indicates a difference of approximately 22 points between actual leadership performance and employee expectations in the creative style category.

The Creative style is the leadership style showing the most significant difference, with employees rating the Creative attributes very highly, at a score of 70, while their leaders score below 50.

The high score (70) indicates that, in contrast with the Strategic and Task styles, European employees place a high value on leaders who are characterized by the Creative leadership competencies.

The employees show a strong preference for a creative, inspiring, and courageous leader, scoring higher on this leadership style than on the other seven. This translates into a strong demand among European employees for a leader of vision and innovation who is prepared to deal with the increasing complexity of the business environment and who sees creativity and continuous improvement as the keys to success. European employees seek a leader who acts as a source of inspiration, motivating the workforce and taking courageous business decisions. These expectations, however, are significantly above the requirements their leaders need to meet in order to achieve business excellence.

Comments on the Specialist

There was a difference of approximately 15 points between the leaders' profile and the employees' ideality profile.

The employees' low score on the Specialist leadership style (below 35) can be seen as the mirror image of the high value they place on the Captain and Creative styles. The solitary nature of the Specialist leader, his lack of "people" skills and ability to inspire, are the direct antithesis of the Captain's and the Creative leader's attributes. The Specialist style of leadership is clearly not appreciated or regarded by employees as being of great value.

European leaders, whose Specialist score was significantly above the employee rating for that style, place a greater value on this leadership style than their employees do.

2.8. Conclusions

In seeking to achieve business excellence, European leaders may encounter resistance among their employees. Of crucial significance in this regard is the fact that European employees place a markedly lower value on the Team Builder and Strategic competencies than is required for business excellence. By contrast, their "ideal" leader is heavily characterized as being creative, inspiring, and an active problem solver.

The clear findings from this research study were that the five crucial drivers of business excellence are the Team Builder, the Captain, the Strategic, the Creative, and the Impulsive leadership styles (Fig. 4). Leaders trying to achieve business excellence must therefore view the high-level attainment of these sets of leadership competencies as their paramount objective.

It is important to remember, however, that this must not be done at the cost of neglecting other leadership competencies. As the Excellent Leadership profile demonstrates, the other leadership styles may be of less importance to achieving business excellence than the five leadership styles mentioned above, but this does not mean that they should be neglected altogether. The overall balance of the ELP requires the other leadership styles to be maintained at levels within the ELP interval. Maintaining a certain focus on these competencies is therefore still an important aspect of excellent leadership.

3. MONITORING THE IMPLEMENTATION OF THE SUCCESS CRITERIA FOR BUSINESS EXCELLENCE

Section 2 showed how it is possible to measure and hence to understand the softer parts of TQM (the intangibles). Remember that Dr. Deming talked about "the most important numbers being unknown and unknowable," i.e., measures of the qualitative world. Section 2 shows an example of how it is possible to measure the mind-set of people by using statistical methods. This section gives an overview on how to monitor and improve tangibles (things, processes, etc.) as well as intangibles. Business excellence can be achieved only if continuous improvements are focused on both areas. Such a focus is an important element of the leadership part of Figures 1 and 2.

3.1. The Plan–Do–Check–Action Cycle for Business Excellence

The problem with leadership is that most managers are confused about how to practice leadership. They need one or more simple models from which they can learn what their main leadership tasks are and how to integrate those tasks in the strategic planning process, a process that generates the yearly business plan and also longer term plan for the company (3–5 year plan) each year. The European model for business excellence may help managers to solve that problem. Both the yearly business plan and the long-term strategic plan can be designed by using the nine criteria of the model; i.e., the plan should comprise the result criteria of the model (*what* you want to achieve) and the enabler criteria as well (*how* you decide to work, i.e., *how* you plan to use intangibles). Figure 9 gives an overview of this Plan–Do–Check-Action (PDCA) approach.

It is seen from Figure 9 that action consists of a yearly self-assessment of *what* you have achieved and *how* you achieved the results. Such a yearly self-assessment is invaluable as input to the next year's strategic planning process.

During the year the plan is implemented with the help of people in the company's processes, and the results on people satisfaction, customer satisfaction, impact on society, and business results come in. This implementation may be visualized as a deployment of the plan to the Do and Check levels as shown in Figure 10.

Figures 9 and 10 give the guidelines or the overall framework for finding a way to business excellence. The guidelines are monitored by the PDCA cycle in which Study and Learn (Check) constitute the crucial precondition for continuous improvement of the strategic planning process. The framework has been linked to the European model for TQM and business excellence.

As was pointed out in Section 2, the European model for business excellence is not explicit enough on creativity and learning. For that reason, and also because companies outside Europe may wish to apply other models (e.g., the Malcolm Baldridge model), a more general model is proposed. We call the model the PDCA-leadership cycle for business excellence. This

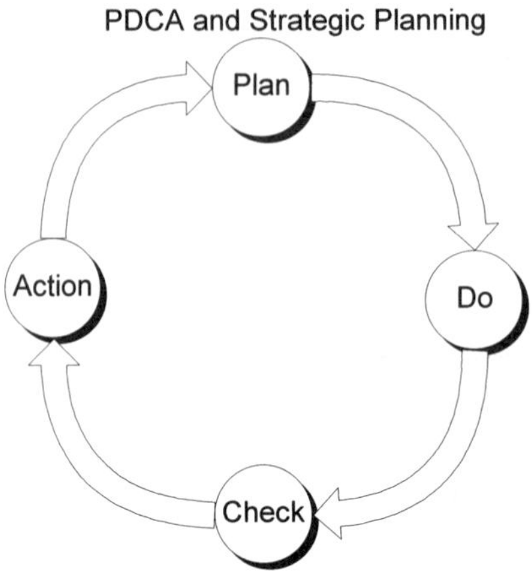

Plan:
How? 1. Leadership
2. People Management
3. Policy and Strategy
4. Resources
5. Processes
What? 6. People Satisfaction
7. Customer Satisfaction
8. Impact on Society
9. Business Results
Action: 10. Self-Assessment

Figure 9 The elements of Plan in relation to the yearly strategic planning process (items 1–10).

model, which contains the key leadership elements for business excellence, is shown in Figure 11.

It is seen from Figure 11 that the Plan component contains the vision, mission, and goals of the company together with the business plan, which contains goals for both tangibles and intangibles. In the Do phase the plan has to be deployed through policy deployment. Two other elements are crucial for an effective implementation of the business plan: (1) the leadership style of all managers and (2) education and training. The Check phase of the PDCA–leadership cycle comprises two elements: (1) Gaps between goals and results have to be identified, and (2) the gaps have to be studied for learning purposes. Once we understand why the gaps came up we are ready for Action. This phase should result in new ideas for improvement of people, processes, and products and new ideas for motivation of the people.

PDCA and Implementation

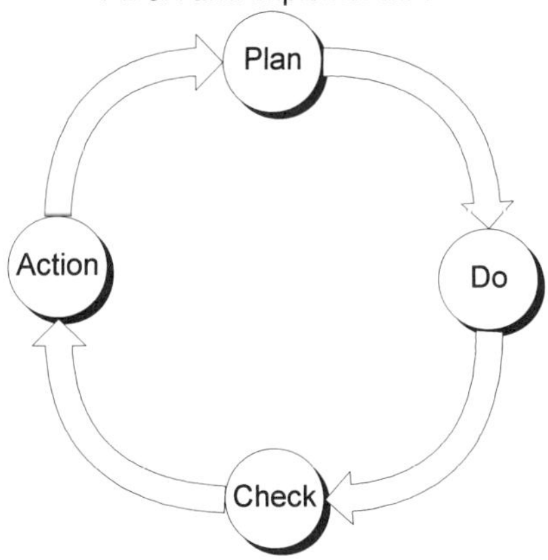

Plan: 1. Leadership
2. People Management
3. Policy and Strategy
4 Resources
Do: 5. Processes
Check: 6. People Satisfaction
7. Customer Satisfaction
8. Impact on Society
9 Business Results
Action: 10. Self-Assessment

Figure 10 Deployment of the plan to the Do and Check levels.

With this raw material the company has strong input for the next PDCA-leadership cycle for business excellence.

Let us look more specifically at education and training in the Do phase.

3.2. Education and Training for Business Excellence

The overall purpose of education and training is to build quality into people so that it is possible to practice real empowerment for business excellence. This can be achieved only if education and training are part of an overall leadership process where improvements in both tangibles and intangibles support each other as natural elements of the strategic planning process. Tangible world class results are evidence of business excellence, but the precondition for the tangible results are the intangible results such as recognition, achievement, and self-realization. The intangible results are a pre-

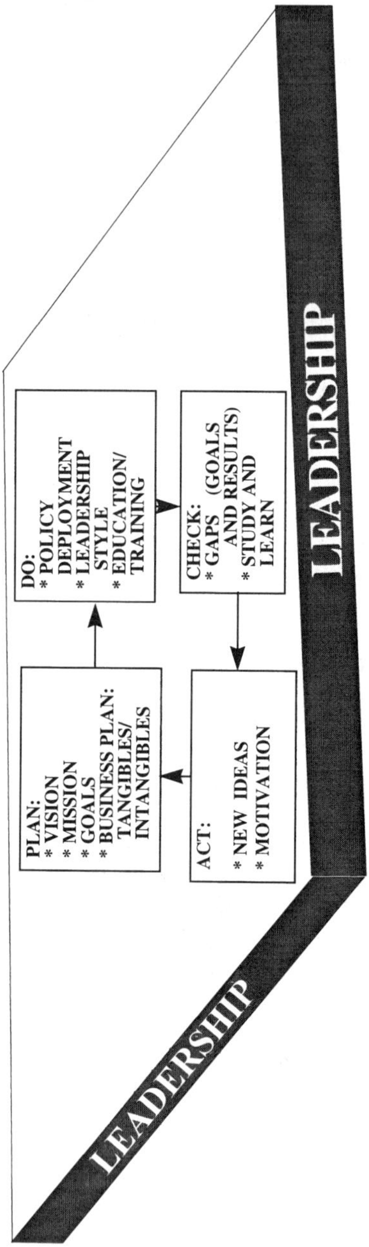

Figure 11 The PDCA-leadership cycle for business excellence.

condition for building values into the processes, i.e., value building of intangible processes, which again will improve the tangible results. Figure 12 shows how this process is guided by the principles of the TQM pyramid supplemented by education and training.

If we look at Education and Training (Fig. 12), we see that it forms the foundation of a temple and that its aim, quality of people, is the roof of the temple. The pillars of the temple are the main elements of Education and Training: (1) learning, (2) creativity, and (3) team building. Training in team building is a necessary element to support and complement creativity and learning. The importance of team building was also clearly demonstrated in Section 2 of this chapter (see Figs. 4 and 7).

The main elements of the three pillars are shown in Figures 13–15. It is seen that the elements of each pillar are subdivided into a logic part and a nonlogic part. The logic part of each pillar contains the tools to be used for improvement of tangibles (things, processes, etc.), and the nonlogic part contains the models, principles, and disciplines that are needed to improve intangibles such as the mind-set of people (mental models, etc.). Learning and applying the tools from the logic part of the three pillars may also gradually have an indirect positive effect on intangibles.

Most of the methods presented in this volume are related to the logic part of the three pillars. To build quality into people and to achieve business excellence, logic is not enough. Education and training should also comprise the nonlogic part of the pillars, which is a precondition for effective utilization of the well-known logical tools for continuous improvement. It is a common learning point of world class companies that managers are the most important teachers and coaches of their employees. That is the main reason why education and training are integrated in the PDCA-leadership cycle for business excellence.

4. CONCLUSION

In this chapter the role of statistical methods in monitoring the implementation of TQM and business excellence has been discussed. It has been argued that in order to achieve business excellence it is necessary to continuously improve tangibles (things, processes) as well as intangibles (e.g., the mind-set of people). Improving the mind-set of people is the same as building quality into people. Improvement of tangibles requires education and training on the well-known statistical tools such as statistical process control. Improvement of intangibles requires education and training on nonlogical models, principles, and disciplines. Both types of education and training are needed to achieve business excellence.

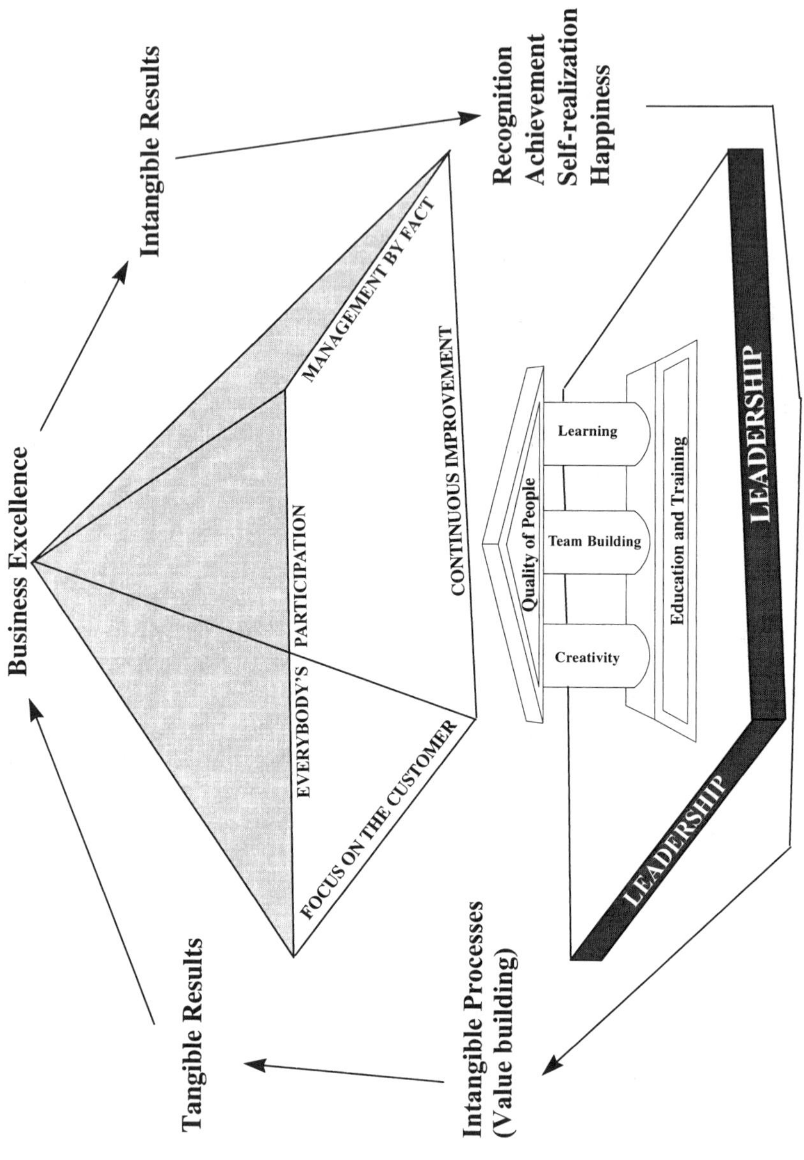

Figure 12 The continuous improvement process for business excellence.

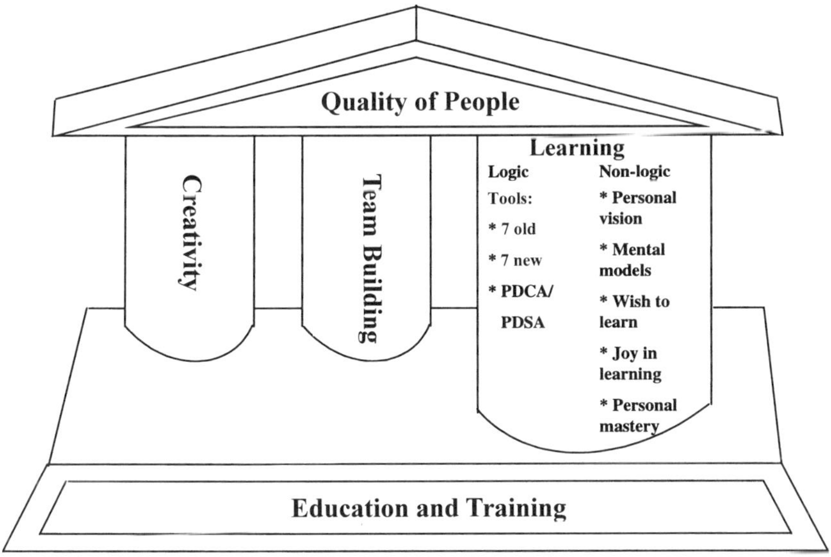

Figure 13 The logic and nonlogic parts of Learning in Education and Training.

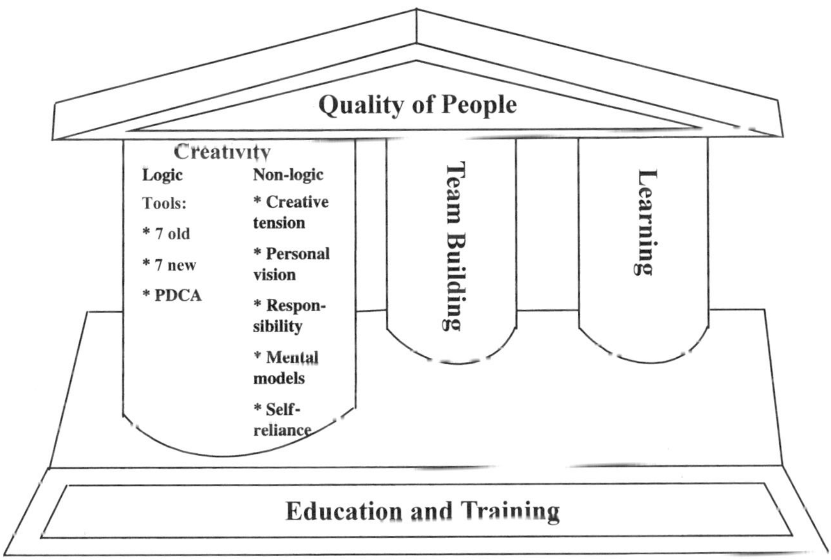

Figure 14 The logic and nonlogic parts of Creativity in Education and Training.

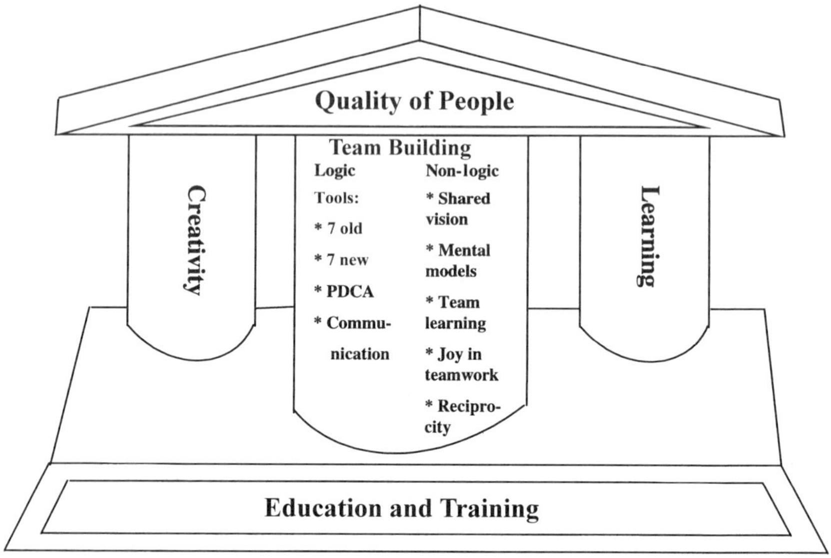

Figure 15 The logic and nonlogic parts of Team Building in Education and Training.

Section 2 showed how it is possible to use statistical methods to understand and improve the soft part of TQM implementation: leadership styles. Without understanding the effects of leadership styles it is impossible to practice effective leadership. As leadership is both the foundation of the TQM pyramid in Figure 1 and the first enabler criterion of the European model for business excellence (see Fig. 2), it is obvious that the first step on the journey to business excellence should be to try to assess and improve the different leadership styles of the company's managers.

Section 3 showed how leadership can be practiced and monitored by a simple PDCA leadership cycle. It was shown that in this cycle the implementation of the company's business plan is accomplished by people working in the different processes that are running day by day. These people need education and training in the well-known statistical tools for improvement of tangibles (things, processes, etc.) as well as education and training in models, principles, and disciplines for improvement of intangibles (the mind-set). It was argued that the company's business plan should contain improvement goals for tangibles as well as intangibles. Only in this way can business excellence be achieved.

RECOMMENDED READINGS

Dahlgaard JJ, Kristensen K. Vejen til Kvalitet (in Danish). Centrum, Aarhus, Denmark, 1992.

Dahlgaard JJ, Kristensen K, Kanji GK. The Quality Journey—A Journey without an End. Carfax, Abingdon, UK, 1994, and Productivity Press, India, 1996.

Dahlgaard JJ, Kristensen K, Kanji GK. Total Quality Leadership and Employee Involvement (The TQM pyramid), Proceedings of ICQ'96, Yokohama, JUSE, 1996.

Dahlgaard JJ, Kristensen K. Milliken Denmark. A case study included in the European Way to Excellence Case Study Project, EU DGIII and EFQM, 1997.

Dahlgaard JJ, Kristensen K, Kanji GK. The Fundamentals of Total Quality Management. Chapman & Hall, London, UK, 1997.

Park Dahlgaard SM, Kondo Y. Quality Motivation (in Danish). Centrum, Aarhus, Denmark, 1994.

Fritz R. Corporate Tides: Redesigning the Organization. Butterworth-Heinemann, Oxford, UK, 1994.

Imai M. Kaizen—The key to Japan's competitive success. Kaizen Institute, Tokyo, Japan, 1986.

Kanji GK, Asher M. Total Quality Management Process—A Systematic Approach. Carfax, Abingdon, UK, 1993.

Kanji GK. Quality and statistical concepts. In: Kanji GK, ed. Proceedings of the First World Congress, Sheffield, UK. Chapman & Hall, Cornwall, UK, 1995.

Marjaro S. Managing Ideas for Profit: The Creative Gap. McGraw-Hill, London, UK, 1992.

Morgan A. Imaginization: The Art of Creative Management. Sage Publ., London, UK, 1986.

Senge PM. The Fifth Discipline—The Arts & Practice of the Learning Organization Doubleday Currency, New York, 1990.

Urban GL. A second industrial revolution. Across the Board 32(2), 1995.

5

A Methodological Approach to the Integration of SPC and EPC in Discrete Manufacturing Processes

Enrique Del Castillo
Pennsylvania State University, University Park, Pennsylvania

Rainer Göb and Elart von Collani
University of Wuerzburg, Wuerzburg, Germany

1. INTRODUCTION

The control of industrial manufacturing processes has long been considered from two different points of view. *Statistical process control* (SPC), which traces back to the work of Walter Shewhart in the 1920s, was originally developed for discrete manufacturing industries (industries concerned with the production of discrete items). On the other hand, continuous process industries, chemical industries for instance, used various forms of adjustment strategies administered by automatic controllers. This type of process control became known as *engineering process control* (EPC) or *automatic process control* (APC). Separately, both approaches have received enormous interest in the academic literature.

Interest in SPC and EPC integration originated in the 1950s in the chemical industries. Part of this interest is due to the inertial elements in this type of production process (e.g., raw materials with drifting properties) that result in autocorrelated quality characteristics of the end product. Traditional SPC methods assume instead i.i.d. quality characteristics, and problems of a high number of false alarms and the difficulty in detecting process shifts occur under (positive) autocorrelation at low lags. If the autocorrelation structure can be modeled and a compensatory variable can be found to modify the quality characteristic, then an EPC scheme is put into place to compensate for such drifting behavior. However, abrupt, large shifts in the quality characteristic indicate major failures or errors in the process that cannot generally be compensated for by the EPC controller. For this reason, many authors have suggested that an SPC chart be added at the output of an EPC-controlled production process to detect large shifts. There is no clear methodology, however, that models such integration efforts in a formal and general way.

In contrast, interest in SPC–EPC integration in discrete part manufacturing is more recent. In this type of production process, elements that induce autocorrelation are not common. However, drifting behavior of a process that "ages" is common. A typical example of this is a metal machining process in which the performance of the cutting tool deteriorates (in many cases, almost linearly) with time. Many years ago, when market competition was not so intense, specifications were wide enough for a production process to drift without producing a large proportion of nonconforming product. With increasing competition, quality specifications have become more rigorous, and drifting behavior, rather than being tolerated, is actively compensated for by simple EPC schemes.

Academic interest in the area of SPC–EPC integration has occurred as a natural reaction to the requirements of industrial practices. However, most of the approaches suggested during the discussion on this problem argued from the point of view of practical necessities alone. Proponents of either side admit that many control problems in modern manufacturing processes cannot be solved by either SPC or EPC alone. As a consequence, methods from each field are recommended as auxiliary tools in a scheme originally developed either for SPC or for EPC applications alone. None of these approaches have been really successful from a methodological point of view. The models used were originally designed for either proper SPC or EPC applications but not for an integration of the two. The practical necessity of an integrating approach to industrial control problems is obvious, but a rigorous mathematical model to reflect this need is still missing. As a reaction to this methodological gap, the present chapter establishes a simple model that integrates the positions of SPC and EPC.

2. MODELS PROPOSED IN THE LITERATURE FOR SPC–EPC INTEGRATION

Although diverse authors have discussed the different aims and strategies of SPC and EPC (e.g., Barnard, 1963; MacGregor, 1988, 1990; Box and Kramer, 1992; Montgomery et al., 1994), few specific models have been proposed for the integration of these fields. Among these models we find algorithmic statistical process control (ASPC) and run-to-run control procedures.

2.1. ASPC

Vander Weil et al. (1992) (see also Tucker et al., 1993) model the observed quality characteristic ξ_t of a batch polymerization process at time t as

$$\xi_t = \mu I_{t \geq t_0} + bx_{t-1} + \frac{1 - \theta\mathcal{B}}{1 - \phi\mathcal{B}}\, \varepsilon_t \tag{1}$$

where the first term on the right represents a shift of magnitude μ that occurs at time t_0, x_t is the compensatory variable, and the noise term is a stationary ARMA(1,1) stochastic process. In what the authors refer to as algorithmic statistical process control (ASPC), process shifts are monitored by a CUSUM chart, whereas the ARMA noise is actively compensated for by an EPC scheme. Using a similar approach, Montgomery et al. (1994) presented some simulation results. Clearly, ASPC is focused on continuous production processes.

A basic weakness of the APSC approach is that there is no explicit stochastic model for the time t_0 of shift occurrence.

2.2. Run-to-Run Process Control

Sachs et al. (1995) (see also Ingolfsson and Sachs, 1993) assume instead a simple linear regression model with no dynamic effects for controlling certain semiconductor manufacturing processes. The model is

$$\xi_t = \mu + bx_{t-1} + \varepsilon_t \tag{2}$$

By using a control chart on the residuals of model (2), called "generalized SPC" by the authors, their method applies two different types of EPC schemes: an Exponentially Weighted Moving Average (EWMA)-based controller if the observed deviation from target is "small" (called "gradual control" by the authors) and a Bayesian controller that determines the moment and magnitude of larger deviations in case they occur. Other authors (Butler and Stefani, 1994; Del Castillo and Hurwitz, 1997) extended model (2) to the case where deterministic trends and ARMA(1,1) noise exist.

A basic weakness of the run-to-run models is that the classical rationale of SPC applications is a shift, stochastic in time of occurrence and/or in magnitude. Again, this is not reflected by the run-to-run models.

3. BASIC FEATURES OF MODELS IN SPC AND EPC

3.1. Process Changes in SPC and EPC

Any approach to process control needs a model of *process changes*, i.e., a model for the changes in the process parameters (output mean, output variance, output proportion nonconformance) that occur throughout production time. On this topic, the traditional approaches to SPC and EPC differ significantly, corresponding to their origins in different types of industries.

The overwhelming majority of SPC models identify process changes as brupt shifts of the process parameters due to assignable causes, disturbances or shocks that affect the manufacturing facilities. These shift models generally share four basic assumptions:

1. The magnitude of shifts is large relative to the process noise variance.
2. Shifts are rare events; the period up to the occurrence of the first shift or between two successive shifts is large.
3. Shifts can result from a variety of assignable causes. Detection of a specific assignable cause requires expert engineering knowledge of the production process, and it is time-consuming and expensive.
4. Control actions to remove assignable causes of variation are time-consuming and expensive, requiring skilled staff, machinery, and material. These actions result in rearranging the process parameters to the "in-control" or target values, e.g., recentering the process mean. That is, these actions are *corrective* in nature.

Under these assumption, a constant automatic adjustment strategy (i.e., an EPC) obviously is not the appropriate remedy for process variation.

Engineering process control models that originate in continuous process industries consider process changes in the form of *continuous drifts*. In contrast to the shift models of SPC, the basic assumptions of the EPC approach are that

1. The drift is slow. Measured over a short time interval, the drift effect is small relative to the noise variance.
2. The drift permanently continues throughout the production time.
3. The drift is an inherent property of the production process. Expert engineering knowledge of the production process provides knowledge of the structure of the drift process. To a certain degree, the drift effect can be estimated and predicted.
4. The control actions taken to counteract the drift effect are minor in terms of time and expense. They follow a repetitive procedure

or algorithm that can be left to automatic controllers. These control actions have no effect on the process parameters but rather compensate for observed deviations of the quality characteristic. That is, the causes of the drifting behavior are not "corrected" but only *compensated* for.

Under these assumptions, constant automatic adjustments are a reasonable control strategy.

3.2. Open-Loop and Closed-Loop Behavior of a Process

The mathematical models used by SPC and EPC reflect different ideas about process changes and process control. We shall explain the differences and similarities for the simple situation of a process that, at successive discrete time points $0, 1, 2, \ldots$, produces output with a single quality characteristic $\xi_0, \xi_1, \xi_2, \ldots$. An essential aspect is the distinction between the *open-loop* behavior (behavior without control actions) and *closed-loop* behavior (behavior in the presence of control actions) of such a process.

In SPC, detection of an assignable cause and subsequent corrective action occurs only rarely. If it occurs, it amounts to a complete *renewal* of the manufacturing process. Hence it is useful to split the entire production run into the periods (*renewal cycles*) between two successive corrective actions (renewals) and to consider each renewal cycle along a separate time axis $0, 1, 2, \ldots$ with corresponding output quality characteristics $\xi_0, \xi_1, \xi_2, \ldots$. The effect of control actions is not reflected in the output model.

In standard EPC, control actions are taken regularly at each time point. Without these permanent compensatory actions the process would exhibit a completely different behavior. A model of the process behavior without control is indispensable for the design and evaluation of control rules. Thus we have the open-loop output quality characteristics $\xi'_0, \xi'_1, \xi'_2, \ldots$ of the process without control (left alone) and the closed-loop output quality characteristics $\xi_0, \xi_1, \xi_2, \ldots$ of the process subject to control actions.

3.3. Process Changes in SPC Models

Statistical process control is designed for manufacturing processes that exhibit discrete parameter shifts that occur at random time points. Thus in SPC models the most general form of the output process $(\xi'_t)_{\mathbb{N}_0}$ is the sum of a *marked point process* and a *white noise* component. This approach is expressed by the model

$$\xi_t' = \mu_t + \varepsilon_t \tag{3}$$

In this formula $(\mu_t)_{\mathbb{N}_0}$ is a marked point process,

$$\mu_t = \mu^* + \sum_{i=1}^{N_t} \delta_i \tag{4}$$

with a target μ^*, with marks $\delta_1, \delta_2, \ldots$ representing the sizes of shifts $1, 2, \ldots$ and a counting process $(N_t)_{\mathbb{N}_0}$ that gives the number of shifts in time interval $[0; t)$. $(\varepsilon_t)_{\mathbb{N}_0}$ in Eq. (3) is a white noise process independent of $(\mu_t)_{\mathbb{N}_0}$. The white noise property is expressed by

$$E[\varepsilon_t] = 0, \qquad V[\varepsilon_t] = \sigma^2, \qquad E[\varepsilon_t \varepsilon_{t+k}] = 0 \qquad \text{for all } t \in \mathbb{N}_0;\ k \in \mathbb{N} \tag{5}$$

A simple and popular instance of a marked point process is one with shifts occurring according to a *Poisson process* $(N_t)_{\mathbb{N}_0}$. Most investigations on control charts use further simplifications. For instance, deterministic absolute values $|\delta_i| = \Delta\,(\Delta > 0)$ of the shifts are frequently assumed. Many approaches assume a *single* shift of a given absolute value Δ that occurs after a random (often assumed to be exponentially distributed) time ν. In this case we have

$$\xi_t' = \begin{cases} \mu^* + \varepsilon_t & \text{if } t \le \nu \\ \mu^* + \gamma\Delta + \varepsilon_t & \text{if } t > \nu \end{cases} \tag{6}$$

where the random variable γ is the sign of the deviation from target with

$$P(\gamma = 1) = p, \qquad P(\gamma = -1) = 1 - p, \qquad p \in [0; 1]$$

In the case of one-sided shifts, we have $p = 0$ or $p = 1$; in the case of two-sided shifts, it is usually assumed that $p = 0.5$.

3.4. Process Changes in EPC Models

Engineering process control is designed for manufacturing processes that exhibit continuous parameter drifts. Some typical instances of open-loop output sequences $(\xi_t')_{\mathbb{N}_0}$ in EPC models are as follows.

ARMA Models

An important family of models used to characterize drifting behavior occurring due to autocorrelated data is the family of ARMA(p, q) models (Box and Jenkins, 1976):

$$\xi_t' = \phi_1 \xi_{t-1}' + \phi_2 \xi_{t-2}' + \cdots + \phi_p \xi_{t-p}' - \lambda_1 \varepsilon_{t-1} - \lambda_2 \varepsilon_{t-2} - \cdots - \lambda_q \varepsilon_{t-q} + \varepsilon_t \tag{7}$$

where $(\varepsilon_t)_{\mathbb{N}_0}$ is a white noise sequence [see Eq. (5)]. By introducing the backshift operator $\mathcal{B}^k f_t = f_{t-k}$, Eq. (7) can be written as

$$(1 - \phi_1 \mathcal{B} - \phi_2 \mathcal{B}^2 - \cdots - \phi_p \mathcal{B}^p)\xi_t' = (1 - \lambda_1 \mathcal{B} - \lambda_2 \mathcal{B}^2 - \cdots - \lambda_q \mathcal{B}^q)\varepsilon_t$$

or as

$$\xi_t' = \frac{\lambda_q(\mathcal{B})}{\phi_p(\mathcal{B})}\, \varepsilon_t$$

where $\lambda_q(\mathcal{B})$ and $\phi_p(\mathcal{B})$ are stable polynomials in $\mathcal{B}$. Sometimes, nonstationary ARIMA(p, d, q) models of the form

$$\xi_t' = \frac{\lambda_q(\mathcal{B})}{(1 - \mathcal{B})^d \phi_p(\mathcal{B})}\, \varepsilon_t$$

have been used instead to model drifting behavior in continuous production processes.

Deterministic Drift

If the drifting behavior is caused by aging of a tool (see, e.g., Quesenberry, 1988), a simple regression model of the form

$$\xi_t' = T + dt + \varepsilon_t \tag{8}$$

is sufficient to model most discrete manufacturing processes. Here, T is a target value and dt is a deterministic time trend.

Unit Root Trend

Alternatively, a "unit root" process can be used to model linear drifts by using

$$\xi'_t - \xi'_{t-1} = (1 - \mathcal{B})\xi'_t = d + \lambda_q(\mathcal{B})\varepsilon_t \tag{9}$$

For example, if $\lambda_q(\mathcal{B}) = 1$, then (9) is a random walk with drift d that has behavior similar to that given by (8) but with variance that increases linearly with time.

3.5. Common Structure of SPC and EPC Models

Analyzing the SPC models of Section 3.3 and the EPC models of Section 3.4, we can point out a common structure that is helpful in developing an approach for an integrating model.

We shall decompose the output ξ'_t into two components. One of these components is a function of the white noise variables ε_s alone and represents completely uncontrollable random variation. The other component, θ_t, represents the effective state of the process, which is subject to an inherent drift or to a shift due to an assignable cause. Formally, this means to consider the output equations (3), (7), (8), and (9) as special cases of the model

$$\xi'_t = \mathcal{F}_t(\theta_t, (\varepsilon_s)_{s \le t}) \tag{10}$$

In many cases θ_t is a deterministic function of t that coincides with the output mean, i.e., $E[\xi'_t] = \theta_t$. The argument $(\varepsilon_s)_{s \le t}$ is required to allow for possible cumulative effects of the white noise variables on the process output; see (13) or (14) below.

Let us rewrite the models (3), (7), (8), and (9) in terms of (10).

As to the shift model (3), let

$$\mathcal{F}_t(\theta_t, (\varepsilon_s)_{s \le t}) = \theta + \varepsilon_t \qquad \text{with} \qquad \theta_t = \mu^* + \sum_{i=1}^{N_t} \delta_i \tag{11}$$

For the deterministic drift model (8), let

$$\mathcal{F}_t(\theta_t, (\varepsilon_s)_{s \le t}) = \theta_t + \varepsilon_t \qquad \text{with} \qquad \theta_t = T + dt \tag{12}$$

For the random walk with drift model [see (9) with $\lambda_q(\mathcal{B}) = 1$], let

$$\mathcal{F}_t(\theta_t, (\varepsilon_s)_{s \le t}) = \theta_t + \varepsilon_t \qquad \text{with} \qquad \theta_t = \xi'_{t-1} + dt \tag{13}$$

Finally, for the ARMA(p, q) model (7), we can identify

$$\mathcal{F}_t(\theta_t, (\varepsilon_s)_{s \le t}) = \theta_t + \sum_{i=0}^{q} b_i \varepsilon_t \tag{14}$$

where $b_i = -\lambda_i \mathcal{B}^i$ and where we also identify

$$\theta_t = \sum_{i=0}^{p-1} a_i \xi_t' \tag{15}$$

with $a_i = -\phi_i \mathcal{B}^i$.

4. MODELING THE INTEGRATION OF SPC AND EPC

Simultaneous application of SPC and EPC procedures to the same manufacturing process makes sense only if the process exhibits both kinds of changes considered in Section 3: discrete and abrupt variation by shift, which represents the position of SPC, and continuous variation by drift, which represents the position of EPC. Consequently, an integrative model for SPC and EPC should contain components corresponding to the two types of process variation models given in Section 3: a marked point process component to justify the use of SPC (see Section 3.3) and a drift component to justify the use of EPC (see Section 3.4).

In view of the common structure of SPC and EPC models formulated by (10), a unifying model for SPC and EPC can be expressed by the following model for the uncontrolled (open-loop) process output ξ_t':

$$\xi_t' = \mathcal{F}_t\Big((\mu_s^{(1)})_{s \le t}, \ldots, (\mu_x^{(K)})_{s \le t}, (\vartheta_s^{(1)})_{s \le t}, \ldots, (\vartheta_s^{(M)})_{s \le t}, (\varepsilon_s)_{s \le t}\Big) \tag{16}$$

where $(\mu_s^{(1)})_{\mathbb{N}_0}, \ldots, (\mu_s^{(K)})_{\mathbb{N}_0}$ are K different marked point processes representing the effect of shifts to be treated by SPC [see Eq. (4)], $(\vartheta_s^{(1)})_{\mathbb{N}_0}, \ldots, (\vartheta_s^{(M)})_{\mathbb{N}_0}$ are M different drift processes that represent the effect of continuous drifts to be treated by EPC [see Eqs. (12), (13), and (15)], and $(\varepsilon_s)_{\mathbb{N}_0}$ is a white noise sequence [see Eq. (5)]. For some applications it is necessary to choose all past values $(\mu_s^{(i)})_{s \le t}, (\vartheta_s^{(j)})_{s \le t}, (\varepsilon_s)_{s \le t}$ as the arguments of the functions $\mathcal{F}_t$ to allow for possible cumulative effects of $\mu_t^{(i)}, \mu_{t-1}^{(i)}, \ldots, \vartheta_t^{(j)}, \vartheta_{t-1}^{(j)}, \ldots, \varepsilon_t, \varepsilon_{t-1}, \ldots$ on ξ_t' (see Section 4.2, random walk drift model).

Equation (16) gives a generic framework for a process model that integrates the positions of SPC and EPC. Let us now consider three important examples with one drift component, i.e., with $M = 1$.

4.1. Additive Disturbance

In many cases an abrupt shift can be modeled as a translation of the output value ξ_t'. To express this situation in the terms of model (16), we choose

$$\xi_t' = \mathcal{F}_t\big((\mu_s)_{s\le t}, (\vartheta_s)_{s\le t}, (\varepsilon_s)_{s\le t}\big) = \mu_t + \vartheta_t + \mathcal{G}_t\big((\varepsilon_s)_{s\le t}\big)$$

where $(\mu_s)_{\mathbb{N}_0}$ is a shift process of the type introduced by Eq. (4), $(\vartheta_s)_{\mathbb{N}_0}$ is a process that represents the effect of continuous drifts [see Eqs. (12), (13), and (15)], and $(\varepsilon_s)_{\mathbb{N}_0}$ is a white noise sequence [see Eq. (5)]. In many cases, we simply have $\mathcal{G}_t((\varepsilon_s)_{s\le t}) = \varepsilon_t$ [see Eqs. (11) and (12)]. For examples of functions $\mathcal{G}_t((\varepsilon_s)_{s\le t})$ that express a cumulative effect of the white noise variables, see Eq. (14).

4.2. Shift in Drift Parameters

Usually the models for drift processes $(\vartheta_t)_{\mathbb{N}_0}$ that are used in EPC depend on parameters. These parameters can be subject to shifts during production. Engineering controllers, however, are designed for fixed and known parameter values and cannot handle such sudden parameter shifts. Even *adaptive* EPC schemes have the fundamental assumption that the changes in the parameters are slow compared to the rate at which observations are taken (Åström and Wittenmark, 1989). Thus supplementary SPC schemes are required to detect these abrupt changes (Basseville and Nikiforov, 1993). Let us consider two simple models that will be investigated in some detail in Section 5.

Shift in Trend Parameter—Deterministic Drift Model

In the original parametric model, see (12), let the drift component $(\vartheta_t)_{\mathbb{N}_0}$ be described by a deterministic trend,

$$\vartheta_t = T + dt$$

i.e., by the recursion

$$\vartheta_t - \vartheta_{t-1} = d, \qquad \vartheta_0 = T \tag{17}$$

with a parameter d and a target value T. However, the drift parameter d may be subject to abrupt shifts, as may occur, e.g., when a cutting tool starts to fail. Thus the parameter value at time t should be considered as a random variable μ_t, where $(\mu_t)_{\mathbb{N}_0}$ is a marked point process as given by (4) with target $\mu^* = d$. Replacing d by μ_t in (17) we obtain the output equation

$$\xi_t' = T + \sum_{i=1}^{t} \mu_i + \varepsilon_t = \sum_{i=1}^{t} (\mu_i - d) + \vartheta_t + \varepsilon_t \tag{18}$$

In the scheme of Eq. (16) we have

$$K = 1 = M, \qquad \mathcal{F}_t\big((\mu_s)_{s\le t}, (\vartheta_s)_{s\le t}, (\varepsilon_s)_{s\le t}\big) = \sum_{i=1}^{t} (\mu_i - d) + \vartheta_t + \varepsilon_t$$

Shift in Trend Parameter—Random Walk with Drift Model

In the original parametric model the drift component $(\vartheta_t)_{\mathbb{N}_0}$ is the same as in the deterministic drift model [compare Eqs. (12) and (13)]. Again, the parameter d may be subject to abrupt shifts. Thus the value of the parameter d at time t should be considered as a random variable μ_t, where $(\mu_t)_{\mathbb{N}_0}$ is a marked point process as given by (4) with target $\mu^* = d$. To calculate the effect on the output we have to insert μ_t for d in the difference equation (9) with $\lambda_q(\mathcal{B}) = 1$. We obtain the output equation

$$\xi_t' = T + \sum_{i=1}^{t} \mu_i + \sum_{i=1}^{t} \varepsilon_i = \sum_{i=1}^{t} (\mu_i - d) + \vartheta_t + \sum_{i=1}^{t} \varepsilon_i \tag{19}$$

Equation (19) constitutes a special case of Eq. (16) with

$$K = 1 = M, \qquad \mathcal{F}_t\big((\mu_s)_{s\le t}, (\vartheta_s)_{s\le t}, (\varepsilon_s)_{s\le t}\big) = \sum_{i=1}^{t} (\mu_i - d) + \vartheta_t + \sum_{i=1}^{t} \varepsilon_i$$

4.3. Additive Disturbance and Shift in Drift Parameters

As a generalization we can consider a combination of the models of Sections 4.1 and 4.2: an additive disturbance component $(\mu_t^{(1)})_{\mathbb{N}_0}$ and a shift component $(\mu_t^{(2)})_{\mathbb{N}_0}$ in the drift parameter. Let us sketch this approach for the deterministic trend and the random walk with drift models.

Additive Disturbance and Shift in Trend Parameter—Deterministic Trend Model

Consider the deterministic trend model under the assumption that there are possible shifts of the drift parameter d represented by a marked point process $(\mu_t^{(2)})_{\mathbb{N}_0}$ with target $\mu_2^* = d$ and that there is an additive disturbance represented by a marked point process $(\mu_t^{(1)})_{\mathbb{N}_0}$ with target $\mu_1^* = 0$. Then we obtain the output equation

$$\xi_t' = \mu_t^{(1)} + T + \sum_{i=1}^{t} \mu_i^{(2)} + \varepsilon_t = \mu_t^{(1)} + \sum_{i=1}^{t} (\mu_i^{(2)} - d) + \vartheta_t + \varepsilon_t \tag{20}$$

Equation (2) constitutes a special case of Eq. (16) with

$$K = 2, \qquad M = 1,$$

$$\mathcal{F}_t\Big((\mu_s^{(1)})_{s\le t}, (\mu_s^{(2)})_{s\le t}, (\vartheta_s)_{s\le t}\Big) = \mu_t^{(1)} + \sum_{i=1}^{t} (\mu_i^{(2)} - d) + \vartheta_t + \varepsilon_t$$

Additive Disturbance and Shift in Trend Parameter—Random Walk with Drift Model

Consider the random walk with drift model under the assumption that there are possible shifts of the drift parameter d represented by a marked point process $(\mu_t^{(2)})_{\mathbb{N}_0}$ with target $\mu_2^* = d$ and there is an additive disturbance represented by a marked point process $(\mu_t^{(1)})_{\mathbb{N}_0}$ with target $\mu_1^* = 0$. Then we obtain the output equation

$$\xi_t' = \mu_t^{(1)} + T + \sum_{i=1}^{t} \mu_i^{(2)} + \sum_{i=1}^{t} \varepsilon_i = \mu_t^{(1)} + \sum_{i=1}^{t} (\mu_i^{(2)} - d) + \vartheta_t + \sum_{i=1}^{t} \varepsilon_i \tag{21}$$

In the scheme of Eq. (16) we have

$$K = 2, \qquad M = 1,$$

$$\mathcal{F}_t\Big((\mu_s^{(1)})_{s\le t}, (\mu_s^{(2)})_{s\le t}, (\vartheta_s)_{s\le t}\Big) = \mu_t^{(1)} + \sum_{i=1}^{t} (\mu_i^{(2)} - d) + \vartheta_t + \sum_{i=1}^{t} \varepsilon_i$$

5. ENGINEERING PROCESS CONTROLLERS

If a compensatory variable x_t can be determined in a production system, then a control rule of the form

$$x_t = f(\xi_t, \xi_{t-1}, \ldots; x_{t-1}, x_{t-2}, \ldots) \tag{22}$$

can be devised. Usually, a controller such as Eq. (22) is found by optimizing some performance (or cost) index J. A common index is

$$J_1 = E\left[\frac{1}{N}\sum_{t=1}^{N}(\xi_t - T)^2\right] \tag{23}$$

where T denotes the process target and N is the total number of observations the process is going to be run. Minimization of J_1 results in a *minimum mean square error* (MMSE) controller (Box and Jenkins, 1976), which is also called a *minimum variance* controller by Åström (1970) if ξ_t denotes deviation from target, in which case $T = 0$ in (23). From the principle of optimality of dynamic programming, it can be shown that the minimizing criterion (23) is equivalent to minimizing each $E[(\xi_t - T)^2]$ separately (Söderström, 1994, p. 313).

Other cost indices have been proposed for quality control applications. The following cost index was proposed by Box and Jenkins (1963) for their "machine tool" problem:

$$J_2 = E\left[\frac{1}{N}\sum_{t=1}^{N}[a(\xi_t - T)^2 + c\delta(x_t - x_{t-1})]\right] \tag{24}$$

where $\delta(x) = 0$ if $x = 0$ and $\delta(x) = 1$ if $x \neq 0$. This is a function with quadratic off-target cost and a fixed adjustment cost independent of the magnitude of the adjustment $x_t - x_{t-1}$. With this cost structure, the authors showed that it is optimal to wait until the process is sufficiently far from target in order to perform an adjustment, a policy that resembles an SPC control chart. However, the width of the "adjustment limits" is a function of the relative adjustment cost c/a and is not based on statistical considerations (Box and Jenkins, 1963; Crowder, 1992).

Fixed adjustment costs may be common in certain production processes. However, if x_t represents a setting of some machine (i.e., a setpoint for an automatic controller included with the equipment), then the adjustment cost c is practically zero and J_2 reduces to an MMSE controller.

We now investigate two simple EPC controllers for the drift processes of Eqs. (8) (deterministic trend) and (9) (random walk with drift) in the general framework of Section 4.3. The simpler situations of Sections 4.1 and 4.2 are obtained as special cases of the general scheme. Control rules will be designed according to the J_1 criterion (MMSE). We will assume that the effect of the sequence $(x_t)_{\mathbb{N}_0}$ of compensatory variables on the output process $(\xi_t)_{\mathbb{N}_0}$ is expressed by

$$\xi_t = \xi_t' + x_{t-1} \tag{25}$$

which implies that the full effect of the compensatory variable is felt immediately on the quality characteristic. Furthermore, we assume as before that the noise variables $(\varepsilon_t)_{\mathbb{N}_0}$ form a white noise sequence. These two assumptions guarantee that the closed-loop variables $\xi_0, \xi_1, \ldots$ are all independent. This makes it easier to see how the MMSE criterion (23) is equivalent to requiring that each square deviation be minimized separately without recourse to dynamic programming techniques.

5.1. Control of Deterministic Trend

We consider the deterministic trend model of Section 4.3 with a possible shift in the trend parameter d and an additive disturbance. By (20) and (25), the equation of the output of the controlled process is

$$\xi_t = \mu_t^{(1)} + T + \sum_{i=1}^{t} \mu_i^{(2)} + \varepsilon_t + x_{t-1} \tag{26}$$

It is clear that the control rule has to be designed for the case where the shift components $\mu_t^{(l)}$ are on their targets μ_l^*, i.e., for the case

$$\mu_t^{(1)} = 0, \qquad \mu_t^{(2)} = d$$

By (22), x_{t-1} is independent of ε_t; hence

$$E\left[(\xi_t - T)^2\right] = E\left[(dt + \varepsilon_t + x_{t-1})^2\right] \geq E[\varepsilon_t^2] = \sigma^2$$

Obviously, equality is obtained for

$$x_{t-1} = -dt \tag{27}$$

and at the "current" time t we implement the control action,

$$x_t = -d(t+1) \tag{28}$$

Hence the MMSE controller as defined by (28) corresponds to a pure "feedforward" controller (i.e., the observation ξ_{t-1} is not "fed back" into the control equation, but rather the anticipated disturbance is used). Controller (28) is equivalent to rule d_1 in Quesenberry (1988) if the sample size k of that paper equals 1.

Under the effect of the shift components $(\mu_t^{(l)})_{\mathbb{N}_0}$, the effect of control rule (28) on the output can, by (26), obviously be expressed as

$$\xi_t = \mu_t^{(1)} + T + \sum_{i=1}^{t} \mu_i^{(2)} - dt + \varepsilon_t \tag{29}$$

5.2. Control of Random Walk with Drift

An alternative model for linear drift is the random walk with drift stochastic process. As in the second subsection of Section 4.3, we admit possible additive shifts represented by a process $(\mu_t^{(1)})_{\mathbb{N}_0}$ and possible shifts in the drift parameter represented by a process $(\mu_t^{(2)})_{\mathbb{N}_0}$. By (21) and (25) the equation of the output of the controlled process is

$$\xi_t = \mu_t^{(1)} + T + \sum_{i=1}^{t} \mu_i^{(2)} + \sum_{i=1}^{t} \varepsilon_i + x_{t-1} \tag{30}$$

As in Section 5.1, the control rule has to be designed for the case in which the shift components $\mu_t^{(l)}$ are on their targets μ_l^*, i.e., for the case

$$\mu_t^{(1)} = 0, \qquad \mu_1^{(2)} = \cdots = \mu_t^{(2)} = d$$

By (22), $\xi_{t-1} - T + d + x_{t-1} - x_{t-2}$ is independent of ε_t; hence

$$E\left[(\xi_t - T)^2\right] = E\left[(\xi_{t-1} - T + d + \varepsilon_t + x_{t-1} - x_{t-2})^2\right] \geq E[\varepsilon_t^2] = \sigma^2$$

Obviously, equality is obtained for

$$x_t = T - \xi_t - d + x_{t-1} \tag{31}$$

which defines the MMSE control rule.

It is interesting to contrast control rules (28) and (31). Equation (31) is a "feedback" rule, since the observed value for the quality characteristic (ξ_t) is sent back to the controller to determine the next value of the input variable (x_t). For $x_0 = 0$ we obviously obtain from (31) that

$$x_t = -dt - \sum_{i=1}^{t} (\xi_i - T) \tag{32}$$

The second term on the right-hand side of (32) justifies the name "discrete integral controller" used for this type of control rules.

Finally, let us evaluate the effect of control rule (31) on the output quality characteristic under the effect of the shift components $(\mu_t^{(l)})_{\mathbb{N}_0}$. From (30) and (32) we obtain

$$\xi_t = \begin{cases} \mu_1^{(1)} + T + \mu_1^{(2)} + \varepsilon_1 & \text{if } t = 1 \\ \mu_t^{(1)} - \mu_{t-1}^{(1)} + T + \mu_t^{(2)} - d + \varepsilon_t & \text{if } t \geq 2 \end{cases} \tag{33}$$

Inserting (33) into (32) we obtain

$$x_t = -d - \mu_t^{(1)} - \sum_{i=1}^{t} \mu_i^{(2)} - \sum_{i=1}^{t} \varepsilon_i \tag{34}$$

6. DISCUSSION OF SPC IN THE PRESENCE OF EPC

Consider the output of a manufacturing process under the simple drift controllers of the previous section. The output *without* the presence of parameter shifts is a special case of (29) or (33) with constants $\mu_t^{(1)} = 0$ and $\mu_t^{(2)} = d$. In both cases we obtain

$$\xi_t = T + \varepsilon_t \qquad \text{for all } \; t \in \mathbb{N}$$

In this case, the process output $(\xi_t)_{\mathbb{N}}$ is a sequence of i.i.d. random variables with mean $E[\xi_t] = T$ on target and the minimum possible variance $V[\xi_t] = \sigma^2$.

For a successful SPC–EPC integration, it is necessary to analyze the output of processes under EPC control *with* shift components. For a substantial discussion we need simple instances of shift components.

6.1. Effect of Simple Shifts on EPC-Controlled Processes

For many applications it is appropriate to assume simple structured shifts of the type given by Eq. (6). In the models of Sections 5.1 and 5.2 let us consider the output processes under this type of shift. We assume that

$$\mu_t^{(l)} = \begin{cases} \mu_l^* & \text{if } t \leq \nu_l \\ \mu_l^* + \gamma_l \Delta_l & \text{if } t > \nu_l \end{cases} \tag{35}$$

where $\mu_1^* = 0$, $\mu_2^* = d$ are the target values, $\Delta_l > 0$ is the absolute shift size, ν_l is the random time until occurrence of the shift, and γ_l is the random sign of the shift.

Under these assumptions the output equation (29) of the deterministic trend model becomes

$$\xi_t = T + \gamma_1 \Delta_1 \mathbb{I}_{(\nu_1;+\infty)}(t) + (t - \lfloor \nu_2 \rfloor)\gamma_2 \Delta_2 \mathbb{I}_{(\nu_2;+\infty)}(t) + \varepsilon_t \tag{36}$$

where

$$\mathbb{I}_B(t) = \begin{cases} 1 & \text{if } t \in B \\ 0 & \text{if } t \in \mathbb{R} \backslash B \end{cases} \tag{37}$$

is the indicator function of a set $B \subset \mathbb{R}$.

Applying the same assumptions to the output (33) of the random walk with drift model, we obtain for $t \geq 2$

$$\xi_t = T + \gamma_1 \Delta_1 \mathbb{I}_{(\nu_1;\nu_1+1)}(t) + \gamma_2 \Delta_2 \mathbb{I}_{(\nu_2;+\infty)}(t) + \varepsilon_t \tag{38}$$

For the control variable of the random walk with drift model we obtain, by inserting (35) into (34),

$$x_t = -(t+1)d - \gamma_1 \Delta_1 \mathbb{I}_{(\nu_1;+\infty)}(t) - \gamma_2 \Delta_2 (t - \lfloor \nu_2 \rfloor) \mathbb{I}_{(\nu_2;+\infty)}(t) - \sum_{i=1}^{t} \varepsilon_i \tag{39}$$

The equations for the simpler models with only one possible shift (either additive or in the drift parameter) are obtained from (36) and (38) either by letting $\nu_2 = +\infty$ (only additive shifts) or by letting $\nu_1 = +\infty$ (only shifts in drift parameters).

In the following two sections we discuss (36) and (38) in two practically relevant situations.

6.2. Shifts Occurring During Production Time

In the deterministic trend case, the controller defined by (28) has no feedback from the output and is thus not able to compensate for random shifts. As is obvious from (36), an additive shift takes the process mean away from its target T to $T + \gamma_1 \Delta_1$, but the output at least remains stable in its mean. A shift in the drift parameter is even more harmful. After such a shift, the output mean has a trend component $(t - \lfloor \nu_2 \rfloor)\gamma_2 \Delta_2$. It is obvious that in the presence of possible shifts such a process should be monitored by a supplementary SPC scheme.

The feedback controller of (31) or (32) for the random walk with drift is able to react both on additive shifts and on shifts in the drift parameter. As is obvious from (38), due to the delay of one time period in the controller's action, an additive shift leads to only a single outlier of the output ξ_t at $t = \lfloor \nu_1 + 1 \rfloor$ but remains without effect at further time points. A shift in the drift parameter can be more harmful. After such a shift, the output mean is constantly off target T at $T + \gamma_2 \Delta_2$. However, this shift in the mean has serious consequences only if $|\gamma_2 \Delta_2|$ is large or if the cost of being off target is large. In such cases it is reasonable to monitor the process by supplementary SPC procedures.

6.3. Effect of a Biased Drift Parameter Estimate

In the approach of the model presented in Section 4.2, a biased estimate of the drift parameter d can be interpreted as a shift in the drift parameter that coincides with the setup of the process. This is quite useful as it allows us to study the effect of mistakenly using a biased trend parameter estimate in an EPC scheme. Let $\hat{d}$ be the biased estimate that is used instead of d in the control equations (28) and (31) or (32). Then we can describe the situation by (36) and (38) by letting

$$\Delta_2 = |d - \hat{d}|, \qquad P(\gamma_2 = 1) = 1 \qquad \text{if } d \geq \hat{d}$$

and

$$P(\gamma_2 = -1) = 1 \qquad \text{if } d < \hat{d}$$

The effect of this type of parameter shift in the trend and random walk models is exactly the same as in Section 6.2.

6.4. Effect of Constraints in the Compensatory Variable

An important aspect in practice, usually not addressed in the literature on SPC–EPC integration, is that the compensatory variable must usually be constrained to lie within a certain region of operation, i.e.,

$$A \leq x_t \qquad \text{or} \qquad x_t \leq B \qquad \text{or} \qquad A \leq x_t \leq B$$

for all instants t. In particular, integral controllers such as Eq. (31) can compensate for shifts of any size *provided* that the controllable factor is unconstrained.

It is useful to consider what would happen if the EPC schemes given by Eqs. (28) and (31) were applied to a constrained input process. Since the drift is linear, the control variable x_t moves in the opposite direction than the drift to keep ξ_t on target. However, at some point the controller hits a boundary (either A or B) and remains there afterward. In the control engineering literature this is referred to as "saturation" of the EPC scheme.

Effect of Constraints Under the Deterministic Trend Model

Let us discuss the case of a constrained control variable in the trend model of Section 5.1. For simplicity's sake we discuss only the case of $d > 0$ with a lower bound $A < 0$. The case of $d < 0$ with a corresponding upper bound $B > 0$ is completely analogous.

The relationship between the control variable x_t of Eq. (28) and the constrained control variable $\tilde{x}_t$ is

$$\tilde{x}_t = \begin{cases} x_t = -d(t+1) & \text{if } t \leq -A/d - 1 \\ A & \text{if } t > -A/d - 1 \end{cases} \tag{40}$$

Hence the output $\tilde{\xi}_t$ of the process under constrained control is

$$\begin{aligned} \tilde{\xi}_t &= \mu_t^{(1)} + T + \sum_{i=1}^{t} \mu_i^{(2)} + \tilde{x}_{t-1} + \varepsilon_t \\ &= \begin{cases} \mu_t^{(1)} + T + \sum_{i=1}^{t} \mu_i^{(2)} - dt + \varepsilon_t & \text{if } t \leq \dfrac{-A}{d} \\ \mu_t^{(1)} + T + \sum_{i=1}^{t} \mu_i^{(2)} + A + \varepsilon_t & \text{if } t > \dfrac{-A}{d} \end{cases} \end{aligned} \tag{41}$$

Under the simple shift components of type (35) the output $\tilde{\xi}_t$ satisfies the right-hand side of (36) for $t \leq -A/d$. For $t > -A/d$ we obtain

$$\tilde{\xi}_t = T + \gamma_1 \Delta_1 \mathbb{I}_{(\nu_1;+\infty)}(t) + (t - \lfloor \nu_2 \rfloor)\gamma_2 \Delta_2 \mathbb{I}_{(\nu_2;+\infty)}(t) + dt + A + \varepsilon_t \tag{42}$$

Obviously, the arguments in favor of supplementary application of SPC schemes in the trend model that are put forward in Section 6.2 also hold in the case of constrained controllers.

Effect of Constraints Under the Random Walk Model

In the random walk model we also restrict attention to the case $d > 0$ with a lower bound $A < 0$.

Unlike the situation for the deterministic trend model, the time κ until hitting or falling below the lower bound A is stochastic and is defined by

$$\begin{aligned}\kappa &= \min\{t|x_t \leq A\} \\ &= \min\left\{t|\sum_{i=1}^{t}\varepsilon_i \geq -A-(t+1)d-\gamma_1\Delta_1 \mathbb{I}_{(\nu_1;+\infty)}(t)-\gamma_2\Delta_2(t-\lfloor\nu_2\rfloor)\right. \\ &\qquad \left.\mathbb{I}_{(\nu_2;+\infty)}(t)\right\}\end{aligned} \tag{43}$$

From (43) the distribution of κ can be found by first determining the conditional distribution under ν_i and γ_i and then integrating with respect to the corresponding densities. We shall not investigate this problem here.

The relationship between the control variable x_t of (39) and the constrained control variable $\tilde{x}_t$ is

$$\tilde{x}_t = \begin{cases} x_t & \text{if } t < \kappa \\ A & \text{if } t \geq \kappa \end{cases} \tag{44}$$

Hence the output $\tilde{\xi}_t$ of the process under constrained control is

$$\begin{aligned}\tilde{\xi}_t &= \mu_t^{(1)} + T + \sum_{i=1}^{t}\mu_i^{(2)} + \sum_{i=1}^{t}\varepsilon_i + \tilde{x}_{t-1} \\ &= \begin{cases} \mu_t^{(1)} - \mu_{t-1}^{(1)} + T + \mu_t^{(2)} - d + \varepsilon_t & \text{if } t < \kappa+1 \\ \mu_t^{(1)} + T + \sum_{i=1}^{t}\mu_i^{(2)} + \sum_{i=1}^{t}\varepsilon_i + A & \text{if } t \geq \kappa+1 \end{cases}\end{aligned} \tag{45}$$

Under the simple shift components of type (35) the output $\tilde{\xi}_t$ satisfies the right-hand side of (38) for $2 \leq t < \kappa+1$. For $t \geq 2, t \geq \kappa+1$ we obtain

$$\tilde{\xi}_t = T + \gamma_1\Delta_1\mathbb{I}_{(\nu_1;+\infty)}(t) + td + \gamma_2\Delta_2\max\{0, t-\lfloor\nu_2\rfloor\} + \sum_{i=1}^{t}\varepsilon_i + A \tag{46}$$

In the unconstrained case, supplementary application of SPC in the random walk model makes sense in only special cases (see Section 6.2). From Eq. (46), it is evident that in the constrained case supplementary application of SPC is much more interesting and perhaps indispensable.

6.5. Effect of Using a Wrong Model

We now study what would happen if a wrong drift model is used.

Under a deterministic trend, the relation between the closed-loop output ξ_t and the control variable x_{t-1} is given by (26). If the integral controller defined by (31) is used in this model, the explicit expression for ξ_t is

$$\xi_t = T + \mu_t^{(1)} - \mu_{t-1}^{(1)} + \mu_t^{(2)} - d + \varepsilon_t - \varepsilon_{t-1}$$

Under the simple shifts of type (35) we obtain

$$\xi_t = T + \gamma_1 \Delta_1 \mathbb{1}_{(\nu_1;\nu_1+1]}(t) + \gamma_2 \Delta_2 \mathbb{1}_{(\nu_2;+\infty)}(t) + \varepsilon_t - \varepsilon_{t-1} \tag{47}$$

Whether there are parameter shifts or not, the output exhibits twice as great a variance as in the case of using the correct model. This case occurs in Quesenberry's (1988) d_2 and d_3 rules. If $\mu_t^{(1)} = 0$ and $\mu_t^{(2)} = d$ for all t, then $\xi_t = T + (1 - \mathcal{B})\varepsilon_t$, which is an MA(1) process, an always-stationary time series model (Box and Jenkins, 1976).

If parameter shifts occur, we have the following result. Except for the single outlier for $\nu_1 < t \leq \nu_1 + 1$, ξ_t is permanently off target for $t > \nu_2$ with absolute deviation Δ_2. This, and the uncertainty about the correctness of the assumptions of the model, make it advisable to use SPC methods in addition to the simple EPC schemes.

If, on the contrary, the deterministic trend controller (28) is used in a random walk with drift process, the closed-loop equation is, by (30),

$$\xi_t = T + \mu_t^{(1)} + \sum_{i=1}^{t} \mu_i^{(2)} + \sum_{i=1}^{t} \varepsilon_i - dt$$

In the long run, if we let t grow without bound and we use the inverse of the difference operator, namely,

$$\frac{1}{1 - \mathcal{B}} y_t \equiv \sum_{j=0}^{\infty} y_{t-j}$$

then the closed-loop equation is

$$\xi_t = T + \mu_t^{(1)} + \frac{\mu_t^{(2)}}{1 - \mathcal{B}} - dt + \frac{\varepsilon_t}{1 - \mathcal{B}}$$

If $\mu_t^{(1)} = 0$ and $\mu_t^{(2)} = d$ for all t, then the previous equation reduces to

$$\xi_t = \frac{\varepsilon_t}{1 - \mathcal{B}}$$

which is a nonstationary AR(1) process (Box and Jenkins, 1976) with $\text{var}(\xi_t) \to \infty$ as $t \to \infty$.

For bounded values of t and under the simple shifts of type (35), we obtain

$$\xi_t = T + \gamma_1 \Delta_1 \mathbb{1}_{(\nu_1;+\infty)}(t) + \max\{t - \lfloor \nu_2 \rfloor, 0\} \gamma_2 \Delta_2 + \sum_{i=1}^{t} \varepsilon_i \tag{48}$$

In this case, whether there are parameter shifts or not, the output exhibits variance that increases linearly with time compared with the case of using the correct model. Thus it is evident that using an EPC controller designed for a random walk with drift model is "safer" than using an EPC controller designed for a deterministic trend process in case we selected (by mistake) the wrong drift model.

Taking the shifts into account we have the following result. There is a shift in the mean for $\nu_1 < t$ and a shift that results in a trend for $\nu_2 < t$. Again, given the uncertainty about the correctness of model assumptions it is obviously advisable to use additional SPC methods.

7. SHEWHART CHARTS FOR DETECTION OF SHIFTS IN THE DETERMINISTIC TREND MODEL

In this section we investigate the design of a simple two-sided Shewhart chart with fixed sampling interval for detection of shifts (i.e., abrupt changes) in the trend parameter of the model in Section 5.1. The chart is defined by the triple (n, c, h) of sample size n, control limit width multiple c, sampling distance h (i.e., the number of discrete periods between samples), where $n \in \mathbb{N}$, $c \in (0; +\infty)$, $h \in \mathbb{N}$, $h \geq n$. The control procedure is as follows.

(S1) At time points $h, 2h, 3h, \ldots, kh, \ldots,$ output samples $(\xi_{kh}, \ldots, \xi_{kh+n-1})$ of size n are taken from the process.

(S2) The absolute value $|\bar{\xi}_k - T|$ of the difference of the arithmetic mean

$$\bar{\xi}_k := \frac{1}{n} \sum_{i=0}^{n-1} \xi_{kh+i} \tag{49}$$

of the sample variables from the process target T is compared with the control limit $c\sigma/\sqrt{n}$.

(S3) If $|\bar{\xi}_k - T| \leq c\sigma/\sqrt{n}$, the manufacturing process continues without intervention. If $|\bar{\xi}_k - T| > c\sigma/\sqrt{n}$, the manufacturing process is stopped (giving an *out-of-control signal* or *alarm*) and inspected for the presence of an additive shift or a shift in the trend parameter. If no shift is detected, the manufacturing process continues without further intervention. If a shift is detected, the manufacturing process is *renewed*, i.e., the conditions of the start of the process at time point 0 are restored, e.g., by a repair or by complete overhaul of the production facilities. After the renewal the process is restarted at time point 0 of the next renewal cycle.

From the point of view of the optimality principles of mathematical statistics, there may be better tests for the detection of a shift in the trend model than the test defined by rules (S1), (S2), (S3). Nevertheless, the two following arguments support an investigation of Shewhart charts under our shift model:

1. The simple structure of Shewhart charts simplifies the design of optimum charts in a statistical or economic scheme of optimality.
2. Shewhart charts are widely used in industrial practice. Most often, the charts applied are not designed under a precise statistical and economic model but from a heuristic point of view (sample sizes $n = 3, 5, 7$; 3σ limits as control limits). It is interesting to investigate the behavior of such charts under the trend model of Section 5.1.

Here we investigate Shewhart charts from a statistical point of view. This decision is not supported by principal arguments; it merely reflects an option for simplicity. An economic design is based on variables such as the number of false alarms, length of a renewal cycle, and profit incurred from items during a cycle. It is obvious that for a model that admits both an additive shift and a shift in the drift parameter, the formulas for the distributions and expected values of such variables are rather involved. Thus an investigation into the economic design would lead to mathematical details that far exceed the scope of the present chapter, which is primarily interested in the structure of a fundamental model of SPC–EPC integration and a simple application thereof.

7.1. The Average Run Length

An essential quantity in the statistical design of a Shewhart chart is the *average run length* (ARL), i.e., the expected number of samples until occurrence of an alarm, or, equivalently, the *average time to signal* (ATS), i.e., the expected time until occurrence of an alarm. Run length and time to signal are usually calculated under the simplifying assumption that the process is either stable without a parameter shift or stable at a given parameter shift. In this approach the problem of the time until occurrence of a shift is ignored. Interest concentrates on the question, "How long does it take to obtain an out-of-control signal if the process has entered certain invariant conditions at an arbitrarily fixed time point 0?"

To define the ARL in terms of the model of Section 5.1 we consider as in Section 6.1 fixed absolute shift sizes $\Delta_l \geq 0$ with given signs $z_l \in \{-1, 1\}$. $\Delta_l = 0$ is admitted to express the case that no shift of type i has occurred. Ignoring the times until occurrence of shifts and assuming that the conditions of the process remain fixed from an arbitrarily chosen time point 0 on, we obtain in analogy to (36) the output equation

$$\xi_t = T + z_1\Delta_1 + z_2\Delta_2 t + \varepsilon_t \tag{50}$$

Under the control rules (S1), (S2), (S3), the *run length*, i.e., the number η of samples until occurrence of an alarm, is defined by

$$\eta = \min\{k | k \in \mathbb{N}, |\bar{\xi}_k - T| > c\sigma/\sqrt{n}\} \tag{51}$$

We assume production speed 1; i.e., one item is produced in one time unit. Then the total time until occurrence of an alarm (*time to signal*) is

$$\eta h + n - 1$$

Define $\delta_l = \Delta_l/\sigma$. We use this standardization to avoid the nuisance parameter σ.

The ARL can now be defined as the expected value of η, considered as a function $A(z_1, \delta_1, z_2, \delta_2)$ of the shift amounts δ_l and the signs z_l of shifts:

$$A(z_1, \delta_1, z_2, \delta_2) = E[\eta] \tag{52}$$

The corresponding expected total time until occurrence of an alarm (ATS) is

$$hA(z_1, \delta_1, z_2, \delta_2) + n - 1$$

For explicit calculation of the ARL the noise components $(\varepsilon_t)_{\mathbb{N}}$ are assumed as i.i.d., each ε_t with normal distribution $\mathcal{N}(0; \sigma^2)$. Hence under the output equation (50), the test statistics $(\bar{\xi}_k)_{\mathbb{N}}$ are independent, where $\bar{\xi}_k$ is normally distributed, with parameters

$$E[\bar{\xi}_k] = T + z_1\Delta_1 + z_2\Delta_2\left(kh + \frac{n-1}{2}\right) \tag{53}$$

$$V[\bar{\xi}_k] = \sigma^2/n \tag{54}$$

Hence the alarm probabilities are given by

$$\begin{aligned}\beta_k(z_1, \delta_1, z_2, \delta_2) &= \mathrm{P}(|\,\bar{\xi}_k - T| > c\sigma/\sqrt{n}) \\ &= 1 - \phi(c - z_1\delta_1\sqrt{n} - z_2\delta_2\left(kh + \frac{n-1}{2}\right)\sqrt{n} \\ &\quad + \phi\left(-c - z_1\delta_1\sqrt{n} - z_2\delta_2\left(kh + \frac{n-1}{2}\right)\sqrt{n}\right)\end{aligned} \tag{55}$$

In the case of $\Delta_2 = 0$ (no shift in the drift parameter), the alarm probabilities are constant in the number k of the sample. Hence we have the classical case: The distribution of the run length η is geometric with parameter

$$1 - \phi(c - z_1\delta_1\sqrt{n}) + \phi(-c - z_1\delta_1\sqrt{n}) = \beta(z_1, \delta_1, z_2, 0) = \beta_k(z_1, \delta_1, z_2, 0)$$

Thus, in particular,

$$A(z_1, \delta_1, z_2, 0) = \frac{1}{\beta(z_1, \delta_1, z_2, 0)} \tag{56}$$

In the case of $\Delta_2 > 0$ (shift in the drift parameter), the alarm probabilities vary in the number k of the sample. The distribution of the run length η is determined by the probabilities

$$p(m) = \mathrm{P}(\eta = m) = \beta_m(z_1, \delta_1, z_2, \delta_2)\prod_{k=1}^{m-1}[1 - \beta_k(z_1, \delta_1, z_2, \delta_2)] \tag{57}$$

In this case we obtain no simple expression for the ARL. We have

$$A(z_1, \delta_1, z_2, \delta_2) = \sum_{m=1}^{+\infty} \left\{ m\beta_m(z_1, \delta_1, z_2, \delta_2) \prod_{k=1}^{m-1} [1 - \beta_k(z_1, \delta_1, z_2, \delta_2)] \right\} \tag{58}$$

From a computational point of view, it is better to write Eq. (57) in recursive form:

$$p(m) = p(m-1)\, \frac{\beta_m(z_1, \delta_1, z_2, \delta_2)}{\beta_{m-1}(z_1, \delta_1, z_2, \delta_2)}\, [1 - \beta_{m-1}(z_1, \delta_1, z_2, \delta_2)] \tag{59}$$

with $p(1) = \beta_1(z_1, \delta_1, z_2, \delta_2)$, and thus

$$A(z_1, \delta_1, z_2, \delta_2) = \sum_{m=1}^{+\infty} mp(m) \tag{60}$$

7.2. Example

Chemical mechanical planarization (CMP) is an important process in the manufacturing of semiconductors. A key quality characteristic in a CMP process is the removal rate of silicon oxide from the surface of each wafer. Since the polishing pads wear out with use, a negative tend is experienced in this response, in addition to random shocks or shifts. The removal rate has a target of 1800 and is controlled via a deterministic trend EPC scheme. The errors are normally distributed with mean zero and $\sigma = 60$, and an estimate of the drift $\hat{d}$ is used for control purposes. It is desired not to let the process run for more than an average of 10 samples if a bias in the drift estimate of magnitude $0.01\sigma = 0.6$ exists. In the absence of shifts in the mean or trend, an ARL of 370 is desired. In addition, positive shifts of size $\Delta_1 = 1\sigma$ should be detected, on average, after a maximum of 12 samples if the aforementioned biased trend estimate is (incorrectly) used by the EPC.

Table 1 shows numerical computations for this problem using Eqs. (57)–(60) and varying n from 1 to 10. Clearly, the desired ARL of 370 is obtained with $c = 3$; thus the table shows results for this value of c. From the table, $A(0, 0, -1, 0.01) = 10.86$ for $n = 4$, and $A(1, 1, -1, 0.01) = 13.51$ for $n = 5$. Therefore, the chart design with the smallest sample size that meets the design specifications calls for using $n = 5$ and $c = 3$. The h design parameter ($h \geq n$) should be decided based

Table 1 Average Run Lengths for the Example Problem, $c = 3, h = 10$

n	$A(1, 1, 1, 0.01)$ $= A(-1, 1, -1, 0.01)$	$A(-1, 1, 1, 0.01)$ $= A(1, 1, -1, 0.01)$	$A(0, 0, 1, 0.01)$ $= A(0.0, -1, 0.01)$	$A(1, 1, 0, 0)$ $= A(-1, 1, 0, 0)$
1	9.55	26.62	18.42	43.89
2	5.68	21.37	14.21	17.73
3	4.00	18.15	12.16	9.76
4	3.05	15.64	10.86	6.30
5	2.46	13.51	9.93	4.49
6	2.06	11.67	9.21	3.43
7	1.78	10.06	8.63	2.76
8	1.57	8.66	8.14	2.31
9	1.43	7.46	7.72	2.00
10	1.31	6.43	7.36	1.77

upon economic considerations not discussed in this chapter and was therefore set to 10.

Interestingly, the third and fourth columns in Table 1 indicate that for $n < 9$, a negative drift "masks" positive shifts and vice versa, making it harder to detect a shift [i.e., this occurs when $\text{sign}(z_1) \neq \text{sign}(z_2)$]. Also, we have the relationship

$$A(z_1, \delta_1, z_2, \delta_2) = A(-z_1, \delta_1, -z_2, \delta_2)$$

Figure 1 shows a realization of the controlled sample means for this process ($\bar{\xi}_k$) with no shifts in the mean occurring in the simulated time. The designed chart limits are shown superimposed. In the absence of abrupt shifts in the process, the SPC chart will detect the biased d stimate after an average of 9.93 samples (cf. Table 1, fourth column), although in the figure it was not detected until sample 11. In practice, production will be stopped at the alarm time and corrective action will be taken (e.g., replacing the polishing pad), which will recenter the process.

ACKNOWLEDGMENTS

Dr. Castillo was funded by NSF grants INT 9513444 and DMI 9623669. Drs. Göb and von Collani were funded by DAAD grant 315/PPP/fo-ab.

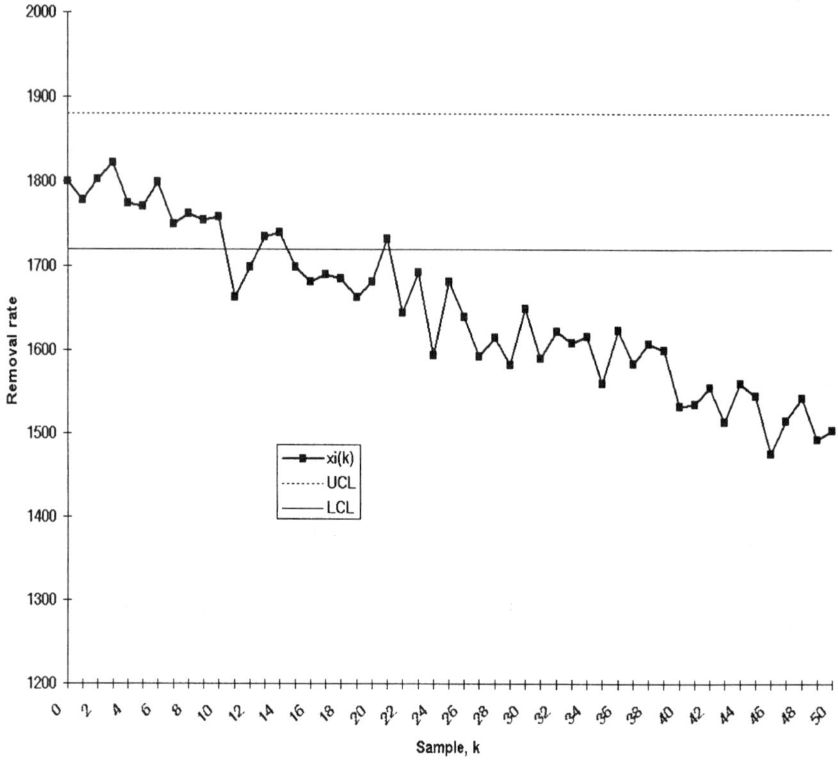

Figure 1 A realization of sample means $\bar{\xi}_k$ for the case of a deterministic trend EPC uses a biased trend estimate. The computed Shewhart limits for detecting such bias are also shown.

REFERENCES

Åström KJ. 1970. Introduction to Stochastic Control Theory. Academic Press, San Diego, CA.

Åström KJ, Wittenmark B. 1989. Adaptive Control. Addison-Wesley, Reading, MA.

Barnard GA. 1959. Control charts and stochastic processes. J Roy Stat Soc Ser B XXI(2): 239–271.

Basseville M, Nikiforov IV. 1993. Detection of Abrupt Changes—Theory and Application. Prentice-Hall, Englewood Cliffs, NJ.

Box GEP, Jenkins G. 1963. Further contributions to adaptive quality control: Simultaneous estimation of dynamics: Nonzero costs. Bull Int Stat Inst 34: 943–974.

Box GEP, Jenkins G. 1976. Time Series Analysis, Forecsting, and Control. Rev. ed. Holden Day, Oakland, CA.

Box GEP, Kramer T. 1992. Statistical process monitoring and feedback adjustment —A discussion. Technometrics 34(3): 251–267.

Butler SW, Stefani JA. 1994. A supervisory run-to-run control of a polysilicon gate etch using in situ ellipsometry. IEEE Trans Semicond Manuf 7(2): 193–201.

Crowder SV. 1992. An SPC model for short production runs: Minimizing expected cost. Technometrics 34: 64–73.

Del Castillo E, Hurwitz A. 1997. Run to run process control: A review and some extensions. J Qual Technol 29(2): 184–196.

Del Castillo E. 1996. Some aspects of process control in semiconductor manufacturing. Proceedings of the 4th Würzburg-Umea Conference in Statistics, pp. 37–52.

Ingolfsson A, Sachs E. 1993. Stability and sensitivity of an EWMA controller. J Qual Technol 25(4): 271–287.

MacGregor JF. 1988. On-line statistical process control. Chem Eng Prog October, pp. 21–31.

MacGregor JF. 1990. A different view of the funnel experiment. J Qual Technol 22: 255–259.

Montgomery DC, Keats JB, Runger GC, Messing WS. 1994. Integrating statistical process control and engineering process control. J Qual Technol 26(2): 79–87.

Quesenberry CP. 1988. An SPC approach to compensating a tool-wear process. J Qual Technol 20(4): 220–229.

Sach E, Hu A, Ingolfsson A. 1995. Run by run process control: Combining SPC and feedback control. IEEE Trans Semicond Manuf 8(1): 26–43.

Söderström T. 1994. Discrete Time Stochastic Systems, Estimation and Control. Prentice-Hall, Englewood Cliffs, NJ.

Tucker WT, Faltin FW, Vander Wiel SA. 1993. ASPC: An elaboration. Technometrics 35(4): 363–375.

Vander Weil SA. Tucker WT, Faltin FW, Doganaksoy N. 1992. Algorithmic statistical process control: Concepts and an application. Technometrics 34(3): 286–297.

6

Reliability Analysis of Customer Claims

Pasquale Erto
University of Naples Federico II, Naples, Italy

1. INTRODUCTION

Reliability theory is substantially the "science of failures," in the same way in which medicine is the "science of diseases," directed toward curing or preventing them. However, because each failure virtually implies the existence of customer dissatisfaction and complaints, reliability is in some ways the science of complaints also and, during the warranty period, the science of claims. Besides, to fully exploit it in the context of quality management, we must always remember that its operative meaning is "the probability that a system possesses and keeps its quality *throughout time*."

In general, reliability is a characteristic of systems that possess and keep during their life the working qualities for which they were designed and realized. In this sense reliability is a *time-oriented quality characteristic* [1] that can also be referred to technical, productive, commercial, and service activities that perform their tasks timely and effectively. Technically, reliability is quantified as the probability of no failures (i.e., of performing the required function) under given environmental and operational conditions and for a stated period of time.

In order to be concrete, let us develop this point of view specifically for the car industry, which constitutes a well-known, crucial, and effective application field.

Today, a new car model must meet a specified reliability level from its initial launching on the market, on pain of obscuring the company image, which will be restored only with difficulty by subsequent improvements. Generally, the reliability targets can be achieved by means of good design, many preliminary life tests on components or subsystems, and a quality control policy. Nevertheless, in the case of mass production, it is essential to verify constantly that these targets are really fulfilled in service. In fact, the in-service reliability level may turn out to be different from the expected one, mainly owing to faults in the production process and/or to unforeseen stresses induced by the real operating environment. Then the manufacturer has a pressing need to collect and analyze field data to detect the causes of a possible discrepancy between the in-service and expected reliability, to be able to immediately adopt the necessary corrective actions.

Obviously, since cars are products with a wide range of operating environments and users, one should have a great many manufactured units under monitoring, over their entire life, to be confident in the measure of their reliability. But the impracticability of such a policy is quite evident. Thus the approach generally undertaken consists in monitoring only the units belonging to homogeneous samples of limited size (e.g., a taxi fleet) and/or controlling the repair operations of manufactured units during the warranty period. Other sources of information, such as the number of spare parts sold, are sometimes used, but they are less informative and are not considered here. The monitoring of a vehicle fleet allows one to collect information about the entire life of the product, taking into account both early and wear-out failures. Moreover, in many cases, these fleets are subjected to more intensive use than normal, and this makes it possible to obtain measures of reliability in a relatively short time. Nevertheless, these measures are generally valid only for the operating environment and use of the particular sample chosen, and it is often difficult to extend them to other situations. Besides, they cannot take into account the impact of subsequent improvements.

The use of warranty data makes timely information available at low cost for reliability evaluations. Obviously, these data are truncated (i.e., limited to the first period of life), so they take into account mainly the impact of early failures, but their use has the advantage of quickly allowing us both to choose corrective actions and to check the effectiveness of those just adopted.

However, to better understand all the information nested in the warranty data, we must consider that usually these data report only the component and failure codes and the mileage interval in which the failure occurred. For instance, with specific reference to the automobile world, no information is directly provided about the number of cars that cover the various mileage intervals without failures, and hence no direct knowledge is available about the population to which the observed failures must be referred. In this situation, only approximate estimation procedures are usually used, that is, procedures that are generally based on some a priori (and subjective) evaluation of car distribution versus mileage intervals.

Instead, using the reliability analysis approach, we can rigorously estimate both the failure and car distributions versus mileage. The method has already been successfully used in real-life cases that are partially reported in an illustrative example included in this chapter.

2. ANALYSES FROM WARRANTY DATA OF CARS

In a modern way of thinking, quality means "customer satisfaction," and it is feasible to realize quality, in this sense, only with the total quality management approach to the management of the whole company. In such a context, the reliability engineers' involvement conforms to this management policy of the car industry too, aiming to involve everyone's commitment to obtain total quality.

However, in order to plan, realize, or control a certain quality level of the cars produced, the availability of an efficacious practical measure of "in-service" quality is first needed. In fact, one of the fundamental rules for the management of total quality consists of turning away from making decisions based exclusively on personal opinions or impressions. Instead, one needs to refer to data that are really representative of the quality as perceived by the customers, such as the warranty data. These data, however, contain only the following information:

1. Vehicle type code
2. Assembly date
3. Component and defect code
4. Mileage to failure

In formal statistical language, the warranty data are failure observations from a sample that is both truncated (at the end of the warranty period) and has items suspended at the number of kilometers effectively covered by the respective customers. Thus, to carry out a reliability analysis, both the number of failures and the number of suspensions for each mileage interval are required. Obviously, the warranty data give no mileage information about those vehicles that are sold and reach the end of each mileage interval without any claim being made. Thus, no direct information is available about the population to which the reported number of failed items must be referred. Therefore, the usual procedures used by the automotive industry [2–4] require the a priori estimation (often arbitrary) of the vehicle distribution versus mileage in order to partition the total number of vehicles under warranty into mileage intervals. Note that this distribution may also be very different from case to case, since it may concern vehicles under special maintenance contract including the warranty period or only a fraction of the vehicles under warranty (e.g., all vehicles from a particular production unit), as well as all vehicles produced in a given period, etc. In Ref. 5 it is stressed that the number of claims at a specific age depends on mileage accumulated, so supplementary information on the mileage accumulated for the population of cars in service is needed.

This chapter shows how one can estimate simultaneously both the mileage and the failure distribution functions without needing any a priori estimation. In the next section, a special case that occurred in a real-life situation, in which the proposed method of analysis was successfully used, is discussed.

3. A REAL-LIFE RELIABILITY ANALYSIS

3.1. The Available Data Set

Failure data normally refer to about 40 different components (or parts) of some car model. The kilometers to failure are typically grouped into equal-width lifetimes, each of 10,000 km, and all vehicles under consideration are sold during the same year in which repairs are made.

In our case from real life [8], 498 cars were sold in the year, and the total number of warranty claims referred to the manufacturer was 70. Furthermore, irrespective of the parts involved, this number of claims were distributed over the lifetimes as shown in Table 1.

The characteristic that makes this case peculiar is that no age (from selling date) distribution and no distribution of covered kilometers are given for the fleet under consideration. Hence, it is not possible to allocate the unfailed units in each lifeime. To overcome this difficulty we can use the reliability analysis approach, introducing an estimation procedure that involves at the same time both the failure and kilometer distributions.

Table 1 Number of Warranty Claims in Each Lifetime

Lifetimes (km/1000)	0–10	10–20	20–30	30–40	> 40
Number of claims	55	11	3	1	0

3.2. Method of Analysis

Let T_f be the random variable (r.v.) representing the "kilometers to failure" and $G(t) = \Pr(T_f \leq t)$ its unknown distribution function. Moreover, let $F(t)$ be the probability that a car sold during the year of observation does not exceed the kilometers t until the end of this year. The experimental context under analysis is equivalent to a sampling (truncated at the end of the warranty period) in which some of the items under life testing have their test randomly suspended before failure. Thus, an r.v. representing the "kilometers to suspension," say T_s, is defined with the unknown distribution function $F(t) = \Pr(T_s \leq t)$. Hence, it follows that the probability that an item fails before t km is

$$\Pr\{(T_f < t) \cap (T_s > T_f)\} = \int_0^t [1 - F(x)]dG(x) \tag{1}$$

and the probability that an item is suspended before t km is

$$\Pr\{(T_s < t) \cap (T_f > T_s)\} = \int_0^t [1 - G(x)]dF(x) \tag{2}$$

T_f and T_s being independent random variables.

Assuming that $G(t)$ and $F(t)$ are exponential functions, that is, $G(t) = 1 - \exp(-at)$ and $F(t) = 1 - \exp(-bt)$, the probability that an item fails before t becomes

$$\frac{a}{a+b}\{1 - \exp[-(a+b)t]\} \tag{3}$$

and, similarly, the probability that an item is suspended before t becomes

$$\frac{b}{a+b}\{1 - \exp[-(a+b)t]\} \tag{4}$$

Some comments on the assumption of the exponential model for $G(t)$ are required. Warranty data are essentially data on early failures; hence, from a theoretical viewpoint, kilometers to failure should have a decreasing failure rate. Then, as an example, a Weibull model with shape parameters less than 1 should be more sound. However, experimental results have shown that in situations similar to the present one (see Refs. 3 and 4), the shape parameter of the Weibull distribution is very close to 1. Hence, it appears that there is no practical advantage to using a more complex model than the exponential one for $G(t)$.

To explain the choice made for $F(t)$, some information given in Ref. 2 about the kilometer distribution versus age (from selling date) for a fleet of European cars can be used. In Ref. 2, information is available on the two 5% tails of the distributions at 3, 6, 9, and 12 months of car age. For each of these distributions a Weibull model that had the same two 5% tails can be assumed. In Figure 1 the kilometer distributions (at 3, 6, 9, and 12 months of age) are reported using a Weibull probability distribution. Then the "compound" kilometer distribution, which corresponds to car ages (from selling dates) uniformly distributed over the range of 12 months, is drawn. As can be seen from the estimates reported in Table 2 (calculated with the maximum likelihood method), this distribution turns out to be very close to the exponential, being close to one its shape parameter. Even if this result cannot be considered decisive proof, the exponential model appears to be at the very least the preferential candidate for $F(t)$ in the present situation.

3.3. Estimation Procedure

In order to estimate the two unknown parameters, a and b, we use a very powerful statistical estimation method, the *maximum likelihood* method. The logic of this method is very simple, and even those who are not statisticians can take advantage of it. It is founded on the idea that the probability law we are looking for is most likely the one—of the hypothesized family—that shows the maximum joint probability density of the collected data

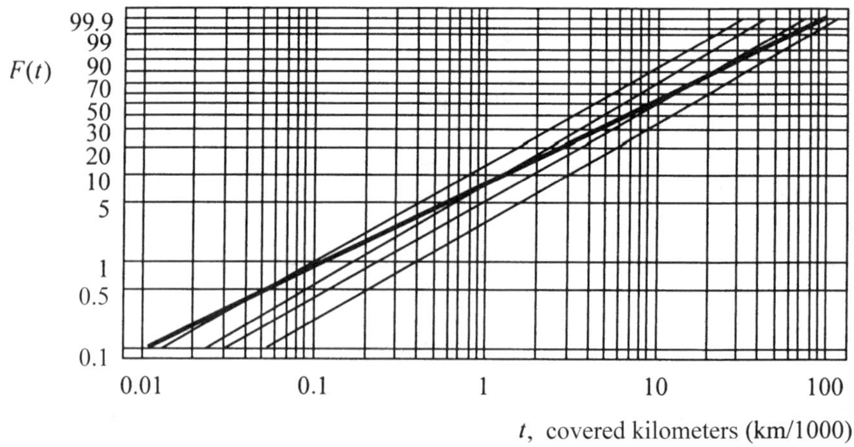

Figure 1 Weibull kilometer distributions at 3, 6, 9, and 12 months and the corresponding compound distribution (heavy line).

Table 2 Parameter Estimates of the Kilometer Distribution Drawn in Figure 1

	3 mo	6 mo	9 mo	12 mo	Compound
Weibull shape parameter	1.13	1.17	1.13	1.14	0.97
Weibull scale parameter	5.85	11.04	17.56	23.25	13.56

sample (called the *likelihood* function, L). So those values of the unknown parameters that maximize the L function are the *maximum likelihood estimates*.

The definition of the probabilities of both failure and suspension allows us to construct the L function for a sample arising from the experimental situation under study [8]. In fact, letting

N = total number of cars under observation
n_i = number of failures in the ith lifetime (T_i, T_{i+1}), $T_1 = 0$
m = number of lifetimes
$n = \sum_{i=1}^{m} n_i$ = total number of failures

the likelihood function, L, is found to be proportional to

$$L \propto \left[\int_0^{\infty} [1 - G(x)]dF(x)\right]^{N-n} \prod_{i=1}^{m} \left[\int_{T_i}^{T_{i+1}} [1 - F(x)dG(x)\right]^{n_i}$$
$$= \left(\frac{b}{a+b}\right)^{N-n} \prod_{i=1}^{m} \left[\frac{a}{a+b} \{\exp[-(a+b)T_i] - \exp[-(a+b)T_{i+1}]\}\right]^{n_i} \quad (5)$$

The values of a and b that maximize this function (given a sample with known N and n_i) are the needed maximum likelihood estimates of the unknown parameters a and b.

Reparameterizing for convenience the likelihood function in terms of a and $c = a + b$ and equating to zero the partial derivatives of $\ln(L)$ yields the equations

$$\frac{a}{c} = \frac{n}{N} \quad (6a)$$

and

$$\sum_{i=1}^{m} n_i \frac{T_{i+1}\exp(-cT_{i+1}) - T_i \exp(-cT_i)}{\exp(-cT_i) - \exp(-cT_{i+1})} = 0 \quad (6b)$$

To solve this last nonlinear equation in c, an iterative method is required (e.g., the Newton–Raphson method). An initial tentative value, c^*, can be easily found by taking into account that the probability that an item fails in the ith interval has the nonparametric estimate n_i/N. Thus, as an example, for the first lifetime ($T_1 = 0$, T_2), (see equation 3),

$$\Pr[(0 < T_f \leq T_2) \cap (T_s > T_f)] = \frac{a}{c}[1 - \exp(-cT_2)] = \frac{n_1}{N} \tag{7}$$

and since from the first likelihood equation, $a/c = n/N$, an initial tentative value follows:

$$c^* = \frac{-\ln(1 - n_1/n)}{T_2} \tag{8}$$

Then the estimates a and $b = c - a$ can be obtained without any further difficulty.

3.4. Practical Example and Comments

The maximum likelihood method was applied to the sample of warranty claims from real life given in Table 1. The following estimates of the unknown parameters a and b were found:

$$a = 2.11 \times 10^{-5}\,\text{km}^{-1} \qquad \text{and} \qquad b = 12.93 \times 10^{-5}\,\text{km}^{-1}$$

The model chosen appears to fit the experimental data with an extremely high degree of accuracy. In Table 3 the observed and the estimated numbers of failures for each lifetime are reported. The estimated average number of kilometers for the fleet under test is $1/b = 7734$ km, which is a very plausible value for a fleet of cars whose ages are distributed over 12 months.

Table 3 Comparison of Collected and Estimated Numbers of Claims

Lifetimes (km/1000)	0–10	10–20	20–30	30–40	> 40
No. of claims, observed	55	11	3	1	0
No. of claims, estimated	54	12	3	1	0

4. SOME CONCLUDING REMARKS

Sometimes, the evaluation of the ability of an item to satisfy customers can be carried out effectively only via reliability analysis. In fact, some stated or implicit customers' needs must be satisfied over time, because they are time-oriented characteristics. So reliability methods conceived to assess the probability of performing required functions for a stated period of time can be the most suitable ones.

The analysis of customer claims presented in this work is only one example of the possible applications of the reliability approach. Another example, not restricted to the warranty period, is given in Ref. 9.

Moreover, it must be pointed out that reliability analysis can be used not only to constantly verify that the quality targets are really fulfilled in service, but also to help improve any time-oriented quality characteristic of a product by estimating its current level in the field. Therefore, collecting field failure data over the entire life of a product and performing a reliability analysis can be an effective policy for achieving continuous improvement of products [6].

In this context, reliability analysis allows one not only to control the failure or degradation of product performance, but also to reduce the variation in the performance over time among copies of the same item [7]. In order to do that, many practical methods for estimating failure distributions are available in the reliability literature, primarily those that integrate the competence of both engineers and statisticians [10, 11], since, as has already been said, only by involving everyone's commitment can total quality be achieved.

REFERENCES

1. Brunelle RD, Kapur KC. Customer-centered reliability methodology. Proceedings of Annual Reliability and Maintainability Symposium, Philadelphia, 1997, pp. 286–292.
2. Toth-Fay R. Evaluation of field reliability on the basis of the information supplied by the warranty. Proceedings of EOQC II European Seminar on Life Testing and Reliability, Torino, Italy, 1971, pp. 71–78.
3. Turpin MP. Application of computer methods to reliability prediction and assessment in a commercial company. Reliab Eng 3: 295–314, 1982.
4. Vikman S, Johansson B. Some experiences with a programmed Weibull routine for the evaluation of field test results. Proceedings of EOQC II European Seminar on Life Testing and Reliability, Torino, 1971, p. 70 (ext vers pp. 1–20).
5. Kalbfleish JD, Lawless JF, Robinson JA. Methods for the analysis and prediction of warranty claims. Technometrics 33(3): 273–285, 1991.

6. Jones JA, Hayes JA. Use of a field failure database for improvement of product reliability. Reliab Eng Syst Safety 55: 131–134, 1997.
7. Wayne KY, Kapur KC. Customer driven reliability: Integration of QFD and robust design. Proceedings of Annual Reliability and Maintainability Symposium, Philadelphia, 1997, pp. 339–345.
8. Erto P, Guida M. Some maximum likelihood reliability estimates from warranty data of cars in users' operation. Proceedings of European Reliability Conference, Copenhagen, 1986, pp. 55–58.
9. Erto, P. Reliability assessments by repair shops via maintenance data. J Appl Stat 16(3): 303–313, 1989.
10. Erto P, Guida M. Estimation of Weibull reliability from few life tests. J Qual Reliab Eng Int 1(3): 161–164, 1985.
11. Erto P, Giorgio M. Modified practical Bayes estimators. IEEE Trans Reliab 45(1): 132–137, 1996.

7

Some Recent Developments in Control Charts for Monitoring a Proportion

Marion R. Reynolds, Jr.
Virginia Polytechnic Institute and State University, Blacksburg, Virginia

Zachary G. Stoumbos
Rutgers University, Newark, New Jersey

1. INTRODUCTION

Control charts are used to monitor a production process to detect changes that may occur in the process. In many applications, information about the process may be in the form of a classification of items from the process into one of two categories, such as defective or nondefective, or nonconforming and conforming. The process characteristic of interest is the proportion p of items that fall in the first category. For convenience in describing the problem being considered, the labels "defective" and "nondefective" will be used in this paper for the two categories. It is usually assumed that the items from the process are independent with probability p of being defective. This would then imply that the total number of defective items in a sample of n items, say T, has a binomial distribution. In most quality control applications the primary objective in using a control chart would be to detect an increase in p, because an increase in p corresponds to a decrease in quality. However, a decrease in p would be of interest if it is important to document an improvement in process quality. Woodall (1997) gives a general review of control charts that can be applied to the problem of monitoring p.

The traditional approach to applying a control chart to monitor p is to take samples of size n at regular intervals and plot the values of the sample proportion defective, T/n, on a Shewhart p-chart. The p-chart usually has control limits set at ± 3 standard deviations from the in-control value p_0 (three-sigma limits). Although the p-chart is relatively easy to set up and interpret, it has a number of disadvantages. These disadvantages are particularly critical when p_0 is close to zero. Monitoring a process with p_0 close to 0 is becoming more and more common with the increasing emphasis on high quality production. Thus, it is important to be aware of the disadvantages of the p-chart and to consider better alternatives.

The distribution of T is discrete, and when p_0 is close to zero the distribution of T is also highly skewed (unless n is very large). This results in the p-chart with 3σ limits having properties very different from what would be expected from a normal distribution with 3σ limits. For example, for many values of n and p_0 that might occur in applications, the calculated lower control limit is negative, so there is, in effect, no lower control limit. This means that the chart will not be able to detect decreases in p. In addition, when $p = p_0$, the discreteness and skewness of T can result in a probability above the upper control limit that is far from the value 0.00135 expected from the normal distribution. If the probability is far above 0.00135, then the false alarm rate will be much higher than expected, whereas if the probability is far below 0.00135, then the false alarm rate will be much lower than expected. A lower than expected false alarm rate is undesirable because it means that process changes will be detected more slowly than necessary.

The p-chart is a Shewhart control chart that plots, at each sampling point, the proportion defective for that sample alone. Information from past samples is not used, and this results in a chart that is not very efficient for detecting small changes in p. In particular, if p_0 is close to 0, then the p-chart requires a very large value of n to detect a small increase in p within a reasonable length of time. The traditional approach to improving the efficiency of a Shewhart chart for detecting small process changes is to use runs rules. For the p-chart, the use of runs rules might also enable decreases in p to be detected when there is no lower control limit. The disadvantages of using large numbers of runs rules are that the chart is more difficult to interpret and the evaluation of the statistical properties of the chart is much more complicated. Most evaluations of the statistical properties of runs rules are based on the symmetrical normal distribution, with regions within the control limits specified in terms of the standard deviation of the statistic being plotted. Applying these runs rules to the p-chart results in the same problem as with the control limits; the discreteness and skewness of the distribution of T can result in runs rules with properties much different than expected.

An additional disadvantage to using a p-chart with runs rules is that using runs rules is not the most efficient way to detect small changes in p. A better approach to obtaining a control chart that will detect small process changes is to use a chart, such as a CUSUM chart, that directly and efficiently uses the past sample data at each sample point. Although the discreteness of the distribution of T will be an issue with a CUSUM chart, the fact that the CUSUM chart is based on a sum will make the discreteness much less of a problem than in the case of the p-chart. In the past, a hindrance to the application of CUSUM charts to monitor p is that it is difficult for the practitioner to determine the CUSUM chart parameters that will give specified properties. Some tables or figures have been published [see, for example, Gan (1993)], but these tables and figures do not include all values of n and p_0 that would be of interest in applications.

Another approach to obtaining more efficient control charts is to use a control chart that varies the sampling rate as a function of the process data. Although a large number of papers have been published in recent years on variable sampling rate control charts [see, e.g., Reynolds (1996a) and Stoumbos and Reynolds (1997b)], only a few papers have been published on the specific problem of monitoring p [see, e.g., Rendtel (1990) and Vaughan (1993)]. The application of variable sampling rate control charts to monitoring p has been hindered by the difficulty of determining the chart parameters that will give specified properties.

The objective of this chapter is to consider three highly efficient control charts for monitoring p that can be used in three different situations. The first control chart is a CUSUM chart, called the Bernoulli CUSUM chart, that can be used in situations in which all items from the process are inspected. The use of 100% inspection is becoming more common as automatic inspection systems are implemented. Also, in the highly competitive global markets of today there is an increasing emphasis on maintaining a very low proportion of product that is defective or that does not meet specifications. The sampling rates that are necessary to discriminate between very low values of p will frequently correspond to 100% inspection. CUSUM charts for this problem have been considered before [see, e.g., Bourke (1991)]. A disadvantage of these CUSUM charts has been that designing a CUSUM chart for a particular application has been difficult unless the values of n and p_0 in the application happen to correspond to values in published tables. A contribution of the current chapter is to show how to design a CUSUM chart for the case of 100% inspection using relatively simple and highly accurate approximations.

The second control chart to be considered is a CUSUM chart that can be applied in situations in which samples of n items are taken from the process at regular intervals. CUSUM charts for this problem have been studied before [see, e.g., Gan (1993)]. As in the case of 100% inspection, a disadvantage of the CUSUM chart in this situation has been that designing one for a particular application has been difficult unless the values of n and p_0 in the application happen to correspond to published results. A contribution of the current chapter is to show how to design a CUSUM chart for the binomial distribution using relatively simple and highly accurate approximations.

The third control chart to be considered here is a chart that can be applied when it is not feasible to use 100% inspection but it is feasible to vary the sample size used at each sampling point depending on the data obtained at that sampling point. The sample size is varied by applying a *sequential probability ratio test* (SPRT) at each sampling point. This SPRT chart for monitoring p is a variable sampling rate control chart, and it is much more efficient than charts that take a fixed-size sample. Methods based on relatively simple and highly accurate approximations are given for designing the SPRT chart.

The remainder of this chapter is organized as follows. Sections 2–5 pertain to the Bernoulli CUSUM chart, Sections 6–8 pertain to the binomial CUSUM chart, and Sections 9–12 pertain the the SPRT chart. For each chart, a description is given, the evaluation of statistical properties is discussed, a design method is explained, and a design example is given. Some general conclusions are given in Section 13.

2. THE BERNOULLI CUSUM CHART WHEN USING 100% INSPECTION

When all items from the process are inspected, the results of the inspection of the ith item can be represented as a Bernoulli observation X_i, which is 1 if the ith item is defective and 0 otherwise. Then p corresponds to $P(X_i = 1)$. The control chart to be considered for this problem is a CUSUM chart based directly on the individual observations $X_1, X_2, \ldots$ without any grouping into segments or samples. This Bernoulli CUSUM chart is defined here for the problem of detecting an increase in p. The problem of detecting a decrease in p, as well as additional details about the Bernoulli CUSUM chart, are given in Reynolds and Stoumbos (1999).

For detecting an increase in p, the Bernoulli CUSUM control statistic is

$$B_i = \max\{0, B_{i-1}\} + (X_i - \gamma), \qquad i = 1, 2, \ldots \tag{1}$$

where $\gamma > 0$ is the reference value. After the inspection of item i this CUSUM statistic adds the increment $X_i - \gamma$ to the previous value as long as the previous value is nonnegative, but resets the cumulative sum to 0 if the previous value drops below 0. The starting value B_0 is frequently taken to be 0 but can be taken to be a positive value if a head start is desired [see Lucas and Crosier (1982) for a discussion of using a head start in a CUSUM chart]. This chart will signal that there has been an increase in p if $B_i \geq h_B$, where h_B is the control limit. The reference value γ can be chosen by using the representation of a CUSUM chart as a sequence of SPRTs. To determine the value of γ it is necessary to specify a value $p_1 > p_0$ that represents an out-of-control value of p that should be detected quickly. For a given in-control value p_0 and a given out-of-control value p_1, define the constants r_1 and r_2 as

$$r_1 = -\log \frac{1 - p_1}{1 - p_0} \qquad \text{and} \qquad r_2 = \log \frac{p_1(1 - p_0)}{p_0(1 - p_1)} \tag{2}$$

Then, from the basic definition of an SPRT (see Section 9), it can be shown that the appropriate choice for γ is

$$\gamma = r_1/r_2 \tag{3}$$

It will usually be convenient if $\gamma = 1/m$, where m is an integer. For example, if $p_0 = 0.005$ and p_1 is chosen to be $p_1 = 0.010$, then this will give $r_1 = 0.00504$, $r_2 = 0.6982$, and $r_1/r_2 = 0.00722 = 1/138.6$. In this case, if p_1 is adjusted slightly from 0.010 to 0.009947, then r_1/r_2 will decrease slightly to 1/139. This means that the possible values of B_i will be integer multiples of 1/139, and this will be convenient for plotting the chart. In general, if p_0 and p_1 are small, then a slight change in p_1 will be sufficient to make $\gamma = 1/m$, where m is an integer. In most cases the precise specification of p_1 will not be critical, so this slight change in p_1 will be of no practical consequence.

3. PROPERTIES OF THE BERNOULLI CUSUM CHART

The performance of a control chart is usually evaluated by looking at the average run length (ARL), which is the expected number of samples required to signal. In the current context of 100% inspection, there may be no natural division of observations into samples or segments, and thus, to avoid confusion, we will use the *average number of observations to signal* (ANOS) instead of the ARL to measure the performance of control charts.

Assuming that the production rate is constant, the ANOS can be easily converted to time units, and for purposes of exposition we will frequently refer to the ANOS as a measure of detection time. When the process is in control ($p = p_0$), it is desirable to have a large ANOS so that the rate of false alarms is low. On the other hand, when there has been a significant change in p, it is desirable to have a small ANOS so that this change in p is detected quickly. In the previous section p_1 was defined as a value of p that should be detected quickly, and thus the ANOS should be small at $p = p_1$. However, in practice, it is usually desirable to consider a range of values of p around p_1 and to have a chart with good performance for all of these values of p.

The ANOS of the Bernoulli CUSUM can be evaluated by formulating the CUSUM as a Markov chain [see Reynolds and Stoumbos (1999)]. This approach gives the exact ANOS when r_1/r_2 is a rational number, but the disadvantage is that a computer program is usually required. The approach to be given here is from Stoumbos and Reynolds (1996) and Reynolds and Stoumbos (1999) and is based on using approximations developed by Wald (1947) and diffusion theory corrections to these approximations obtained by Reynolds and Stoumbos (1999) by extending the work of Siegmund (1985). The approximation for the ANOS that is obtained using this approach will be called the *corrected diffusion* (CD) approximation. The CD approximation will form the basis of a highly accurate and relatively simple design method that requires only a pocket calculator to design the Bernoulli CUSUM for practical applications.

4. A METHOD FOR DESIGNING THE BERNOULLI CUSUM CHART

To design a Bernoulli CUSUM chart for a particular application it will be necessary to specify p_0, the in-control value of p, and p_1, the value of p that the chart is designed to detect. The values of p_0 and p_1 will then determine the reference value γ through Eq. (3). As discussed above, when p_0 and p_1 are small it will usually be convenient to adjust p_1 slightly so that $\gamma = 1/m$, where m is an integer. The design method is presented here for the case in which $0 < p_0 < 0.5$. The case in which $p_0 \geq 0.5$ is discussed in Reynolds and Stoumbos (1999).

In designing the chart it is also necessary to determine the value for the control limit h_B. The value of h_B will determine the false alarm rate and the speed with which the chart detects increases in p. A reasonable approach to determining h_B is to specify a desired value of the ANOS when $p = p_0$ and then choose h_B to achieve approximately this value of the ANOS. It will usually not be possible to achieve exactly a desired value of the ANOS because the Bernoulli distribution is discrete. Once h_B is chosen to achieve approximately the specified ANOS at $p = p_0$, it will be desirable to look at the ANOS at $p = p_1$ and at other values of p to determine whether detection of shifts in p will be fast enough. In practice, it may be necessary to adjust h_B to achieve a reasonable balance between the desire to have a low false alarm rate (achieved by choosing a large h_B) and fast detection of shifts in p (achieved by choosing a small h_B).

The CD approximation to the ANOS of the Bernoulli CUSUM chart uses an adjusted value of h_B, which will be denoted by h_B^*, in a relatively simple formula. This adjusted value of h_B is

$$h_B^* = h_B + \varepsilon(p_0)\sqrt{p_0 q_0} \tag{4}$$

where $\varepsilon(p)$ can be approximated by

$$\varepsilon(p) \approx \begin{cases} 0.410 - 0.0842\log(p) - 0.0391[\log(p)]^3 \\ \quad -0.00376[\log(p)]^4 - 0.000008[\log(p)]^7 & \text{if } 0.01 \le p < 0.5 \\ \dfrac{1}{3}\left(\sqrt{\dfrac{1-p}{p}} - \sqrt{\dfrac{p}{1-p}}\right) & \text{if } 0 < p < 0.01 \end{cases} \tag{5}$$

When $p = p_0$, the CD approximation to the ANOS is

$$\text{ANOS}(p_0) \approx \frac{e^{h_B^* r_2} - h_B^* r_2 - 1}{|r_2 p_0 - r_1|} \tag{6}$$

For given values of r_1 and r_2 and a desired value for the in-control ANOS, Eq. (6) can be used to find the required value of h_B^*, and then (4) and (5) can be used to find the required value of h_B. Finding h_B^* using (6) can be accomplished by simple trial and error.

In most applications it will be desirable to determine how fast a shift from p_0 to p_1 will be detected. The CD approximation to the ANOS when $p = p_1$ is

$$\text{ANOS}(p_1) \approx \frac{e^{-h_B^* r_2} + h_B^* r_2 - 1}{|r_2 p_1 - r_1|} \tag{7}$$

Note that h_B^* uses p_0 even though the ANOS is being approximated at p_1. Approximations to the ANOS for other values of p and a discussion of the accuracy of the CD approximation are given in Reynolds and Stoumbos (1999).

5. AN EXAMPLE OF DESIGNING A BERNOULLI CUSUM CHART

Consider a production process for which it has been possible to maintain the proportion defective at a low level, $p_0 = 0.005$, except for occasional periods in which the value of p has increased above this level. All items from this production process are automatically inspected, and a Shewhart p-chart is currently being used to monitor this process. Items are grouped into segments of $n = 200$ items for purposes of applying the p-chart. If 3σ limits are used with the p-chart, then the upper control limit is 0.01996, and this is equivalent to signaling if $T_j \geq 4$, where T_j is the number of defectives in the jth segment. When $p = p_0 = 0.005$, this results in $P(T_j \geq 4) = 0.01868$, and it was decided that this probability of a false alarm was too high. Thus, the upper control limit of the p-chart was adjusted so that a signal is given if $T_j \geq 5$, and this gives a probability of 0.00355 for a false alarm. There is no lower control limit because giving a signal for $T_j = 0$, the lowest possible value of T_j, would result in $P(T_j = 0) = 0.3670$ when $p = p_0$, and thus the false alarm rate would be unacceptably high. When $p = p_0$, the expected number of segments until a signal is $1/0.00355 = 282.05$. Each segment consists of 200 items, so this corresponds to an in-control ANOS of 56,410 items.

To design a Bernoulli CUSUM chart for this problem, suppose that process engineers decide that it would be desirable to quickly detect any special cause that increases p from 0.005 to 0.010 and that the in-control ANOS should be roughly 56,410 (the value corresponding to the p-chart in current use). From a previous discussion of the case of $p_0 = 0.005$ and $p_1 = 0.010$, it was shown that adjusting p_1 slightly from 0.010 to 0.009947 would give $r_1/r_2 = 1/139$, and thus $m = 139$. Using trial and error to solve (6) to give $\text{ANOS}(p_0) \approx 56{,}410$ results in a value of h_B^* of 6.515 [this value of h_B^* will give an in-control ANOS of 56,408 according to the approximation of Eq. (6)]. Then, using (4) and (5) to convert to h_B gives $\varepsilon(p) = 4.646$, $\varepsilon(p_0)\sqrt{p_0 q_0} = 0.328$, and $h_B = 6.187$. As a point of interest, the exact in-control ANOS using $h_B = 6.187$ can be calculated to be 56,541 by using the methods given in Reynolds and Stoumbos (1999). Thus, in this case the CD approximation gives results that are extremely close to the exact value and certainly good enough for practical applications.

After h_B has been determined, Eq. (7) can be used to determine how fast a shift from p_0 to p_1 will be detected. Using $h_B^* = 6.515$ in (7) gives

ANOS(p_1) $\approx$ 1848. Interestingly, the exact ANOS can be calculated to be 1856, so the CD approximation is also very good at $p = p_1$. At $p = p_1$, the ANOS of the p-chart is 3936 items. Thus, the p-chart would require on average more than twice as long as the Bernoulli CUSUM chart to detect a shift from p_0 to p_1.

6. THE BINOMIAL CUSUM CHART

In many applications 100% inspection of the process output will not be feasible, and thus samples from the output will have to be used for monitoring. In this section, the problem of monitoring p when the data from the process consist of samples of fixed size n that are taken at fixed sampling intervals of length d is investigated. If T_k is used to represent the total number of defectives observed in the kth sample, then the statistics $T_1, T_2, \ldots$ are independent binomial random variables. The control chart to be considered here is a CUSUM chart based on these statistics.

The binomial CUSUM chart uses the control statistic

$$Y_k = \max\{0, Y_{k-1}\} + (T_k - n\gamma), \qquad k = 1, 2, \ldots \tag{8}$$

and signals at sample k if $Y_k \geq h_Y$, where Y_0 is the starting value and γ is given by (3). The reference value of this CUSUM chart is $n\gamma = nr_1/r_2$, and this reference value is appropriate for detecting a shift to p_1.

In the current situation in which samples are taken from the process, the performance of a control chart can be measured by the *average time to signal* (ATS). As in the case of using the ANOS in previous sections, when $p = p_0$ the ATS should be large, and when p shifts from p_0 the ATS should be small. In evaluating the ATS of the binomial CUSUM chart, it wll be assumed for simplicity that the time required to take and plot a sample of n observations is negligible relative to the time d between samples. In this case, the ATS can be expressed as the product of d and the *average number of samples to signal* (ANSS). Gan (1993) discusses Markov chain methods for evaluating the ANSS of the binomial CUSUM chart. Here we use CD approximations to design the binomial CUSUM chart.

When the ATS is used as a measure of the time required to detect a shift in p, the ATS is usually computed assuming that the shift in p occurs when process monitoring starts. However, in many cases the process may run for a while at the in-control value p_0 and then shift away from p_0 at some random time in the future. In this case the detection time of interest is the time from the shift to the signal by the control chart. For control charts such as the CUSUM chart, the computation of this expected time is complicated by the fact that the CUSUM statistic may not be at its starting value when the shift in p occurs. If it is assumed that the CUSUM statistic has reached its stationary or steady-state distribution by the time the shift occurs, then the expected time from the shift to the signal is called the *steady-state* ATS (SSATS). When peforming comprehensive comparisons of different control charts, it is appropriate to consider the SSATS as a measure of detection time. However, for the limited comparisons to be given in the design examples in this paper, the ATS will be used.

7. A METHOD FOR DESIGNING THE BINOMIAL CUSUM CHART

A method for designing the binomial CUSUM can be developed by using CD approximations to the ANSS and the ATS. This method is presented here for the special case in which $p_0 < 0.5$ and $1/n\gamma$ is a positive integer. Extensions of this method to more general cases are currently under development.

The CD aproximation to the ANSS of the binomial CUSUM uses an adjusted value of h_Y, which will be denoted as h_Y^*, in a relatively simple formula. The adjusted value of h_Y is

$$h_Y^* = h_Y + \tfrac{1}{3}(1 - 2p_0), \qquad 0 < p_0 < 0.5 \tag{9}$$

When $p = p_0$, the CD approximation to the ANSS is

$$\text{ANSS}(p_0) \approx \left(\frac{r_2}{n(r_2 p_0 - r_1)}\right)\left(h_Y^* - \frac{n\gamma(e^{h_Y^* r_2} - 1)}{1 - e^{-n\gamma r_2}}\right) \tag{10}$$

When $p = p_1$, the CD approximation to the ANSS is

$$\text{ANSS}(p_1) \approx \left(\frac{r_2}{n(r_2 p_1 - r_1)}\right)\left(h_Y^* - \frac{n\gamma e^{-n\gamma r_2}(e^{h_Y^* r_2} - 1)}{e^{h_Y^* r_2}(1 - e^{-n\gamma r_2})}\right) \tag{11}$$

An approximation to the ATS is obtained by multiplying the ANSS by the sampling interval d.

To design a binomial CUSUM chart, the values of p_0 and p_1 can be specified, and then these values will determine the reference value γ through Eq. (3). To use the CD approximations given above it will be necessary to choose values of n and p_1 such that $1/n\gamma$ is a positive integer. In this case, the possible values of the binomial CUSUM statistic Y_k will be integer multiples of $n\gamma$. Thus, in considering values for h_Y, it is sufficient to look at values that are integer multiples of $n\gamma$. These values of h_Y will correspond to certain values of h_Y^* using (9). Using the approximation (10), the value of h_Y^* can be selected that will give approximately the desired value for the in-control ANSS. Then the h_Y to be used can be obtained from h_Y^* by using (9).

8. AN EXAMPLE OF DESIGNING A BINOMIAL CUSUM CHART

Consider a situation similar to the example in Section 5 in which a Shewhart p-chart is being used to monitor a production process for which the in-control value of p is $p_0 = 0.005$. Instead of using 100% inspection for this process, suppose that it is necessary to take samples from the process output. The value of p_0 is relatively small, and thus it is necessary to take relatively large samples for the p-chart to be able to detect small increases in p above p_0. Suppose that samples of size $n = 200$ items are used (the same as the size of the segments in the example in Section 5). To keep the total sampling effort to a reasonable level, the samples are taken from the process every $d = 4$ hr. As in the previous example, the upper control limit of the p-chart was adjusted so that a signal is given if $T_k \geq 5$, which gives $P(T_k \geq 5) = 0.00355$ when $p = p_0$. This corresponds to an in-control ANSS of $1/0.00355 = 282.05$ and an in-control ATS of $4(282.05) = 1128.2$ hr.

Consider now the design of a binomial CUSUM chart assuming that samples will be taken every $d = 4$ hr as described above. Suppose that process engineers decide that it is important to detect a shift in p from $p_0 = 0.005$ to $p_1 = 0.010$ and that it would be reasonable to have an in-control ANSS of approximately 282 (the same as the value for the p-chart). As in the example in Section 5, adjusting p_1 slightly from 0.010 to 0.009947 will give $\gamma = r_1/r_2 = 1/139$. If n is taken to be 139, then the reference value of the binomial CUSUM chart becomes $n\gamma = 139/139 = 1$. Many practitioners might prefer to have $n = 140$, rather than 139, and this can be achieved by an additional slight adjustment in p_1. If p_1 is adjusted to 0.009820, then this will give $\gamma = r_1/r_2 = 1/140$. Then, taking $n = 140$ gives a reference value of $n\gamma = 140/140 = 1$.

The reference value for the binomial CUSUM chart is 1, so it follows that it is sufficient to look at values for h_Y that are integer multiples of 1. If several values of h_Y are tried, it is found that using $h_Y = 5$ in Eq. (9) gives $h_Y^* = 5.33$, and using this h_Y^* in Eq. (10) gives ANSS$(p_0) \approx 228.7$. As a point of interest, the exact ANSS for this value of h_Y is 228.6. Using $h_Y = 6$ in Eq. (9) gives $h_Y^* = 6.33$, and using this h_Y^* in (10) gives ANSS$(p_0) \approx 471.8$ (the exact value is 471.3). Neither of these ANSS values is extremely close to the desired value of 282, but suppose that it is decided that 228.7 corresponding to $h_Y = 5$ is close enough. Using $h_Y = 5$ will give an in-control ATS of approximately $4 \times 228.7 = 914.8$. Using $h_Y = 5$ and $h_Y^* = 5.33$ in (11) gives ANSS(p_1) ≈ 11.6 (the exact value is 11.9). This corresponds to an ATS at $p = p_1 = 0.0098$ of approximately $4 \times 11.6 = 46.4$ hr. At $p = p_1$, the ATS of the p-chart is 80.3 hr. Thus, the binomial CUSUM chart will detect a shift to p_1 faster than the p-chart will. Note that the p-chart is sampling at a higher rate than the CUSUM chart (200 every 4 hr versus 140 every 4 hr), but the CUSUM chart has a slightly higher false alarm rate.

When the p-chart is being used to detect small increases in p above a small value of p_0, it is necessary to use a large sample size to detect this increase in a reasonable amount of time. This may require that the sampling interval d be relatively long in order to keep the sampling cost to a reasonable level. However, for the binomial CUSUM chart it is not necessary to have n large; it is actually better to take smaller samples at shorter intervals. Thus, as an alternative to taking a sample of $n = 140$ every $d = 4$ hr, consider the option of taking a sample of $n = 70$ every $d = 2$ hr. If the binomial CUSUM chart uses $n = 70$ and $p_1 = 0.009820$, then the reference value will be $n = 70/140 = 0.5$, and the possible values for Y_k will be integer multiples of 0.5. Thus, it is sufficient to look at values for h_Y that are integer multiples of 0.5. If the p-chart has an in-control ATS of 1128 and it is desirable to have approximately the same value for the binomial CUSUM with $d = 2$, then the in-control ANSS should be $1128/2 = 564$. Using $h_Y = 5.5$ in (9) gives $h_Y^* = 5.83$, and using this h_Y^* in (10) gives ANSS(p_0) ≈ 558.5 (the exact value is 557.9). This corresponds to an in-control ATS of approximately $2 \times 558.5 = 1117.0$. Using (11) gives ANSS(p_1) ≈ 24.6 (the exact value is 25.1). This corresponds to an ATS at $p = p_1 = 0.0098$ of approximately $2 \times 24.6 = 49.2$ hr. Compared to the p-chart, this binomial CUSUM chart has almost the same false alarm rate and a lower sampling rate, yet it will detect a shift to p_1 much faster.

As another alternative to taking a sample of $n = 140$ every $d = 4$ hr, consider the option of taking a sample of $n = 35$ every hour. If the binomial CUSUM chart uses $n = 35$ and $p_1 = 0.009820$, then the reference value will be $n\gamma = 35/140 = 0.25$, and the possible values for Y_k will be integer multiples of 0.25. Thus, it is sufficient to look at values for h_Y that are integer multiples of 0.25. If the p-chart has an in-control ATS of 1128 and it is desirable to have approximately the same value for the binomial CUSUM with $d = 1$, then the in-control ANSS should be 1128. Using $h_Y = 5.75$ in (9) gives $h_Y^* = 8.08$, and using this h_Y^* in (10) gives ANSS(p_0) ≈ 1228.2 (the exact value is 1226.6). Because $d = 1$, this corresponds to an in-control

ATS of 1226.2 hr. Using (11) gives ANSS(p_1) $\approx$ 50.7 (the exact value is 51.3). This corresponds to an ATS at $p = p_1 = 0.0098$ of 50.7 hr.

The three binomial CUSUM charts that have been considered here have the same sampling rate of 35 observations per hour. However, their false alarm rates are not exactly the same, so it is difficult to do precise comparisons of the charts. But based on the results given for these charts, it seems clear that taking small samples at frequent intervals would give fast detection of process shifts. If n is reduced to the smallest possible value, 1, then the binomial CUSUM chart reduces to the Bernoulli CUSUM chart discussed previously. Using $n = 1$ might be the best way to apply a CUSUM chart from a statistical point of view, but taking samples of size $n = 1$ might be inconvenient in some applications.

9. THE SPRT CHART

CUSUM charts, such as the binomial CUSUM chart described previously, can be thought of as sequences of SPRTs carried out over successive sampling points. The SPRT chart to be considered in this section is based on using SPRTs in a different way. In particular, the SPRT chart is based on applying a sequential test (an SPRT) to the individual items inspected at each sampling point. A description of the SPRT chart for the case of monitoring a general parameter is given by Stoumbos and Reynolds (1996) and, for the case of monitoring the mean of a normal distribution, in Stoumbos and Reynolds (1997b). More details about the current problem of applying the SPRT chart to monitor p are given in Reynolds and Stoumbos (1998). In the context of hypothesis testing, the SPRT is a general sequential test that can be applied to test a simple null hypothesis against a simple alternative hypothesis. For the case of a test involving the proportion defective p, the SPRT can be used to test the null hypothesis H_0: $p = p_0$ against the alternative hypothesis H_1: $p = p_1$. In the context of monitoring p, p_0 would be the in-control value of p, and p_1 would be a value that should be detected quickly, as defined in previous sections.

Suppose that a sampling interval of length d is used for sampling from the process. At each sampling point items from the process are inspected one by one and an SPRT is applied, with the sample size used at each sampling point being determined by the SPRT. If items can be inspected quickly enough, then the inspection can be done on consecutive items as they come from the process. For example, if an item is produced every 10 sec and the inspection and recording of the result take no more than 10 sec, then inspection can be done as the items are produced. On the other hand, if the inspection rate is slower than the production rate, then inspection could be done after production on items that have been accumulated. Alternatively, inspection could be done on items as they come from production, with some items skipped. For example, if an item is produced every 10 sec but inspection requires between 30 and 40 sec, then every third or fourth item could be inspected during inspection periods.

If the SPRT applied at sampling point k accepts $H_0\colon p = p_0$, then the decision is that the process is in control. The process is then allowed to continue to the next sampling point, $k+1$, at which time another SPRT is applied. But if the SPRT applied at sampling point k rejects H_0, then this is taken as a signal that there has been a change in p. Action should then be taken to find and eliminate the cause of this change in p. Thus, the SPRT chart involves applying an SPRT at each sampling point and giving a signal whenever one of these SPRTs rejects H_0.

To define the SPRT that is applied at sampling point k, let the Bernoulli random variable X_{ki} be defined by $X_{ki} = 1$ if the ith item at sampling point k is defective and by $X_{ki} = 0$ otherwise. The statistic used by the SPRT is defined in terms of a log likelihood ratio using the density $f(x; p) = p^x(1-p)^{1-x}$ of X_{ki}. After the jth item is inspected at sampling point k, this log likelihood ratio statistic is

$$S_{kj} = \sum_{i=1}^{j} \log \frac{f(X_{ki}; p_1)}{f(X_{ki}; p_0)} = \sum_{i=1}^{j} (r_2 X_{ki} - r_1) = r_2 T_{kj} - r_1 j \tag{12}$$

where the constants r_1 and r_2 are defined by (2), and

$$T_{kj} = \sum_{i=1}^{j} X_{ki} \tag{13}$$

is the total number of defective items in the first j items inspected at sampling point k.

The SPRT chart requires the specification of two constants a and b, $b < a$, and uses the following rules for sampling and making decisions.

1. At sampling point k, if $b < S_{kj} < a$, then continue sampling.
2. At sampling point k, if $S_{kj} \geq a$, then stop sampling and signal that p has changed.
3. At sampling point k, if $S_{kj} \leq b$, then stop sampling at sampling point k and wait until sampling point $k+1$ to begin applying another SPRT.

The inequality

$$b < S_{kj} < a \tag{14}$$

determines when the SPRT continues sampling and is usually called the *critical inequality* of the SPRT. In some applications it may be more convenient to carry out the SPRT by dividing S_{kj} by r_2 to obtain an equivalent critical inequality. If $p_1 > p_0$, then this equivalent critical inequality is

$$g < T_{kj} - \gamma j < h \tag{15}$$

where $g = b/r_2$, $h = a/r_2$, and γ is given by (3). Thus, after inspecting the jth item at sampling point k, the SPRT is carried out by determining T_{kj}, subtracting γj, and comparing the result to g and h. If (15) holds, then inspection is continued at this point; if $T_{kj} - \gamma j \geq h$, then sampling is stopped and a signal is given; and if $T_{kj} - \gamma j \leq g$, then sampling is stopped until the time for sample $k + 1$ is reached.

As in the cases of the Bernoulli CUSUM and the binomial CUSUM, it will usually be convenient to have $\gamma - 1/m$, where m is a positive integer, so that the SPRT statistic $T_{kj} - \gamma j$ in (15) will take on values that are integer multiples of γ. It will usually be possible to make $\gamma = 1/m$ by a slight adjustment of p_1. When $\gamma = 1/m$, m a positive integer, the acceptance limit g in (15) can be chosen to be an integer multiple of $1/m$, and this will ensure that the SPRT statistic $T_{kj} - \gamma j$ will exactly hit g when the test accepts H_0. In the development of the SPRT and the SPRT chart that follows it is assumed that $\gamma = 1/m$ and that g is an integer multiple of γ. If $T_{kj} - \gamma j$ is an integer multiple of γ, then it follows that the rejection limit h can also be taken to be an integer multiple of γ, although $T_{kj} - \gamma j$ may still overshoot h when the test rejects H_0

10. THE PROPERTIES OF THE SPRT CHART

When evaluating any hypothesis test, a critical property of the test is determined by either the probability that the test accepts the null hypothesis or the probability that the test rejects the null hypothesis, expressed as functions of the value of the parameter under consideration. Following the convention in sequential analysis, we work with the *operating characteristic* (OC) function, which is the probability of accepting H_0 as a function of p. For most hypothesis tests the sample size is fixed before the data are taken, but for a sequential test the sample size, say N, depends on the data and is thus a random variable. Therefore, for a sequential test the distribution of N must be considered. Usually, $E(N)$, called the *average sample number* (ASN), is used to characterize the distribution of N.

Each SPRT either accepts or rejects H_0, and thus the number of SPRTs until a signal has a geometric distribution with parameter $1 - \mathrm{OC}(p)$. Because each SPRT corresponds to a sample from the process, the expected number of SPRTs until a signal is the ANSS. For the SPRT chart, the ANSS for a given p, say ANSS(p), is thus the mean of the geometric distribution, which is

$$\mathrm{ANSS}(p) = \frac{1}{1 - \mathrm{OC}(p)} \tag{16}$$

When there is a fixed time interval d between samples and the time required to take a sample is negligible, then the ATS is the product of d and the ANSS. Thus, the ATS at p, say ATS(p), is

$$\mathrm{ATS}(p) = d\,\mathrm{ANSS}(p) = \frac{d}{1 - \mathrm{OC}(p)} \tag{17}$$

When $p = p_0$, then $1 - \mathrm{OC}(p_0) = \alpha$, where α is the probability of a type I error for the test. The ATS is then

$$\mathrm{ATS}(p_0) = \frac{d}{\alpha} \tag{18}$$

When $p = p_1$, then $\mathrm{OC}(p_1) = \beta$, where β is the probability of a type II error for the test. The ATS is then

$$\mathrm{ATS}(p_1) = \frac{d}{1 - \beta} \tag{19}$$

Exact expressions for the OC and ASN functions of the SPRT for p can be obtained by modeling the SPRT as a Markov chain [see Reynolds and Stoumbos (1998)]. These expressions, however, are relatively complicated, and thus it would be convenient to have simpler expressions that could be used in pratical applications. The remainder of this section is concerned with presenting some simple approximations to the OC and ASN functions. These approximations to the OC and ASN functions are presented here for the case in which $0 < p_0 < 0.5$. The case in which $p_0 \geq 0.5$ is discussed in Reynolds and Stoumbos (1998).

When the SPRT is used for hypothesis testing, it is usually desirable to choose the constants g and h such that the test has specified probabilities for type I and type II errors. The CD approximations to the OC and ASN functions use an adjusted value of h, which will be denoted by h^*, in a relatively simple formula. The adjusted value of h^* is

$$h^* = h + \tfrac{1}{3}(1 - 2p_0), \qquad 0 < p_0 < 0.5 \tag{20}$$

It is shown in Reynolds and Stoumbos (1998) that choices for g and h^* based on the CD approximations are

$$h^* \approx \frac{1}{r_2} \log\left(\frac{1-\beta}{\alpha}\right) \tag{21}$$

and

$$g \approx \frac{1}{r_2} \log\left(\frac{\beta}{1-\alpha}\right) \tag{22}$$

If nomial values are specified for α and β, then g and h^* can be determined by using Eqs. (21) and (22), and then the value of h can be obtained from h^* by using Eq. (20).

The CD approximation to the ASN at p_0 and p_1 can be expressed simply in terms of α and β [see Reynolds and Stoumbos (1998)]. For $p = p_0$, this expression is

$$\text{ASN}(p_0) \approx \frac{\alpha \log[(1-\beta)/a] + (1-\alpha)\log[\beta/(1-\alpha)]}{r_2 p_0 - r_1} \tag{23}$$

and for $p = p_1$ the expression is

$$\text{ASN}(p_1) \approx \frac{(1-\beta)\log[(1-\beta)/\alpha] + \beta \log[\beta/(1-\alpha)]}{r_1 p_1 - r_1} \tag{24}$$

Thus, for given α and β, evaluating the ASN at p_0 and p_1 is relatively easy.

11. A METHOD FOR DESIGNING THE SPRT CHART

To design the SPRT chart for practical applications it is necessary to determine the constants g and h used in each SPRT. In many applications it is desirable to specify the in-control average sampling rate and the false alarm rate and design the chart to achieve these specifications. Specifying the sampling interval d and ASN(p_0) will determine the in-control average sampling rate, and specifying $\mathrm{ATS}(p_0)$ will determine the false alarm rate. Once these quantities are specified, the design proceeds as follows.

The value of α is determined by using Eq. (18) and the specified values of d and $\mathrm{ATS}(p_0)$. Then, using (23), the value of β can be determined from the specified value of $\mathrm{ASN}(p_0)$ and the value of α just determined. Expression (23) cannot be solved explicitly for β in terms of α and $\mathrm{ASN}(p_0)$, so the solution for β will have to be determined numerically. Once α and β are determined, Eqs. (21), (22), and (20) can be used to determine g and h.

12. AN EXAMPLE OF DESIGNING AN SPRT CHART

To illustrate the design and application of the SPRT chart, consider an example similar to the examples of Sections 5 and 8 in which the objective is to monitor a production process with $p_0 = 0.005$. Suppose that the current procedure for monitoring this process is to take samples of $n = 200$ every $d = 4$ hr and use a p-chart that signals if five or more defectives are found in a sample. Suppose that items are produced at a rapid rate and an item can be inspected in a relatively short time. In this case, process engineers are willing to use a sequential inspection plan in which items are inspected one by one and the sample size at each sampling point depends on the data at that point. In this example the time required to obtain a sample is short relative to the time between samples, so neglecting this time in computations of quantities such as the ATS seems to be reasonable.

As in the example in Section 5, suppose that p_1 is specified to be 0.010 and then adjusted slightly to 0.009947, so that $\gamma = 1/139$. For the first phase of the example, suppose that it is decided that the SPRT chart should be designed to have the same sampling interval, the same in-control average sampling rate, and the same false alarm rate as the p-chart. Then d can be taken to be 4, the target for $\mathrm{ASN}(p_0)$ can be taken to be 200, and the target for $\mathrm{ATS}(p_0)$ can be taken to be 1128 hr.

First consider the problem of finding g and h in critical inequality (15) of the SPRT. Using the specifications decided upon for the chart, Eq. (18) implies that α should be 0.003545. Then, solving (23) numerically for β gives $\beta = 0.7231$. Then, using Eqs. (21) and (20) gives $h^* = \log(0.2769/0.003545)/0.6928 = 6.2906$ and $h = 5.9606$, and using Eq. (22) gives $g = \log(0.7231/0.996455)/0.6928 = -0.4628$. Rounding g and h to the nearest multiple of $1/139$ gives $g = -64/139 = -0.4604$ and $h = 828/139 = 5.9568$. Thus, the SPRT chart can be applied in this case by using the critical inequality

$$-0.4604 < T_{kj} - (j139) < 5.9568 \quad (25)$$

The in-control ASN of this chart should be approximately 200 (the exact value is 198.97), and the in-control ATS should be approximately 1128 hr (the exact value is 1128.48 hr). Using Eq. (24), this chart's ATS at $p = p_1 = 0.009947$ should be approximately $d/(1-\beta) = 4/\ (1-0.7231) =$ 14.45 hr (the exact value is 14.51 hr). Thus, compared to the value of 78.73 hr for the p-chart, the SPRT chart will provide a dramatic reduction in the time required to detect the shift from p_0 to p_1.

The value chosen for p_1 is really just a convenient design device for the SPRT chart, so this value of p would usually not be the only value that should be detected quickly. Thus, when designing an SPRT chart in practice, it is desirable to use the CD approximation (or the exact methods) given in Reynolds and Stoumbos (1998) to find the ATS for a range of values of p around p_1. For the evaluation to be given here, exact ATS values for the SPRT chart were computed and are given in column 3 of Table 1. ATS values for the p-chart are given in column 2 of Table 1 to serve as a basis of comparison. Comparing columns 2 and 3 shows that, except for large shifts in p, the SPRT chart is much more efficient than the p-chart. When considering the binomial CUSUM in Section 8, it was argued that it is better to take small samples at more frequent intervals than to take large samples at long intervals. To determine whether this is also true for the SPRT chart, an SPRT chart was designed to have an approximate in-control ASN of 50 and a sampling interval of $d = 1$ hr. This would give the same sampling rate of 50 observations per hour as in columns 2 and 3. The ATS values of this second SPRT chart are given in column 4 of Table 1. Comparing columns 3 and 4 shows that using a sampling interval of $d = 1$ with ASN $= 50$ is better than using a sampling interval of $d = 4$ with ASN $= 200$, especially for detecting large shifts.

In some applications, the motivation for using a variable sampling rate control chart is to reduce the sampling cost required to produce a given detection ability [see Baxley (1996), Reynolds (1996b), and Reynolds and Stoumbos (1998)]. Because the SPRT chart is so much more efficient than the p-chart, it follows that the SPRT chart could achieve the detection ability of the p-chart with a much smaller average sampling rate. To illustrate this point, the design method given in Section 11 was used to design some SPRT charts with lower average sampling rates. Columns 5 and 6 of Table 1 contain ATS values of two SPRT charts that have an in-control average sampling rate of approximately half the value for the p-chart (approximately 25 observations per hour). The SPRT chart in column 5 uses $d = 2.0$ and has ASN$(p_0) \approx 50$, and the SPRT chart in column 6 uses $d = 1.0$ and has ASN$(p_0) \approx 25$. Although these two SPRT charts are sam-

Table 1 ATS Values for the p-Chart and the SPRT Charts When $p_0 = 0.005$

	p-Chart	SPRT charts					
	obs/hr = 50,	obs/hr = 50		obs/hr ≈ 25		obs/hr ≈ 12.5	
p	$n = 200$, $d = 4.0$	ASN = 199.0 $d = 4.0$	ASN = 49.8 $d = 1.0$	ASN = 51.1 $d = 2.0$	ASN = 24.1 $d = 1.0$	ASN = 25.6 $d = 2.0$	ASN = 11.4 $d = 1.00$
0.005	1128.2	1128.5	1131.6	1100.7	1172.2	1100.1	1233.8
0.008	172.8	32.3	30.7	53.8	56.6	91.7	102.5
0.010	77.3	14.3	12.7	23.8	24.7	43.2	48.0
0.015	21.8	7.3	5.6	10.5	10.6	18.9	20.7
0.020	10.8	5.7	3.9	7.3	7.2	12.9	13.9
0.030	5.6	4.7	2.6	5.0	4.7	8.5	8.8
0.050	4.1	4.2	1.8	3.4	3.0	5.4	5.4

pling at half the rate of the p-chart, they are still faster at detecting shifts in p. Columns 7 and 8 of Table 1 contain ATS values of two SPRT charts that have an in-control average sampling rate of approximately one-fourth the value for the p-chart (approximately 12.5 observations per hour). The SPRT chart in column 7 uses $d = 2.0$ and has $\mathrm{ASN}(p_0) \approx 50$, and the SPRT chart in column 6 uses $d = 1.0$ and has $\mathrm{ASN}(p_0) \approx 25$. Comparing columns 5 and 6 to column 2 shows that the SPRT charts with half the sampling rate of the p-chart offer faster detection than the p-chart. Columns 7 and 8 show that an SPRT chart with about one-fourth the sampling rate of the p-chart will offer roughly the same detection capability as the p-chart.

13. CONCLUSIONS

It has been shown that the Bernoulli CUSUM chart, the binomial CUSUM chart, and the SPRT chart are highly efficient control charts that can be applied in different sampling situations. Each of these charts is much more efficient than the traditional Shewhart p-chart. The design methods based on the highly accurate CD approximations provide a relatively simple way for practitioners to design these charts for practical applications. Although the design possibilities for these charts are limited slightly by the discreteness of the distribution of the inspection data, this discreteness is much less of a problem than for the p-chart.

The SPRT chart is a variable sampling rate control chart that is much more efficient than standard fixed sampling rate charts such as the p-chart. The increased efficiency of the SPRT chart can be used to reduce the time required to detect process changes or to reduce the sampling cost required to achieve a given detection capability.

REFERENCES

Baxley RV Jr. (1995). An application of variable sampling interval control charts. J Qual Technol 27: 275–282.

Bourke PD. (1991). Detecting a shift in fraction nonconforming using run-length control charts with 100% inspection. J Qual Technol 23: 225–238.

Gan FF. (1993). An optimal design of CUSUM control charts for binomial counts. J Appl Stat 20: 445–460.

Lucas JM, Crosier RB. (1982). Fast initial response for CUSUM quality control schemes: Give your CUSUM a head start. Technometrics 24. 199–205.

Rendtel U. (1990). CUSUM schemes with variable sampling intervals and sample sizes. Stat Papers 31: 103–118.

Reynolds MR Jr. (1996a). Variable sampling interval control charts with sampling at fixed times. IIE Trans 28: 497–510.
Reynolds MR Jr. (1996b). Shewhart and EWMA variable sampling interval control charts with sampling at fixed times. J Qual Technol 28: 199–212.
Reynolds MR Jr, Stoumbos ZG. (1999). A CUSUM chart for monitoring a proportion when inspecting continuously. J. Qual Technol 31: 87–108.
Reynolds MR Jr, Stoumbos ZG. (1998). The SPRT chart for monitoring a proportion. IIE Trans 30: 545–561.
Siegmund D. (1985). Sequential Analysis. Springer-Verlag, New York.
Stoumbos ZG, Reynolds MR Jr. (1996). Control charts applying a general sequential test at each sampling point. Sequent Anal 15: 159–183.
Stoumbos ZG, Reynolds MR Jr. (1997a). Corrected diffusion theory approximations in evaluating properties of SPRT charts for monitoring a process mean. Nonlinear Analysis 30: 3987–3996.
Stoumbos ZG, Reynolds MR Jr. (1997b). Control charts applying a sequential test at fixed sampling intervals. J Qual Technol 29: 21–40.
Wald A. (1947). Sequential Analysis. Dover, New York.
Woodall WH. (1997). Control charts based on attribute data: Bibliography and review. J Qual Technol 29: 172–183.
Vaughan TS. (1993). Variable sampling interval *np* process control chart. Commun Stat: Theory Methods 22: 147–167.

8
Process Monitoring with Autocorrelated Data

Douglas C. Montgomery
Arizona State University, Tempe, Arizona

Christina M. Mastrangelo
University of Virginia, Charlottesville, Virginia

1. INTRODUCTION

The standard assumptions when control charts are used to monitor a process are that the data generated by the process when it is in control are normally and independently distributed with mean μ and standard deviation σ. Both μ and σ are considered fixed and unknown. An out-of-control condition is created by an assignable cause that produces a change or shift in μ or σ (or both) to some different value. Therefore, we could say that when the process is in control the quality characteristic at time t, x_t, is represented by the model

$$x_t = \mu + \varepsilon_t, \qquad t = 1, 2, \ldots \tag{1}$$

where ε_t is normally and independently distributed with mean zero and standard deviation σ. This is often called the Shewhart model of the process.

When these assumptions are satisfied, one may apply either Shewhart, CUSUM, or EWMA control charts and draw reliable conclusions about the state of statistical control of the process. Furthermore, the statistical properties of the control chart, such as the false alarm rate with 3σ control limits, or the average run length, can be easily determined and used to provide guidance for chart interpretation. Even in situations where the normality assumption is violated to a slight or moderate degree, these control charts will still work reasonably well.

The most important of these assumptions is that the observations are independent (or uncorrelated), because conventional control charts do not perform well if the quality characteristics exhibit even low levels of correlation over time. Specifically, these control charts will give misleading results in the form of too many false alarms if the data are autocorrelated. This point has been made by numerous authors, including Berthouex et al. (1978), Alwan and Roberts (1988), Montgomery and Friedman (1989), Alwan (1992), Harris and Ross (1991), Montgomery and Mastrangelo (1991), Yaschin (1993), and Wardell et al. (1994).

Unfortunately, the assumption of uncorrelated or independent observations is not even approximately satisfied in some manufacturing processes. Examples include chemical processes in which consecutive measurements on process or product characteristics are often highly correlated and automated test and measurement procedures in which every quality characteristic is measured on every unit in time order of production. The increasing use of on-line data acquisition systems is shrinking the interval between process observations. As a result, the volume of process data collected per unit time is increasing dramatically [see the discussion in Hahn (1989)]. All manufacturing processes are driven by inertial elements, and when the interval between samples becomes small relative to these forces, the observations on the process will be correlated over time.

It is easy to given an analytical demonstration of this phenomenon. Figure 1 shows a simple system consisting of a tank of volume V, with an input and output material stream having flow rate f. Let w_t be the concentration of a certain material in the input stream at time t and x_t be the corresponding concentration in the output stream at time t. Assuming homogeneity within the tank, the relationship between x_t and w_t is

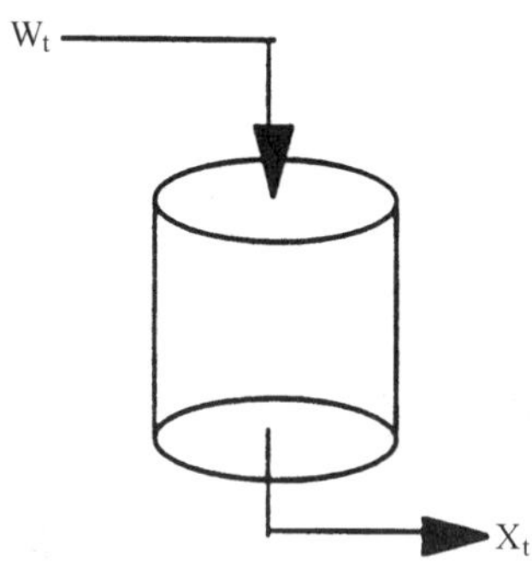

Figure 1 A tank with volume V and input and output material streams.

$$x_t = w_t - T\frac{dx_t}{dt}$$

where $T = V/f$ is often called the *time constant* of the system.

If the input stream experiences a step change of w_0 at time $t = 0$ (say), then the output concentration at time t is

$$x_t = w_0(1 - e^{-t/T})$$

Now, in practice, we do not observe x_t continuously but only at small, equally spaced intervals of time, Δt. In this case,

$$x_t = x_{t-1} + (w_t - x_{t-1})(1 - e^{-\Delta t/T}) = aw_t + (1 - a)x_{t-1}$$

where $a = 1 - e^{-\Delta t/T}$.

The properties of the output stream concentration x_t depend on those of the input stream concentration w_t and the sampling interval. Figure 2 illustrates the effect of the mean of w_t on x_t. If we assume that the w_t are uncorrelated random variables, then the correlation between successive values of x_t (or autocorrelation between x_t and x_{t-1}) is given by

$$\rho = 1 - a = e^{-\Delta t/T}$$

Note that if Δt is much greater than T, then $\rho \simeq 0$. That is, if the interval between samples Δt in the output stream is long, much longer than the time constant T, then the observations on output concentration will be uncorrelated. However, if $\Delta t \leq T$, this will not be the case. For example, if

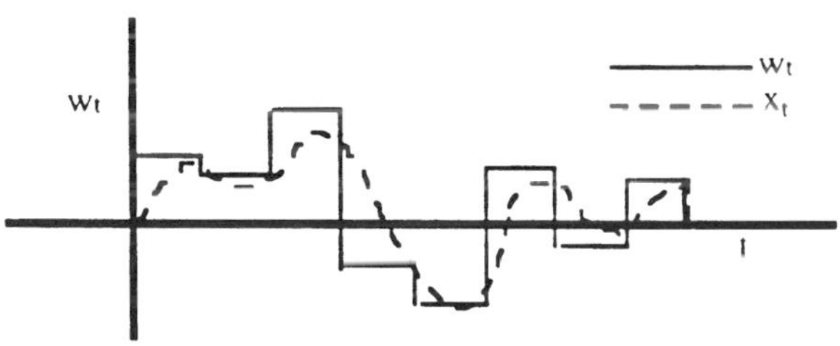

Figure 2 The effect of the process on x_t.

$$\Delta t/T = 1, \qquad \rho = 0.37$$
$$\Delta t/T = 0.5, \qquad \rho = 0.61$$
$$\Delta t/T = 0.25, \qquad \rho = 0.78$$
$$\Delta t/T = 0.10, \qquad \rho = 0.90$$

Clearly, if we sample at least once per time constant, there will be significant autocorrelation present in the observations. For instance, sampling four times per time constant ($\Delta t/T = 0.25$) results in autocorrelation between x_t and x_{t-1} of $\rho = 0.78$. Autocorrelation between successive observations as small as 0.25 can cause a substantial increase in the false alarm rate of a control chart, so clearly this is an important issue to consider in control chart implementation.

Figure 3 illustrates the foregoing discussion. This is a control chart for individual measurements applied to concentration measurements from a chemical process taken every hour. The data are shown in Table 1. Note that many points are outside the control limits (horizontal lines) on this chart. Because of the nature of the production process and the visual appearance of the concentration measurements in Figure 3, which appear to "drift" or "wander" slowly over time, we would probably suspect that concentration is autocorrelated.

Figure 4 is a scatter plot of concentration at time t (x_t) versus concentration measured one period earlier (x_{t-1}). Note that the points on this graph tend to cluster along a straight line with a positive slope. That is, a relatively low observation of concentration at time $t-1$ tends to be followed by another low value at time t, while a relatively large observation at time $t-1$ tends to be followed by another large value at time t. This type of behavior is indicative of positive autocorrelation in the observations.

It is also possible to measure the level of autocorrelation analytically. The autocorrelation over a series of time-oriented observations is measured by the autocorrelation function

$$\rho_k = \frac{\mathrm{Cov}(x_t, x_{t-k})}{V(x_t)}, \qquad k = 0, 1, \ldots$$

where $\mathrm{Cov}(x_t, x_{t-k})$ is the covariance of observations that are k time periods apart, and we have assumed that the observations (called a time series) have constant variance given by $V(x_t)$. We usually estimate the values of ρ_k with the sample autocorrelation function:

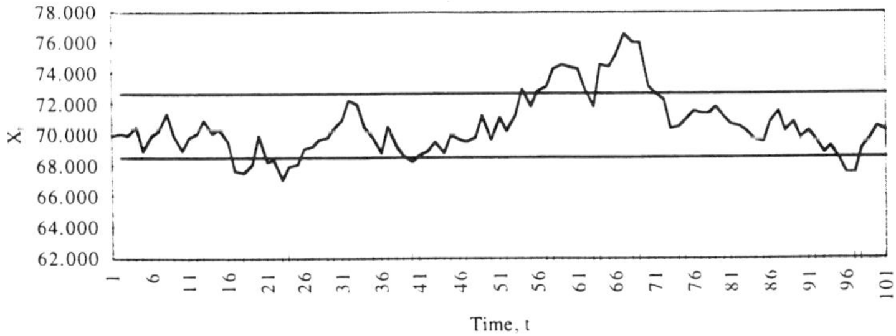

Figure 3 Control chart for individuals.

Table 1 Concentration Data

Time, t	X	Time, t	X	Time, t	X	Time, t	X
1	70.204	26	69.270	51	70.263	76	71.371
2	69.982	27	69.738	52	71.257	77	71.387
3	70.558	28	69.794	53	73.019	78	71.819
4	68.993	29	79.400	54	71.871	79	71.162
5	70.064	30	70.935	55	72.793	80	70.647
6	70.291	31	72.224	56	73.090	81	70.566
7	71.401	32	71.930	57	74.323	82	70.311
8	70.048	33	70.534	58	74.539	83	69.762
9	69.028	34	69.836	59	74.444	84	69.552
10	69.892	35	68.808	60	74.247	85	70.884
11	70.152	36	70.559	61	72.979	86	71.593
12	71.006	37	69.288	62	71.824	87	70.242
13	70.196	38	68.740	63	74.612	88	70.863
14	70.477	39	68.322	64	74.368	89	69.895
15	69.510	40	68.713	65	75.109	90	70.244
16	67.744	41	68.973	66	76.569	91	69.716
17	67.607	42	69.580	67	75.959	92	68.914
18	68.168	43	68.808	68	76.005	93	69.216
19	69.979	44	69.931	69	73.206	94	68.431
20	68.227	45	69.763	70	72.692	95	67.516
21	68.497	46	69.541	71	72.251	96	67.542
22	67.113	47	69.889	72	70.386	97	69.136
23	67.993	48	71.243	73	70.519	98	69.905
24	68.113	49	69.701	74	71.005	99	70.515
25	69.142	50	71.135	75	71.542	100	70.234

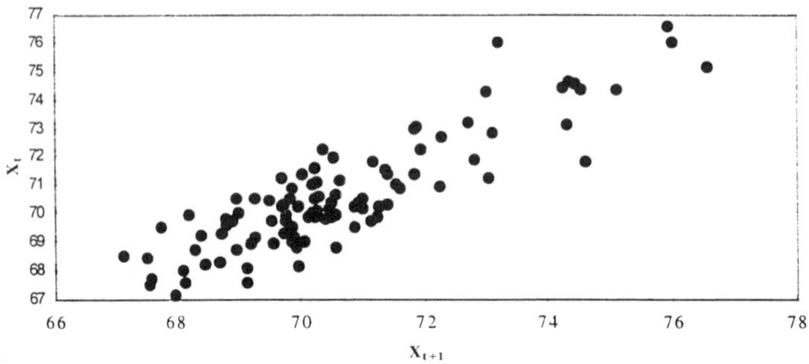

Figure 4 Scatter plot of concentration at time $t(x_t)$ versus concentration measured one period earlier (x_{t-1}).

$$r_k = \frac{\sum_{t=1}^{n-k}(x_t - \bar{x})(x_{t-k} - \bar{x})}{\sum_{t=1}^{n}(x_t - \bar{x})^2}, \qquad k = 0, 1, \ldots, K \tag{2}$$

As a general rule, we need to compute values of r_k for a few values of k, $k \leq n/4$. Many software programs for statistical data analysis can perform these calculations.

The sample autocorrelation function for the concentration data is shown in Figure 5. The dashed line on the graph is the upper two-standard deviation limit on the autocorrelation parameter ρ_k at lag k. The lower limit (not shown here) would be symmetrical. These limits are useful in detecting nonzero autocorrelations; in effect, if a sample autocorrelation exceeds its two-standard deviation limit, the corresponding autocorrelation parameter ρ_k is likely nonzero. Note that there is a strong positive correlation at lag 1;

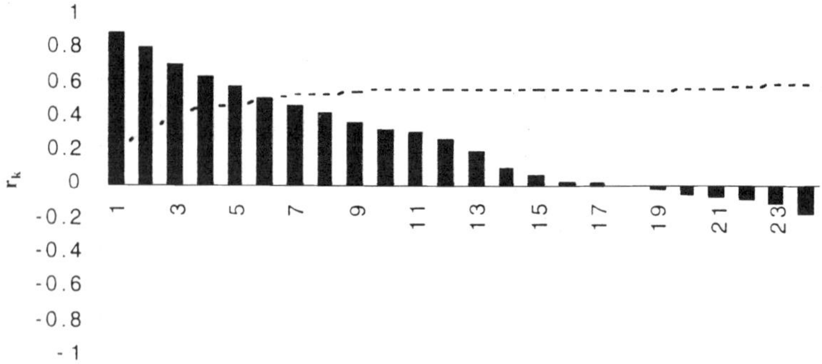

Figure 5 Autocorrelation function for the concentration data.

that is, concentration observations that are one period apart are positively correlated with $r_1 = 0.88$. This level of autocorrelation is sufficiently high to distort greatly the performance of a Shewhart control chart. In particular, because we know that positive correlation greatly increases the frequency of false alarms, we should be very suspicious about the out-of-control signals on the control chart in Figure 3.

Several approaches have been proposed for monitoring processes with autocorrelated data. Just as in traditional applications of SPC techniques to uncorrelated data, our objective is to detect assignable causes so that if the causes are removed, process variability can be reduced. The first is to sample from the process less frequently so that the autocorrelation is diminished. For example, note from Figure 5 that if we only took every 20th observation on concentration, there would be very little autocorrelation in the resulting data. However, since the original observations were taken every hour, the new sampling frequency would be one observation every 20 hr. Obviously, the drawback of this approach is that many hours may elapse between the occurrence of an assignable cause and its detection.

The second general approach may be thought of as a *model-based approach*. One way that this approach is implemented involves building an appropriate model for the process and control, charting the residuals. The basis of this approach is that any disturbances from assignable causes that affect the original observations will be transferred to the residuals. Model-based approaches are presented in the following subsection. The *model-free approach* does not use a specific model for the process; this approach is discussed in Section 3.

2. MODEL-BASED APPROACHES

2.1. ARIMA Models

An approach to process monitoring with autocorrelated data that has been applied widely in the chemical and process industries is to directly model the correlative structure with an appropriate time series model, use that model to remove the autocorrelation from the data, and apply control charts to the residuals. For example, suppose we could model the quality characteristic x_t as

$$x_t = \xi + \phi x_{t-1} + \varepsilon_t \tag{3}$$

where ξ and ϕ $(-1 < \phi < 1)$ are unknown constants and ε_t is normally and independently distributed with mean zero and standard deviation σ. Note how intuitive this model is for the concentration data from examining

Figure 4. Equation (3) is called a first-order autoregressive model; the observations x_t from such a model have mean $\xi/(1-\phi)$ and standard deviation $\sigma/(1-\phi^2)^{1/2}$, and the observations that are k periods apart (x_t and x_{t-k}) have correlation coefficient ϕ^k. That is, the autocorrelation function should decay exponentially just as the autocorrelation function of the concentration data did in Figure 5. Suppose that $\hat{\phi}$ is an estimate of ϕ obtained from analysis of sample data from the process and $\hat{x}$ is the fitted value of x_t. Then the residuals

$$e_t = x_t - \hat{x}_t$$

are approximately normally and independently distributed with mean zero and constant variance. Conventional control charts could now be applied to the sequence of residuals. Points out of control or unusual patterns on such charts would indicate that the parameter ϕ had changed, implying that the original variable x_t was out of control. For details of identifying and fitting time series models such as this one, see Montgomery et al. (1990) and Box et al. (1994).

The parameters in the autoregressive model. Eq. (3), may be estimated by the method of least squares, that is, by choosing the values of ξ and ϕ that minimize the sum of squared errors ε_t. Many statistical software packages have routines for fitting these time series models. The fitted value of this model for the concentration data is

$$x_t = 8.38 + 0.88x_{t-1}$$

We may think of this as an alternative to the Shewhart model for this process.

Figure 6 is an individuals control chart of the residuals from the fitted first-order autoregressive model. Note that now no points are outside the control limits. In contrast to the control chart on the individual measurements in Figure 3, we would conclude that this process is in a reasonable state of statistical control.

Other Time Series Models

The first-order autoregressive model used in the concentration example [Eq. (3)] is not the only possible model for time-oriented data that exhibits a correlative structure. An obvious extension to Eq. (3) is

$$x_t = \xi + \phi_1 x_{t-1} + \phi_2 x_{t-2} + \varepsilon_t \tag{4}$$

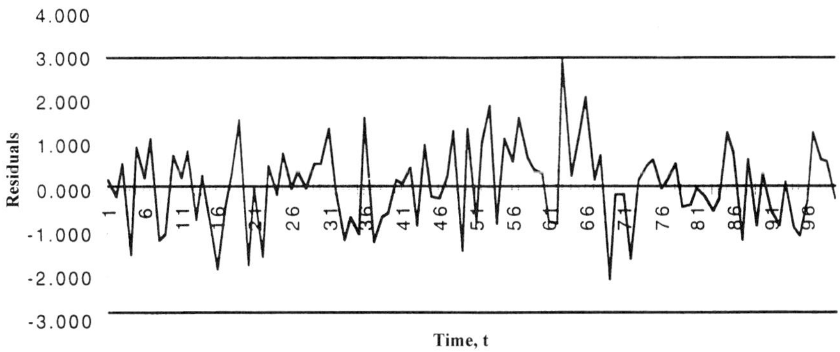

Figure 6 Control chart for individuals applied to the residuals from the AR(1) model.

which is a *second-order autoregressive model.* In general, in autoregressive-type models, the variable x_t is directly dependent on previous observations x_{t-1}, x_{t-2}, and so forth. Another possibility is to model the dependence through the random component ε_t. A simple way to do this is

$$x_t = \mu + \varepsilon_t - \theta\varepsilon_{t-1} \tag{5}$$

This is called a *first-order moving average model.* In this model, the correlation between x_t and x_{t-1} is $\rho_1 = -\theta/(1+\theta^2)$ and is zero at all other lags. Thus, the correlative structure in x_t extends backward for only one time period.

Sometimes combinations of autoregressive and moving average terms are useful. A *first-order mixed model* is

$$x_t = \xi + \phi x_{t-1} + \varepsilon_t - \theta\varepsilon_{t-1} \tag{6}$$

This model often occurs in the chemical and process industries. The reason is that if the underlying process variable x_t is first-order autoregressive and a random error component is added to x_t, the result is the mixed model in Eq. (6). In the chemical and process industries, first-order autoregressive process behavior is fairly common. Furthermore, the quality characteristic is often measured in a laboratory (or by an on-line instrument) that has measurement error, which we can usually think of as random or uncorrelated. The reported or observed measurement then consists of an autoregressive component plus random variation, so the mixed model in Eq. (6) is required as the process model.

We also encounter the *first-order integrated moving average* model

$$x_t = x_{t-1} + \varepsilon_t - \theta\varepsilon_{t-1} \tag{7}$$

in some applications. Whereas the previous models are used to describe stationary behavior (that is, x_t wanders around a "fixed" mean), the model in Eq. (7) describes nonstationary behavior (the variable x_t "drifts" as if there were no fixed value of the process mean). This model often arises in chemical and process plants when x_t is an "uncontrolled" process output, that is, when no control actions are taken to keep the variable close to target value.

The models we have been discussing in Eqs. (3)–(7) are members of a class of time series models called *autoregressive integrated moving average* (ARIMA) models. Montgomery et al. (1990) and Box et al. (1994) discuss these models in detail. While these models appear very different from the Shewhart model [Eq. (1)], they are actually relatively similar and include the Shewhart model as a special case. Note that if we let $\phi = 0$ in Eq. (3), the Shewhart model results. Similarly, if we let $\theta = 0$ in Eq. (5), the Shewhart model results.

Average Run Length Performance for Residuals Control Charts

Several authors have pointed out that residuals control charts are not sensitive to small process shifts [e.g., see Wardell et al. (1994)]. The average run length for the residuals chart from an AR(1) model is

$$\mathrm{ARL}_{\mathrm{RES}} = \frac{1 - P_1 + P}{P} \tag{8}$$

where P_1 is the probability that the run has length 1, that is, the probability that the first residual exceeds ± 3,

$$\begin{aligned} P_1 &= \Pr(\text{run length} = 1) \\ &= 1 - \Phi(3 - \delta) + \Phi(-3 - \delta) \end{aligned} \tag{9}$$

$\Phi(.)$ is the cumulative distribution function of the standard normal distribution. The probability that any subsequent observation will generate an alarm is the probability that e_t exceeds ± 3,

$$P = 1 - \Phi(3 - \delta(1 - \phi)) + \Phi(-3 - \delta(1 - \phi)) \tag{10}$$

See Willemain and Runger (1996) for the complete derivation.

Table 2 ARLs for Residuals Chart

Correlation			Shift, δ/c		
σ	0	0.5	1	2	4
0.00	370.38	152.22	43.89	6.30	1.19
0.25	370.38	212.32	80.37	13.59	1.32
0.50	370.38	280.33	152.69	37.93	2.00
0.90	370.38	364.51	345.87	260.48	32.74
0.99	370.38	368.95	362.76	312.00	59.30

Note: ARLs measured in observations.

Table 2 shows ARL_{RES} for representative values of the autocorrelation coefficient ϕ and shift δ. Note the poor performance of the residuals chart when the correlation is high ($\phi = 0.90$ or $\phi = 0.99$). This problem arises because the AR(1) model responds to the change in the mean level and partially incorporates the shift in the mean into its forecasts, as seen in (13).

Using an Exponentially Weighted Moving Average (EWMA) with Autocorrelated Data

The time series modeling approach illustrated in the concentration example can be time-consuming and difficult to apply in practice. Typically, we apply control charts to several process variables, and developing an explicit time series model for each variable of interest is potentially time-consuming. Some authors have developed automatic time series model building to partially alleviate this difficulty [see Yourstone and Montgomery (1989) and the references therein]. However, unless the time series model is of intrinsic value in explaining process dynamics (as it sometimes is), this approach will frequently require more effort than may be justified in practice.

Montgomery and Mastrangelo (1991) suggested an approximate procedure based on the EWMA. They use the fact that the EWMA can be used in certain situations where the data are autocorrelated. Suppose that the process can be modeled by the integrated moving average model of Eq. (7). It can be easily shown that the EWMA with $\lambda = 1 - \theta$ is the optimal one-step-ahead forecast for this process. That is, if $\hat{x}_{t+1}(t)$ is the forecast for the observation in period $t + 1$ made at the end of period t, then

$$\hat{x}_{t+1}(t) = z_t$$

where $z_i = \lambda x_t + (1 - \lambda) z_{t-1}$ is the EWMA. The sequence of one-step-ahead prediction errors,

$$e_t = x_t - \hat{x}_t(t-1) \tag{11}$$

is independently and identically distributed with mean zero. Therefore, control charts could be applied to these one-step-ahead prediction errors. The parameter λ (or equivalently, θ) would be found by minimizing the sum of squares of the errors e_t.

Now suppose that the process is not modeled exactly by Eq. (7). In general, if the observations from the process are positively autocorrelated and the process mean does not drift too quickly, the EWMA with an appropriate value for λ will provide an excellent one-step-ahead predictor. The forecasting and time series analysis field has used this result for many years; for examples, see Montgomery et al. (1990). Consequently, we would expect many processes that obey first-order dynamics (that is, follow a slow "drift") to be well represented by the EWMA.

Consequently, under the conditions just described, we may use the EWMA as the basis of a statistical process monitoring procedure that is an approximation of the exact time series model approach. The procedure would consist of plotting one-step-ahead EWMA prediction errors (or model residuals) on a control chart. This chart could be accompanied by a run chart of the original observations on which the EWMA forecast is superimposed. Our experience indicates that both charts are usually necessary, as operational personnel feel that the control chart of residuals sometimes does not provide a direct frame of reference to the process. The run chart of original observations allows process dynamics to be visualized.

Figure 7 presents a control chart for individuals applied to the EWMA prediction errors for the concentration data. For this chart, $\lambda = 0.85$. This is the value of λ that minimizes the sum of squares of the EWMA prediction

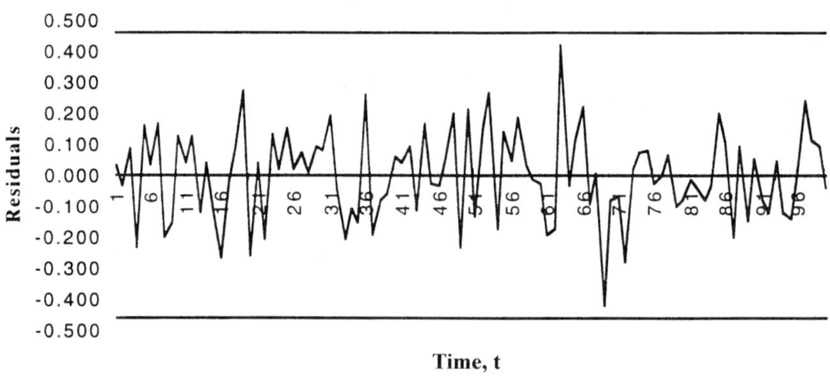

Figure 7 EWMA prediction errors with $\lambda = 0.85$ and Shewhart limits.

errors. This chart is slightly different from the control chart of the exact autoregressive model residuals shown in Figure 6, but not significantly so. Both indicate a process that is reasonably stable, with a period around $t = 62$ where an assignable cause may be present.

Montgomery and Mastrangelo (1991) point out that it is possible to combine information about the state of statistical control and process dynamics on a single control chart. If the EWMA is a suitable one-step-ahead predictor, then one could use z_t as the centerline on a control chart for period $t + 1$ with upper and lower control limits at

$$\mathrm{UCL}_{t+1} = z_t + 3\sigma$$

and

$$\mathrm{LCL}_{t+1} = z_t - 3\sigma \tag{12}$$

and the observation $x_t + 1$ would be compared to these limits to test for statistical control. We can think of this as a *moving centerline EWMA control chart*. As mentioned above, in many cases this would be preferable from an interpretation standpoint to a control chart of residuals and a separate chart of the EWMA, as it combines information about process dynamics and statistical control in one chart.

Figure 8 is the moving centerline EWMA control chart for the data, with $\lambda = 0.85$. It conveys the same information about statistical control as

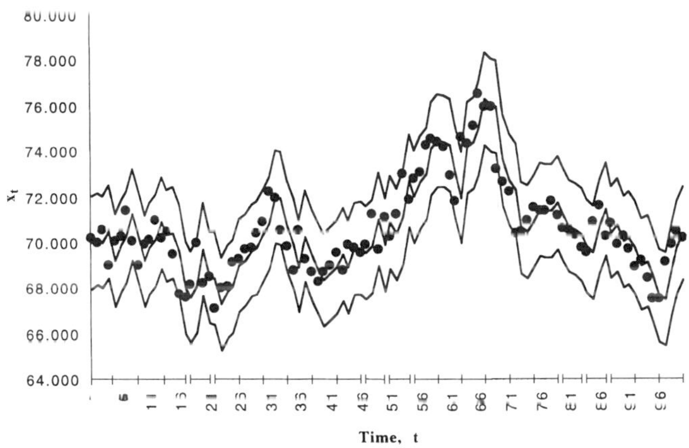

Figure 8 Moving centerline EWMA control chart applied to the concentration data.

the residual or EWMA prediction error control chart in Figure 7, but operating personnel often feel more comfortable with this display.

ARL Performance. Because the EWMA-based procedures presented above are very similar to the residuals control chart, they will have some of the same problems in detecting process shifts. Also, Tseng and Adams (1994) note that because the EWMA is not an optimal forecasting scheme for most processes [except the IMA(1,1) model], it will not completely account for the autocorrelation, and this can affect the statistical performance of control charts based on EWMA residuals or prediction errors. Montgomery and Mastrangelo (1991) suggest the use of supplementary procedures called tracking signals combined with the control charts for residuals. There is evidence that these supplementary procedures considerably enhance the performance of residuals control charts. Furthermore, Mastrangelo and Montgomery (1995) show that if an appropriately designed tracking signal scheme is combined with the EWMA-based procedure we have described, good in-control performance and adequate shift detection can be achieved.

Estimating and Monitoring σ. The standard deviation of the one-step-ahead errors or model residuals σ can be estimated in several ways. If λ is chosen as suggested above over a record of n observations, then dividing the sum of the squared prediction errors for the optimal λ by n will produce an estimate of σ^2. This is the method used in many time series analysis computer programs.

Another approach is to compute the estimate of σ as is typically done in forecasting systems. The mean absolute deviation (MAD) could be used in this regard. The MAD is computed by applying an EWMA to the absolute value of the prediction error,

$$\Delta_t = \alpha|e_t| + (1-\alpha)\Delta_{t-1}, \qquad 0 < \alpha \leq 1$$

Since the MAD of a normal distribution is related to the standard deviation by $\sigma \cong 1.25\Delta_t$ [see Montgomery et al. (1990)], we could estimate the standard deviation of the prediction errors at time t by

$$\sigma \cong 1.25\Delta_t \tag{13}$$

Another approach is to directly calculate a smoothed variance,

$$\hat{\sigma}_t^2 = \alpha e_t^2 + (1-\alpha)\hat{\sigma}_{t-1}^2, \qquad 0 < \alpha \leq 1 \tag{14}$$

MacGregor and Ross (1993) discuss the use of exponentially weighted moving variance estimates in monitoring the variability of a process. They show how to find control limits for these quantities for both correlated and uncorrelated data.

The Weighted Batch Means Control Chart

While control charting the residuals from a time series model is one way to cope with autocorrelation, there is another way to exploit these models. This is the weighted batch means chart, introduced by Runger and Willemain (1995).

Bischak et al. (1993) derived a way to eliminate autocorrelation among the averages of successive data values in discrete-event simulation. Their findings have value for statistical process control, since a way to cancel autocorrelation in subgroups maps the problem of autocorrelated data into the familiar problem of using independent subgroups to monitor process means.

Starting with a stationary ARIMA or ARMA model, Bischak et al. (1993) derived the weights needed to eliminate autocorrelation between batch means as a function of the batch size and the model parameters. Designating the batch size by b and forming the jth batch from consecutive data values $X_{(j-1)b+i}$, the jth weighted batch mean is

$$Y_j = \sum_{i=1}^{b} w_i x_{(j-1)b+i}, \qquad j = 1, 2, \ldots \tag{15}$$

The batch size b can be selected to tune performance against a specified shift δ.

The weights w_i must sum to unity for Y_j to be an unbiased estimate of the process mean μ. For AR(p) processes, the optimal weights are identical in the middle of the batch but differ in sign and magnitude for the first and last values in the batch. For the AR(1) model, the weights are

$$w_1 = \frac{\phi}{(b-1)(1-\phi)} \tag{16a}$$

$$w_i = \frac{1}{b-1}, \qquad i - 2, \ldots, b-1 \tag{16b}$$

$$w_b = \frac{1}{(b-1)(1-\phi)} \tag{16c}$$

For example, with $b = 64$ and $\phi = 0.99$, the middle weights are all 0.016, and the first and last weights are -1.57 and 1.59, respectively.

Given normal data and any bath size $b > 1$, the optimal weights produce batch means that are i.i.d. normal with mean

$$E(Y_j) = \mu \tag{17}$$

and variance

$$\mathrm{Var}(Y_j) = \frac{1}{(1-\phi)^2(b-1)} \tag{18}$$

Given (17) and (18), the standardized value of a shift from μ to $\mu + \delta$ is

$$\Delta_\delta = \delta(1-\phi)(b-1)^{1/2} \tag{19}$$

To adjust the on-target ARL to equal ARL_{ON}, one computes the control limit by solving for Δ_0 in

$$\mathrm{ARL}_{ON} = \frac{b}{1 - \Phi(\Delta_{ON}) + \Phi(-\Delta_{ON})} \tag{20}$$

where b in the numerator accounts for the fact that each batch is b observations long. Then the average run length for the weighted batch means (WBM) chart (measured in individual observations) can be computed as

$$\mathrm{ARL}_{\mathrm{WBM}} = \frac{b}{1 - \Phi(\Delta_{ON} - \Delta_\delta) + \Phi(-\Delta_{ON} - \Delta_\delta)} \tag{21}$$

Table 3 compares $\mathrm{ARL}_{\mathrm{WBM}}$ against $\mathrm{ARL}_{\mathrm{RES}}$ for a range of values of batch size b, shift δ, and autocorrelation ϕ. The proper choice of batch size b results in superior performance for the WMB chart. In general, the WBM chart is more sensitive than the residuals chart for shift $\delta \leq 3$ and autocorrelation $0 \leq \phi \leq 0.99$.

The WBM chart achieves its superiority by, in effect, using larger subgroups of residuals. It is well known that for independent data, larger subgroups provide greater sensitivity to small shifts. Runger and Willemain (1995) show that a form of this conclusion applies to autocorrelated data as well.

Table 3 ARLs of Residuals and Weighted Batch Means Charts

Correlation		Shift d/s				
f	b	0	0.5	1	2	4
0	RES	370.38	155.22	43.89	6.30	1.19
	2	370.38	170.14	53.42	9.21	2.25
	4	370.38	86.03	19.33	4.88	4.00
	8	370.38	48.47	12.57	8.01	8.00
	16	370.38	34.33	16.53	16.00	16.00
	32	370.38	37.32	32.00	32.00	32.00
	64	370.38	64.30	64.00	64.00	64.00
	128	370.38	128.00	128.00	128.00	128.00
	256	370.38	256.00	256.00	256.00	256.00
0.25	RES	370.38	212.32	80.37	13.59	1.32
	2	370.38	226.37	94.01	20.02	3.41
	4	370.38	135.90	37.84	7.70	4.02
	8	370.38	82.97	21.21	8.40	8.00
	16	370.38	56.18	19.72	16.00	16.00
	32	370.38	49.57	32.22	32.00	32.00
	64	370.38	67.61	64.00	64.00	64.00
	128	370.38	128.07	128.00	128.00	128.00
	256	370.38	256.04	256.00	256.00	256.00
0.5	RES	370.38	280.33	152.69	37.93	2.00
	2	370.38	290.61	170.14	53.42	9.21
	4	370.38	215.06	86.03	19.33	4.88
	8	370.38	152.45	48.47	12.57	8.01
	16	370.38	108.52	34.33	16.53	16.00
	32	370.38	85.47	37.32	32.00	32.00
	64	370.38	87.30	64.30	64.00	64.00
	128	370.38	132.01	128.00	128.00	128.00
	256	370.38	256.04	256.00	256.00	256.00
0.9	RES	370.38	364.51	345.87	260.48	32.74
	2	370.38	366.45	355.10	315.45	214.22
	4	370.38	360.48	333.49	254.77	123.84
	8	370.38	351.66	304.84	195.75	74.03
	16	370.38	339.49	270.84	147.13	50.25
	32	370.38	324.61	237.03	115.86	45.99
	64	370.38	310.51	213.34	108.42	66.32
	128	370.38	306.20	216.17	141.32	128.02
	256	370.38	331.71	281.96	256.62	256.00

Table 3 (continued)

Correlation		Shift d/s				
f	b	0	0.5	1	2	4
0.99	RES	370.38	368.95	362.76	312.00	59.30
	2	370.38	370.34	370.22	369.75	367.86
	4	370.38	370.28	369.97	368.76	363.98
	8	370.38	370.18	369.60	367.26	358.19
	16	370.38	370.04	369.04	365.08	350.02
	32	370.38	369.86	368.30	362.20	339.72
	64	370.38	369.66	367.50	359.16	329.50
	128	370.38	369.56	367.13	357.81	325.79
	256	370.38	369.88	368.40	362.73	343.39

Note: ARLs measured in observations.
Source: Runger and Willemain 1995.

3. A MODEL-FREE APPROACH: THE BATCH MEANS CONTROL CHART

Runger and Willemain (1996) proposed an unweighted batch means (UBM) control chart as an alternative to the weighted batch means (WBM) chart for monitoring autocorrelated process data.

The UBM chart differs from the WBM chart by giving equal weights to every point in the batch. let the jth unweighted batch mean be

$$v_j = (b^{-1}) \sum_{i=1}^{b} x_{(j-1)b+i}, \qquad j = 1, 2, \ldots \tag{22}$$

This expression differs from (15) only in that

$$w_i = \frac{1}{b}, \qquad i = 1, \ldots, b \tag{23}$$

The important implication of (23) is that although one has to determine an appropriate batch size b, one does not need to construct an ARMA model of the data. This model-free approach is quite standard in simulation output analysis, which also focuses on inference for long time series with high autocorrelation.

A model-free process-monitoring procedure was the objective of the many schemes considered by Runger and Willemain (1996). That work showed that the batch means can be plotted and approximately analyzed on a standard individuals control chart. Distinct from residuals plots, UBM charts retain the basic simplicity of averaging observations to form a point in a control chart. With UBM, the control chart averaging is used to dilute the autocorrelation of the data.

Procedures for determining an appropriate batch size were developed by Law and Carson (1979) and Fishman (1978a, 1978b). These procedures are empirical and do not depend on identifying and estimating a time series model. Of course, a time series model can guide the process of selecting the batch size and also provide analytical insights.

Runger and Willemain (1996) provided a detailed analysis of batch sizes for AR(1) models. They recommend that the batch size be selected so as to reduce the lag 1 autocorrelation of the batch means to approximately 0.10. They suggest using Fishman's (1978a) procedure, which starts with $b = 1$ and doubles b until the lag 1 autocorrelation of the batch means is sufficiently small. This parallels the logic of the Shewhart chart in that larger batches are more effective for detecting smaller shifts; smaller batches respond more quickly to larger shifts.

Though a time series model is not necessary to construct a UBM chart, Table 4 shows the batch size requirements for the AR(1) model for various values of ϕ (Kang and Schmeiser, 1987). The lower values of σ_{UBM} imply greater sensitivity.

Table 4 Minimum Batch Size Required for UBM Chart for AR(1) Data

ϕ	b	$\sigma(UBM)/\sigma$	$\sigma(UBM)/\sigma$
0.00	1	1.0000	n/a
0.10	2	0.7454	1.1111
0.20	3	0.6701	0.8839
0.30	4	0.6533	0.8248
0.40	6	0.6243	0.7454
0.50	8	0.6457	0.7559
0.60	12	0.6630	0.7538
0.70	17	0.7405	0.8333
0.80	27	0.8797	0.9806
0.90	58	1.2013	1.3245
0.95	118	1.6827	1.8490
0.99	596	3.7396	4.0996

Note: Batch size chosen to make lag-1 autocorrelation of batch means 0.10.

Table 5 Performances of the Unweighted Batch Means Control Chart

ϕ	b	Method	Shift, δ/σ 0	1	2
0.9	60	Approximation	370	199	98
		Monte Carlo	371 ± 2	206 ± 3	100 ± 1
0.95	120	Approximation	370	296	201
		Monte Carlo	371 ± 6	303 ± 4	206 ± 2

Note: ARLs measured in batches. Monte Carlo results based on 5 sets of 5,000 alarms. Uncertainties are 95π confidence intervals.

Table 6 Comparison of Shewhart Charts ARLs for AR(1) Data

ϕ	Method	b	Shift, δ/σ 0.00	0.5	1	2	4
0	REs	1	10000	2823	520	34	2
0.25	RES	1	10000	4360	1183	116	3
	WBM	4	10000	2066	320	23	4
	UBM	4	10000	1279	149	11	4
	WBM	23	10000	233	34	23	23
	UBM	23	10000	210	32	23	23
0.50	REs	1	10000	6521	2818	506	17
	WBM	8	10000	2230	378	33	8
	UBM	8	10000	1607	225	20	8
	WBM	43	10000	397	66	43	43
	UBM	43	10000	367	63	43	43
0.90	RES	1	10000	9801	9234	7279	1828
	WBM	58	10000	6119	2548	548	96
	UBM	58	10000	5619	2133	423	81
	WBM	472	10000	2547	823	476	472
	UBM	472	10000	2504	809	476	472
0.99	RES	1	10000	9995	9974	6977	4508
	WBM	596	10000	9691	8868	6605	3238
	UBM	596	10000	9631	8670	6178	2847
	WBM	2750	10000	9440	8129	6605	3238
	UBM	2750	10000	9420	8074	5434	3225

Note: ARLs measured in observations.

Runger and Willemain (1996) use the following approximation to estimate the performance of the UBM chart:

$$\mathrm{ARL} = \frac{b}{1 - \Phi(\Delta_0 - \delta/\sigma_{\mathrm{UBM}}) + \Phi(-\Delta_0 - \delta/\sigma_{\mathrm{UBM}})} \tag{24}$$

This approximation, which assumes that the batch means are i.i.d. normal with mean μ and standard deviation σ_{UBM} as given in Table 4, was confirmed by Monte Carlo analysis (Table 5).

Since estimating ARLs with (27) is simpler than extensive Monte Carlo analysis, the approximation is used in Table 6. Table 6 compares this ARL with the ARLs of the other two charts for selected values of the autocorrelation parameter ϕ. The batch sizes b were chosen by using Table 3 to provide a WBM chart sensitive to a shift $\delta = 1$. The comparison was made with the in-control $\mathrm{ARL}_0 = 10{,}000$. Table 6 shows that both batch means charts outperform the residuals chart in almost all cases shown, with the UBM chart performing best of all.

REFERENCES

Alwan LC. (1992). Effects of autocorrelation on control chart performance. Commun Stat: Theory Methods 21: 1025–1049.

Alwan LC, Roberts HV. (1988). Time series modeling for statistical process control. J Bus Econ Stat 6(1): 87–95.

Berthouex TM, Hunter, WG, Pallesen L. (1978). Monitoring sewage treatment plants: Some quality control perspectives. J Qual Technol 10: 139–148.

Bischak DP, Kelton WD, Pollock SM. (1993). Weighted batch means for confidence intervals in steady-state simulations. Manag Sci 39: 1002–1019.

Box GEP, Jenkins GM, Reinsel GC. (1994). Time Series Analysis, Forecasting and Control. 3rd ed. Prentice-Hall, Englewood Cliffs, NJ.

Fishman GS. (1978a). Grouping observations in digital simulation. Manage Sci 24: 510–521.

Fishman GS. (1978b). Principles of Discrete Event Simulation. Wiley, New York.

Hahn GJ. (1989). Statistics-aided manufacturing: A look into the future. Am Stat 43: 74–79.

Harris TJ, Ross WH. (1991). Statistical process control procedures for correlated observations. Can J Chem Eng 69: 48–57.

Kang K, Schmeiser B. (1987). Properties of batch means from stationary ARMA time series, Operations Research Letters 6: 19–24.

Law A, Carson JS. (1979). A sequential procedure for determining the length of steady-state. Op Res 29: 1011.

MacGregor JF, Ross TJ. (1993). The exponentially weighted moving variance. J Qual Technol 25: 106–118.

Mastrangelo CM, Montgomery DC. (1995). Characterization of a moving centerline exponentially weighted moving average. Qual Reliab Eng Int 11(2): 79–89.

Montgomery DC, Friedman DJ. (1989). Statistical process control in a computer-integrated manufacturing environment. In: Statistical Process Control in Automated Manufacturing. Keats JB, Hubele NF, eds. Marcel Dekker, New York.

Montgomery DC, Johnson LA, Gardiner JS. (1990). Forecasting and Time Series Analysis. 2nd ed. McGraw-Hill, New York.

Montgomery DC, Mastrangelo CM. (1991). Some statistical process control methods for autocorrelated data. J Qual Technol 23(3): 179–193.

Runger GC, Willemain TR. (1995) Model-based and model-free control of autocorrelated processes. J Qual Technol 27: 283–292.

Runger GC, Willemain TR. (1996). Batch-means control charts for autocorrelated data. IIE Transactions 28: 483–487.

Tseng, Adams BM. (1994). Robustness of forecast-based monitoring schemes. Technical Report, Department of Management Science and Statistics, University of Alabama.

Wardell DG, Moskowitz H, Plante RD. (1994). Run-length distributions of special-cause control charts for correlated processes and discussion. Technometrics 36(1): 3–27.

Willemain TR, Runger, GC. (1996). Designing control charts using an empirical reference distribution. J Qual Technol 28(1): 31.

Yaschin E. (1993). Performance of CUSUM control schemes for serially correlated observation. Technometrics 35: 37–52.

Yourstone S, Montgomery DC. (1989). Development of a real-time statistical process control algorithm. Qual Reliab Eng Int 5: 309–317.

9

An Introduction to the New Multivariate Diagnosis Theory with Two Kinds of Quality and Its Applications

Gongxu Zhang
Beijing University of Science and Technology, Beijing, People's Republic of China

1. MULTIOPERATION AND MULTI-INDEX SYSTEM

In factories multioperation and multi-index systems are very common. A multioperation system is a system in which its product is processed by a production line consisting of two or more operations. A multi-index system is one in which at least one operation has two or more indices, such as a technical index and/or a quality index. For example, a printed circuit production line consists of 17 operations, with at least two indices and at most 27. Again, for analgin (a kind of drug) a production line consists of six operations, with at least two and at most six indices. Such examples exist indeed everywhere.

2. PROBLEMS ENCOUNTERED IN IMPLEMENTING QUALITY CONTROL AND DIAGNOSIS IN A MULTIOPERATION, MULTI-INDEX SYSTEM

In a multioperation, multi-index system, if we want to implement quality control and diagnosis, there are three major problems:

1. In a multioperation production line, the processing of the preceding operating will in general influence the current operation. Since the preceding influence is synthesized with the processing of the current operation,

how do we differentiate one from the other? If we cannot differentiate them, we cannot distinguish their quality responsibility, and then we cannot implement scientific quality control. Evidently, in a multioperation production line, we need to diagnose the preceding influence.

2. In a multi-index production line, there is the problem of correlations among indices. For example, in the operation of etching a printed circuit, the quality index of etching has correlations with the technical indices: NaOH, Cl^-, Cu^{2+}. When the etching index is abnormal, we need to diagnose which technical index or indices induced this abnormality.

3. In a multioperation, multi-index production line, there are both the preceding influence and the correlations among indices, making the problem more complex.

3. HOW TO DIAGNOSE THE PRECEDING INFLUENCE IN A MULTIOPERATION PRODUCTION LINE

In a multioperation production line, we need to use the diagnosis theory with two kinds of quality proposed by Zhang (1982a) to diagnose the preceding influence. The basis of this theory is the concept of two kinds of quality.

3.1. Two Kinds of Quality

According to the different ranges involved in different definitions of quality, there are two kinds of product quality:

1. *Total quality* is the product quality contributed by the current operation and all the preceding operations. It is simply product quality in the usual sense and is felt directly by the customer.
2. *Partial quality* is the quality specifically resulting from the current operation and does not include the influence of the preceding operations. Obviously, it reflects the work quality of the current operation.

These two kinds of quality exist in any operation. Total quality consists of two parts: the partial quality and the preceding influence on it; hence, partial quality is only part of total quality.

3.2. Importance of the Concept of Two Kinds of Quality

The concept of two kinds of quality is very important, as can be seen from the following facts:

1. The two kinds of quality exist at each operation.
2. The concept of two kinds of quality is very general and exists in all processes of production, service and management as well as many other processes.

3.3. Fundamental Thinking of the Diagnosis Theory with Two Kinds of Quality

The so-called diagnosis is always obtained through a comparison of a measured value with the standard value. For example, in order to diagnose the preceding influence, we can take the partial quality (which has no relationship with the preceding influence) of the current operation as the standard value, and the corresponding total quality (which consists of both the partial quality and the preceding influence) as the measured value. Comparing these two kinds of quality, we can diagnose the preceding influence of the current operation. The greater the difference between these two kinds of quality, the more serious the preceding influence.

Here, the key problem is how to measure these two kinds of quality. If we use a control chart to measure them, we can use the Shewhart control chart to measure the total quality and the cause-selecting Shewhart control chart proposed by Zhang (1980) to measure the partial quality. We refer to this as diagnosis with two kinds of control charts. If we use the process capability index to measure the two kinds of quality, we can use the total process capability index (which is just the process capability index in the usual sense), denoted by C_{pt}, to measure the total quality and the partial process capability index, denoted by C_{pp}, which is a new kind of process capability index proposed by Zhang (1982a), to measure the partial quality. We refer to this as diagnosis with two kinds of process capability indices. The former is a realtime diagnosis, and the latter is a diagnosis over time. See Zhang (1989, 1990).

3.4. Steps in Diagnosis with Two Kinds of Control Charts

The steps in diagnosis with two kinds of control charts are as follows.

Step 1. Construct the diagnosis system between adjacent operations with technical relations as shown in Figure 1. In Figure 1, the connection between operations 1 and 2 is the total quality of operation 1, and there exist two kinds of quality at operation 2, i.e., total quality and partial quality. Suppose the total qualities of operations 1 and 2 are measured with two Shewhart charts, and the partial quality is measured with the cause-selecting

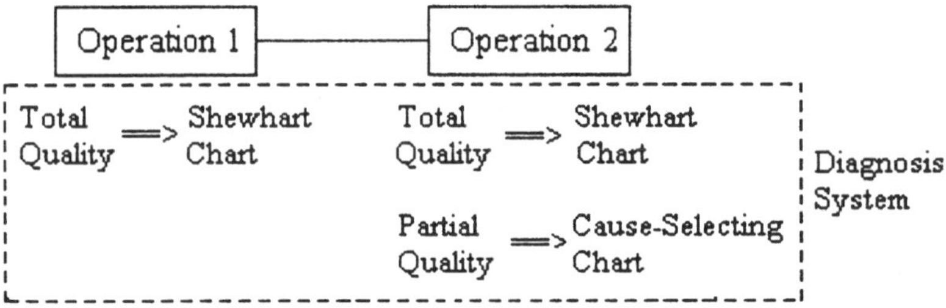

Figure 1 Diagnosis system between adjacent operations.

Shewhart chart; then the diagnosis system can also be referred to as a three-chart diagnosis system.

Step 2. Diagnose the diagnosis system according to the typical case diagnosis table, Table 1. Since each control chart has two states, i.e., the normal state and the abnormal state, the three-chart diagnosis system has eight typical diagnosis cases (see Table 1). Comparing the three charts of the diagnosis system with the three-chart cases of Table 1, we can diagnose the diagnosis system.

From Table 1 we can see that if we do not have the diagnosis theory with two kinds of quality and use only Shewhart charts at each operation, we may get a false alarm or alarm missing for cases II, III, VI, and VII. This is already verified by experience in factories. It is not a fault of the Shewhart chart itself; in fact, it is due to our misunderstanding of the Shewhart chart. The Shewhart chart can be used to reflect total quality only; thus it includes the preceding influence. Using the Shewhart chart as if it reflects the partial quality only and has no relation with the preceding influence is wrong; see Zhang (1992b, p. 173).

3.5. Characteristics of Diagnosis with Two Kinds of Control Charts

In Table 1, the diagnosis of each typical case is derived only from ordinary logical deduction; we did not use probability and statistics. Thus there are no two kinds of errors. Table 1 also considers the connection between preceding and succeeding operations.

The Shewhart chart used in Table 1 can be replaced by some other all-control chart, for example, the CUSUM chart or the T^2 chart. But, at the same time, the cause-selecting chart should be replaced by the corresponding cause-selecting CUSUM chart, the cause-selecting T^2 chart, etc.

Table 1 Typical Case Diagnosis Table

Typical cases	Shewhart chart for operation 1	Shewhart chart for operation 2	Cause-selecting chart for operation 2	Diagnosis
I	+	+	+	The partial quality is abnormal. The preceding influence is also abnormal.
II	+	+	–	The partial quality is normal. The preceding influence is abnormal.
III	+	–	+	The partial quality is abnormal. The preceding influence is also abnormal. But the one offsets the effect of the other.
IV	+	–	–	The preceding influence is abnormal, but the partial quality offsets its effect and makes the total quality of operation 2 to be normal.
V	–	+	+	The partial quality is abnormal. The preceding influence is normal.
VI	–	+	–	The partial quality and the preceding influence are both normal, but their total effect is to make the total quality of operation 2 become abnormal.
VII	–	–	+	The partial quality is abnormal. But the preceding influence offsets it to make the total quality of operation 2 become abnormal.
VIII	–	–	–	The partial quality, the preceding influence, and the total quality are all normal

Note:"–" means normal and "+" means abnormal.

4. HOW TO DIAGNOSE THE CORRELATION AMONG INDICES FOR A MULTI-INDEX OPERATION

For a multi-index operation, we need to use the multivariate diagnosis theory with two kinds of quality proposed by Zhang (1996b, 1997). The fundamental thinking of this theory is similar to that of the diagnosis theory with two kinds of control charts. But since this is the multivariate case, we use the multivariate T^2 control chart and the cause-selecting T^2 control chart instead of the corresponding Shewhart chart and the cause-selecting Shewhart chart in the three-chart diagnosis system. Since the statistics of the T^2 control chart include the covariance matrix of each variable (assuming that the number of variables is p),

$$\begin{bmatrix} s_{11} & s_{12} & \cdots & s_{1p} \\ s_{21} & s_{22} & \cdots & s_{2p} \\ \vdots & \vdots & \ddots & \vdots \\ s_{p1} & s_{p2} & \cdots & s_{pp} \end{bmatrix}$$

where S_{ij}, $i \neq j$, is the covariance, the T^2 control chart can consider completely the correlation among variables.

The multivariate T^2 control chart was proposed by Hotelling in 1947 and was well used in Western countries for multivariate cases. Its merits are that (1) it considers the correlations among variables and (2) it can give us exactly the probability of the first kind of error, α. But its greatest drawback is that it cannot diagnose which variable induced the abnormality when the process is abnormal. On the other hand, the best merit of the diagnosis theory with two kinds of quality is that it can be used to diagnose the cause of abnormality in the process. Hence Zhang proposed a new multivariate diagnosis theory with two kinds of quality to combine the above-stated theories together so that we can concentrate their merits and at the same time avoid their drawbacks.

5. HOW TO SIMULTANEOUSLY DIAGNOSE THE PRECEDING INFLUENCE AND THE CORRELATION AMONG INDICES IN A MULTIOPERATION, MULTI-INDEX SYSTEM

From the preceding discussions it is evident that we need to use the diagnosis theory with two kinds of quality in order to diagnose the preceding influence, and we also need to use the multivariate diagnosis theory with two kinds of quality in order to diagnose the correlated indices. In such a complex system, it is not enough to depend on the technology only; we must consider statistical process control and diagnosis (SPCD) too. Besides the diagnosis theories of Western countries always diagnose all variables simultaneously. Suppose the number of variables is p and the probability of the first kind of error in diagnosing a variable is α, then the probability of no first kind of error in diagnosing p variables is

$$P_0 = (1 - \alpha)^p \approx 1 - p\alpha$$

Thus, the probability of the first kind of error in diagnosing p variables is

$$P_1 = 1 - P_0 \approx p\alpha$$

i.e. it is proportional to the number of variables. In the case of a great number of variables, the value of P_1 may become intolerable. To solve this problem, Zhang and his Ph.D. candidate Dr. Huiyin Zheng (Zheng, 1995) proposed the multivariate stepwise diagnosis theory in 1994.

5.1. Fundamentals of the Multivariate Stepwise Diagnosis Theory

If we tested that the population of all variables concerned with the problem is abnormal, we want to identify the abnormal variable. Instead of diagnosing each variable contained in this population, we need only diagnose the most probable assignable variable each time, for by so doing we can decrease the number of steps of diagnosis needed. The steps of the multivariate stepwise diagnosis theory are as follows:

Step 1. Test the abnormality of the population of all variables. If it is normal, the diagnosis stops; otherwise proceed to step 2.

Step 2. Select the most probable assignable variable and test whether it is abnormal or not.

Step 3. Test the remaining population of variables. If it is normal, then the diagnosis stops, otherwise return to step 2.

Repeat steps 1–3 until we can ascertain each variable to be normal or abnormal.

In practice, in general, it takes only one to three steps to complete the multivariate diagnosis process.

5.2. Compiling the Windows Software DTTQ2000

We have compiled the Windows software DTTQ2000 (DTTQ = diagnosis theory with two kinds of quality), which combines the diagnosis theory with two kinds of quality, the multivariate diagnosis theory with two kinds of quality, and the multivariate stepwise diagnosis theory. So far we have diagnosed the multioperation, multi-index production lines of eleven factories more than 40 times using DTTQ2000. All results of these diagnoses have been in accordance with practical production.

5.3. Necessity of Application of the Multivariate Diagnosis Theory with Two Kinds of Quality and Its DTTQ2000 Software

Today's society has developed into an era of high quality and high reliability. The percent defective of some electronic products is as low as the parts per million or even parts per billion level, so production technology at the worksite must be combined with statistical process control and diagnosis (SPCD) to guarantee product quality. In fact, the requirements of SPCD with respect to product quality are more severe than those of technology. For example, the control limits of control charts are, in general, situated within the specification limits. In addition, we consider significant variations in product quality and nonrandom arrangements of points plotted between control limits on the control chart to be abnormal and take action to eliminate such abnormalities. But, on the other hand, technology does not pay attention to such facts.

At the worksite, technicians in general take one of the following actions whenever there is a need of multivariate control: (1) Put all parameters of the current operation to be within the specification limits or (2) adopt the Shewhart control chart to control each parameter of the current operation. In fact, these two actions are virtually the same; both oversimplify the multivariate problem and resolve it into several univariate problems. Here, unless all the variables are independent, otherwise we must consider the correlations among variables. For example, in the printed circuit production line there are altogether 27 indices at the operation of the factory Desmear/PTH. If we supervise this process with 27 x–R_s control charts supervising each of the 27 indices individually, then we can supervise 27 averages and 27 standard deviations, i.e.,

$$\mu_i, \sigma_i, \qquad i = 1, 2, ..., 27$$

But we cannot supervise the correlations among variables, i.e., the covariances among indices, with such a univariate x–R_s control chart. There are altogether 351 [$= 27(27 - 1)/2$] covariance parameters or coefficients of correlation to be supervised. Only by using multivariate diagnosis theory with two kinds of quality can we supervise all 405 ($= 27 + 27 + 351$) process parameters and implement the SPCD. Using the DTTQ2000 software we have diagnosed eleven factories in China, and all the diagnostic results have been in fairly good agreement with the actual production results.

Using the DTTQ2000 software with a microcomputer, it takes only about 1 min to perform one diagnosis; thus, it saves much time on the spot. Not only is the diagnosis correct, but it also avoids the subjectivity of the working personnel.

In a factory, it always takes a long time to train an experienced engineer in quality control and diagnosis. If we use the DTTQ2000 software, the training time is much reduced.

5.4. How to Establish SPCD in a Multioperation, Multi-index System

In a multioperation, multi-index system, in order to establish the SPCD we must consider three principles:

Principle 1. A multioperation production line must consider the preceding influence.

1. If there is no preceding influence, the partial quality will be equal to the total quality at the current operation, and we can use the Shewhart control chart (which is only a kind of all-control chart) to control it.
2. If there is a preceding influence, there exist two kinds of quality, total quality and partial quality at the current operation. Total quality can be controlled by the all-control chart, and partial quality can be controlled with the cause-selecting control chart.
3. Except for the first operation or the above-stated case 1, we can construct a three-chart diagnosis system as shown in Figure 1. Then we can diagnose this diagnosis system according to the typical case diagnosis table, Table 1.

Principle 2. A multi-index production line must consider the correlation among indices.

1. If the indices are not related, we can use a univariate all-control chart to control each index individually.
2. If the indices are related, we need to use a multivariate all-control chart to control the whole index system.

Principle 3. In a multioperation, multi-index system, we need to consider both the preceding influence and the correlation among indices, which makes the problem of implementing the SPCD more complex. The multivariate diagnosis system with two kinds of quality is a method for solving this complex problem, and its implementations show that the theory is in good accordance with actual practice.

6. APPLICATIONS OF THE MULTIVARIATE DIAGNOSIS THEORY WITH TWO KINDS OF QUALITY

Here, we show some practical examples of the multivariate diagnosis theory with two kinds of quality as follows.

Example 1

Operations 4 and 5 of a production line for the drug analgin have five indices, three of which belong to the preceding operation; the other two belong to the succeeding operation. Their data are as follows (see group 51 data in Table 2):

$$\text{Preceding operation: } x_1 = 8.80, \; x_2 = 97.71, \; x_3 = 89.11$$
$$\text{Succeeding operation: } x_4 = 95.67, x_5 = 4.37$$

Using the DTTQ2000 software, we know that the T^2 value is 18.693, greater than the upper control limit (UCL) of 13.555 of the T^2 control chart (Fig. 2), which means that the process is abnormal. Then, by diagnosing with DTTQ2000, we know that index x_5 is abnormal.

Example 2

Using the DTTQ2000 Windows software to diagnose the same desmear/ PTH operations of three printed circuit factories, A, B, C, we obtained Figure 3. Compare and criticize these three factories.

Table 2 Data for Operations 4 and 5 of Analgin Production Line

Group No.	x_1	x_2	x_3	x_4	x_5	T^2	Diagnosis
27	11.70	96.08	84.84	93.88	1.35	11.988	Normal
28	9.70	95.85	86.55	93.51	2.18	11.311	Normal
29	9.70	95.85	86.55	95.24	1.32	3.765	Normal
30	7.66	98.61	91.06	95.34	1.39	4.050	Normal
⋮	⋮	⋮	⋮	⋮	⋮	⋮	⋮
47	9.00	98.42	89.57	95.89	1.17	2.942	Normal
48	8.00	97.24	89.34	95.67	2.98	5.544	Normal
49	8.00	97.24	89.34	95.14	1.77	1.148	Normal
50	8.80	97.71	89.11	95.90	1.25	1.369	Normal
51	8.80	97.71	89.11	95.67	4.37	19.214	Abnormal

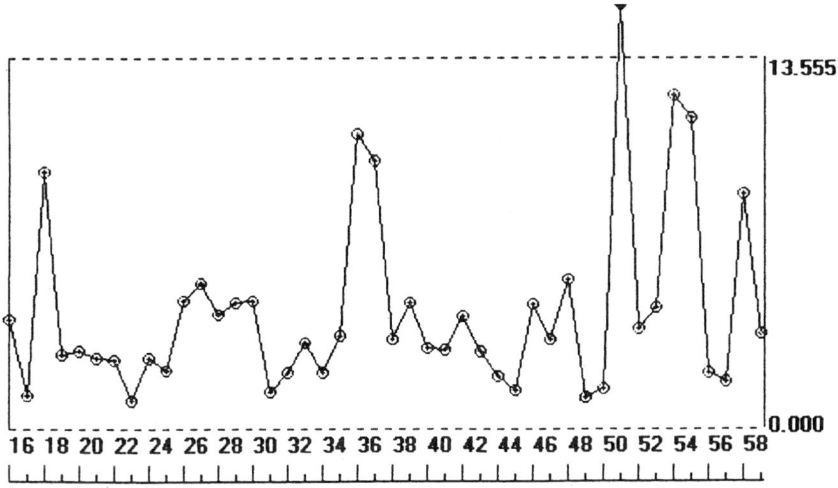

Figure 2 T^2 control chart.

From Figure 3 we see that the desmear/PTH operation of factory A (Fig. 3a) is under statistical control; but the desmear/PTH operation of factory B (Fig. 3b) has a record of an average of 1.0 point per month plotted outside the UCL of the T^2 control chart; and the same operation in factory C (Fig. 3c) has an average of 1.3 points per month plotted outside the UCL of the T^2 control chart. Hence, factories A, B, and C are in descending order according to the work quality of the desmear/PTH operation. Thus, the multivariate diagnosis theory with two kinds of quality can be used to give us an objective evaluation of the quality of each factory. This method can also be used to point out their direction of quality improvement.

7. CONCLUSION

1. According to what has been stated above, we can see that the multivariate diagnosis theory with two kinds of quality and its DTTQ2000 Windows software have prospects of being applied to the field of multioperation, multi-index systems. Its greatest merit is that it considers the multivariate characteristics of the multioperation, multi-index system and can control all objects that should be controlled by the system.
2. The implementation of this theory at eleven factories in China shows that production practices are in fair agreement with the theory.

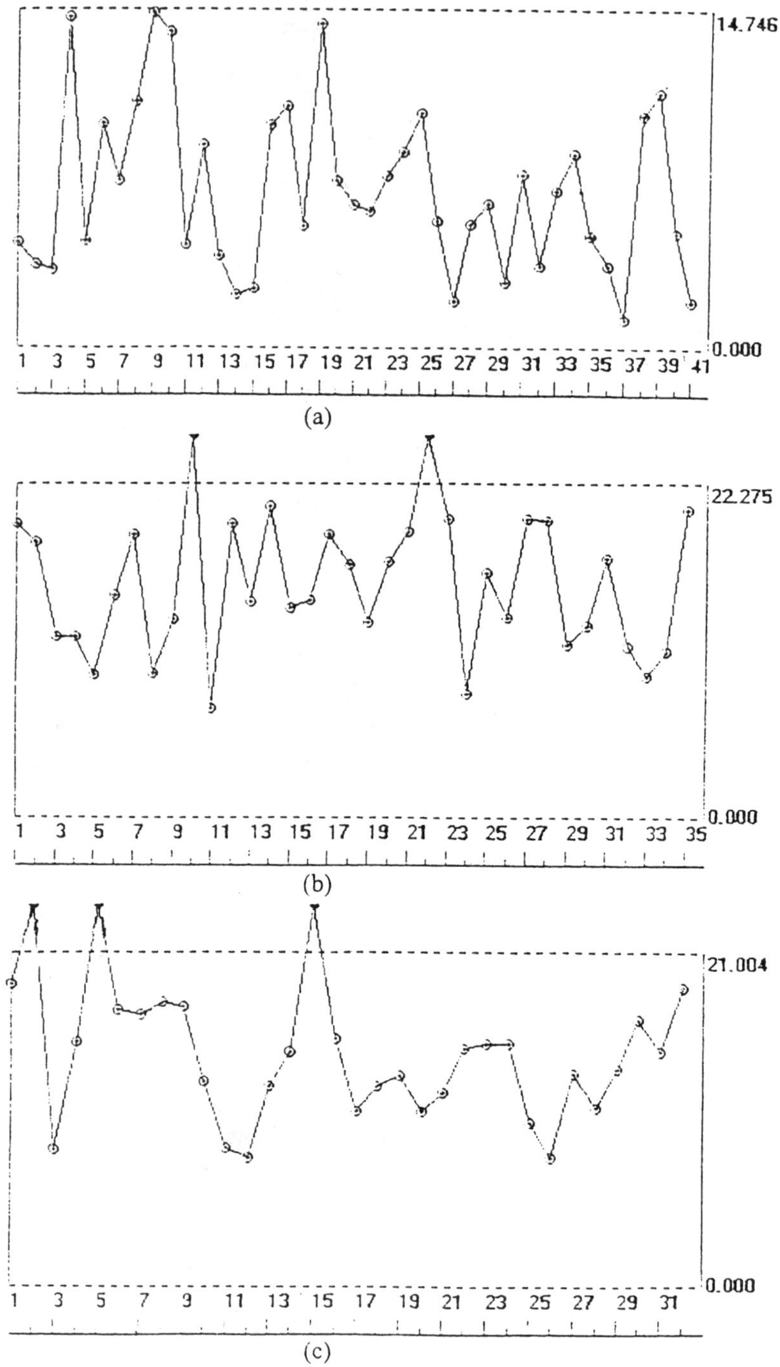

Figure 3 Comparison of T^2 charts of Desmear/PTH operations of three factories A, B, C.

REFERENCES

Chen Z.Q., Zhang G. (1996a). Fuzzy control charts. China Qual, October 1996.

Zhang G. (1980). A new type of quality control chart allowing the presence of assignable cause—the cause-selecting control charts. Acta Electron Sin 2:1–10.

Zhang G. (1982a). Control chart design and a diagnosis theory with two kinds of quality. Second Annual Meeting of CQCA, February 1982, Guilin, P. R. China.

Zhang G. (1982b). Multiple cause-selecting control charts. Acta Electronic Sin 3:31–36, May 1982.

Zhang G. (1983). A universal method of cause-selecting—The standard transformation method. Acta Electron Sin 5:1983.

Zhang G. (1984a). Cause-Selecting Control Chart: Theory and Practice. The Publishing House of Post and Telecommunication, Beijing, 1984.

Zhang, G. (1984b). A new type of control charts—cause-selecting control charts and a theory of diagnosis with control charts. Proceedings of World Quality Congress '84, pp 175–185.

Zhang G. (1985). Cumulative control charts and cumulative cause-selecting control charts. J China Inst of Commun 6:31–38.

Zhang G. (1989). A diagnosis theory with two kinds of quality. Proceedings of the 43rd American Quality Congress Transactions, pp 594–599. Reprinted in TQM, UK, No. 2, 1990.

Zhang, G. (1992a). Cause-Selecting Control Chart and Diagnosis. The Aarhus School of Business, Aarhus, Denmark, 1992.

Zhang G. (1992b). Textbook of Quality Management. The Publisher of High Education, 1992.

Zhang G, Dahlgaard JJ, Kristensen K. (1996b). Diagnosing quality: theory and practice. Research Report. MAPP, Denmark, 1996

Zhang G. (1997). An introduction to the multivariate diagnosis theory with two kinds of quality. China Qual, February 1997, pp 36–39.

Zheng H. (1995). Multivariate theory of quality control and diagnosis. Doctoral dissertation, Beijing University of Aeronautics and Astronautics, 1995.

10

Applications of Markov Chains in Quality-Related Matters

Min-Te Chao
Academia Sinica, Taipei, Taiwan, Republic of China

1. INTRODUCTION

To evaluate the performance of a control chart, one of the key elements is the average run length (ARL), which is difficult to calculate. However, if the underlying observations can be embedded into a finite Markov chain, then an exact ARL can be found if the observations are discrete, and approximations are available if they are continuous. In this chapter I provide a systematic review of many quality-related topics in situations in which a finite Markov chain can be employed.

The most fundamental statistical system consists of a set of independent random variables. Although this structure contains all the essential features for statistical analysis, and in fact most ideas for statistical analysis may have their origin traced back to this simple case, it nevertheless lacks the versatility to describe the more complex systems that are often encountered in real-life applications. In this chapter I describe the next simplest case, the Markov chain model, under which various quality-related problems can be vividly described and analyzed.

The most striking advantage of a Markov chain is its versatility. It can be used to describe, e.g., intricate deterioration processes and complex maintenance procedures (Neuman and Bonhomme, 1975) with relative ease.

Many complicated quality-related processes, when properly arranged, can be embedded into a Markov chain of reasonable size. Since the theory of Markov chains, particularly finite and ergodic Markov chains, is well established, for various applications the essential problem is how to find a reasonably sized Markov chain to describe the underlying quality-related process. Once this is done, the rest of the analysis is standard.

In this chapter I consider various well-known procedures and in each case indicate why such a Markov chain can be constructed. Efforts are placed on exploration rather than on original research. We first list some basic facts about a Markov chain (Section 2), and in Section 3 examples are given of where some exact results can be obtained. The exact ARL formula and its sampling distribution are given in Section 4. I then introduce in Section 5 the Brook and Evans (1972) approximation technique and show how it can be applied to various CUSUMs.

We have concentrated our efforts mostly on control charts. The Markov chain method, however, can also be applied to other quality systems. A list of these procedures is presented in Section 6.

2. BASIC FACTS ABOUT MARKOV CHAINS

In this section I briefly describe the necessary background of a finite Markov chain that will be needed for the rest of this chapter.

Given a sequence of random variables X_1, X_2, . . ., the simplest nontrivial structure we can impose is to assume that the X's are independent and identically distributed (i.i.d.). This assumption is often used to describe a sequence of observations of certain quality characteristics for essentially all kinds of control charts. If the X's are correlated, then the probability structure of these X's can be very complicated. One of the simplest nontrivial dependent cases is that of the X's following a Markov process.

Roughly speaking, a sequence $\{X_n, n \geq 0\}$ of random variables is Markovian if it has the ability to forget the past when the present status is given. When we say "present," "past," or "future," we implicitly assume that there is a time element. We shall in this respect consider the subscript n of X_n as a time index. Mathematically, the Markov property can be expressed by

$$P[X_{n+1} \in A | X_n = x, (X_0 X_1, ..., X_n) \in B] = P[X_{n+1} \in A | X_n = x] \quad (1)$$

for all Borel sets $A \subset R$, $B \subset R^{n+1}$. If, in addition to Eq. (1), the X's take values only in a finite set S, which without loss of generality we may assume to be $S = \{1, 2, ..., s\}$, then we say that the X's follow a finite Markov chain.

For a finite Markov chain, the information contained in Eq. (1) can be summarized into, say,

$$P[X_{n+1}=j|X_n=i,(X_0,X_1...,X_n)\in B]=P[X_{n+1}=j|X_n=1]:=p_{n;ij} \quad (2)$$

If, in addition, $p_{n;ij}$ of (2) is independent of n, then the Markov chain is said to have stationary transition probabilities. In this case, let

$$P=(p_{ij})_{s\times s} \quad (3)$$

and let $\boldsymbol{\pi}=(\pi_1,\pi_2...,\pi_s)$, $\pi_i=P[X_0=i]$. It can be shown that for a Markov chain with stationary transition probabilities the knowledge of $\boldsymbol{\pi}$ and the matrix P is sufficient to determine the joint probability distribution of $(X_0, X_1, X_2, ...)$. We call $\boldsymbol{\pi}$ the initial distribution and P the (stationary) transition probability matrix.

In what follows, we shall always assume that the Markov chains under consideration are finite with a certain initial distribution and a stationary transition matrix.

A good technical reason to use a matrix P is that we can employ matrix algebra to simplify various calculations. For example, the kth-order transition probability

$$p_{ij}^{(k)}=P[X_{n+k}=j|X_n=i]$$

is simply the (i,j)th element of P^k, the kth power of the transition matrix P, i.e.,

$$\begin{aligned}P^k &= P\times P\times ...\times P \quad (k\text{ times})\\ &=(p_{ij}^{(k)})\end{aligned} \quad (4)$$

The entries of P are probabilities, so the row sums of P are unity and the entries themselves are all nonnegative. It may happen that some of the entries of P are 0. But if we look at the sequence $P, P^2, P^3, ...$, it may happen that at some $k>0$ all entries of P^k are strictly positive. If this is the case, this means that if one starts from any state i, in k steps it is possible to reach state j, and this holds true for all $1\le i,j\le s$. If $p_{ij}^{(k)}>0$ for some $k>0$ and for all $1\le i,j\le s$, then we say that the Markov chain is irreducible.

Let $f_j^{(n)}$ be the probability that in a Markov chain starting from state j, the first time it goes back to the jth state is at time n, i.e.,

$$f_j^{(n)}=P[X_1\neq j, X_2\neq j, ..., X_{n-1}\neq j, X_n=j|X_0=j]$$

Let

$$\mu_j=\sum_{n=1}^{\infty} nf_j^{(n)} \quad (5)$$

The quantity μ_j is the average time at which a Markov chain starting from state j returns to state j, and it is called the mean recurrence time for state j. If $\mu_j < \infty$, then state j is said to be ergodic. If $\mu_j < \infty$ for all $j \in S$, then we say that the Markov chain is ergodic.

If a Markov chain is irreducible and ergodic, then the limits

$$u_j = \lim_{n\to\infty} p_{ij}^{(n)}, \qquad i,j \in S \tag{6}$$

exist and are independent of the initial state i. Furthermore, $u_j > 0$, $\sum_{j=1}^{s} u_j = 1$, and

$$u_j = \sum_{i=1}^{s} u_i p_{ij}, \qquad j \in S \tag{7}$$

The vector $\mathbf{u} = (u_1, u_2, ..., u_s)$ is called the absolute stationary probability. If $\pi = \mathbf{u}$, then it can be shown that

$$P[X_n = j] = P[X_0 = j] \tag{8}$$

for all $j \in S$ and for all $n \geq 0$, i.e., the Markov chain is stationary (instead of just having a stationary transition probability).

An interesting feature of Eq. (6) is that its rate of convergence is geometric. Let U be an $s \times s$ matrix consisting of identical rows, where each row is $\mathbf{u}$. Then by (7), $PU = UP = U$, so by induction we have

$$P^n - U = (P - U)^n \tag{9}$$

The fact that $(P - U)^n \to 0$ exponentially fast follows from the Perron–Frobenius theorem, and since it is a little bit too technical we shall not pursue it further. This basically explains that for a well-behaved Markov chain, the series in (5) usually converges because it is basically a geometric series. Also, the long-term behavior of an ergodic Markov chain is independent of its initial distribution.

Let A be a subset of S, and let $T = \inf\{n \geq 1, X_n \in A\}$. Then T is the first entrance time to the set A. For a control chart modeled by a Markov chain, the set A may consist of the region where an alarm should be triggered when $X_n \in A$ occurs for the first time. Thus T is the time when the first out-of-control signal is obtained, and $E(T)$ is closely related to the concept of average run length (ARL). When the control charts become more involved, the exact or approximate calculations of ARLs become involved or impossible with elementary methods. However, most (not all) control charts can be properly modeled by a Markov chain, and essentially all methods developed to calculate the ARLs are more or less based on the possibility that one can embed the control scheme into a Markov chain.

3. DISCRETE CASE: EXACT RESULT

In this section we discuss cases for which an exact finite Markov chain can be found to describe the underlying control chart. I first describe a general scenario where such a representation can be arranged and explain why it can be done.

Assume that the basic observations are $X_1, X_2, \ldots$, which are i.i.d. and take values in a finite set A of size k. The key point is that the X's are discrete and the set A is finite. This may be the case when either the X's themselves are discrete or the X's can be discretized.

Most control charts are of "finite memory"; i.e., at time n the decision of whether to flag an out-of-control signal depends on $X_{n-r+1}, X_{n-r+2}, \ldots, X_n$ only. In other words, we may trace back to consult the recent behavior of the observations to decide whether the chart is out of control, but we do it for at most r steps back, $r < \infty$. The case for which we have to trace back to the infinite past is excluded.

Let $Y_n = (X_{n-r+1}, X_{n-r+2}, \ldots, X_n)$. The random vector Y_n can take as many as $s = k^r < \infty$ possible values. It is easy to see that the Y's follow a Markov chain with an $s \times s$ transition matrix. Since at time n, Y_n is used to decide whether the process is out of control, we see that, conceptually at least, for the scenario described above, there exists a finite Markov chain for which the behavior of the control chart can be completely determined.

However, $s = k^r$ can be a very large number, so the $s \times s$ matrix can be too large to have practical value. Fortunately, this matrix is necessarily sparse (i.e., most entries are 0), and if we take a good look at the rules of the control chart, then the chances are we may find some means to drastically reduce the number of states of the Markov chain. Hence, to implement our general observation, we need case-by-case technical works for various control charts.

Example 1. The Standard $\bar{X}$ Chart

If groups of size n is used against $\pm 3\sigma$ limits, define, for each $\bar{X}_n$,

$$X_n = \begin{cases} 1 & \text{if } \bar{X}_n < \mu - 3\sigma \\ 2 & \text{if } \mu - 3\sigma \le \bar{X}_n < \mu + 3\sigma \\ 3 & \text{if } \bar{X}_n > \mu + 3\sigma \end{cases}$$

Note that the X's are the coded values of the $\bar{X}$'s. As long as our only concern is whether the process is under control, the behavior of the $\bar{X}$ chart can be completely determined by the coded X's. The coded X's are still i.i.d., and this is a special kind of Markov chain. Its transition matrix, when the process is under control, consists of three identical rows:

$$\begin{bmatrix} p_1 & p_2 & p_3 \\ p_1 & p_2 & p_3 \\ p_1 & p_2 & p_3 \end{bmatrix}$$

where $p_1 = p_3 = \Phi(-3)$ and $p_2 = \Phi(3) - \Phi(-3)$, where Φ denotes the cumulative distribution function of a standard normal distribution.

We can do similar things for the standard R chart.

Example 2. Shewhart Control Chart with Supplementary Runs

We often include additional run rules on a standard control chart to increase its sensitivity in detecting a mean shift. For example,

> *Rule T45.* If four of the last five observations are between (–3 and –1) or between 1 and 3, then a signal is suggested.

The well-implemented Western Electric Company (1965) rules also fall into this category.

If we want to implement rule T45 (in addition to the standard $\pm 3\sigma$ rules), we first need to divide the real line into five disjoint intervals:

$$I_i = (a_{i-1}, a_i]$$

with $a_5 = -a_0 = \infty$, $a_i = -5 + 2i$, $1 \le i \le 4$. Hence an $s = 5^5 = 3125$-state Markov chain is sufficient to describe this situation. But a 3145×3145 matrix is too large even for today's computers, so well devised tricks are needed to drastically reduce the value of s. It turns out that it is possible to use a 30-state Markov chain, which is of moderate size.

Rule T45 can be replaced by other run rules or some combinations of them. The idea is that in many cases we may drastically cut the size s, and it is possible to find a constructive method to implement such a simplification. Hence this type of problem (evaluate the exact ARLs and run length distributions for Shewhart control charts with various supplementary run rules) is mathematically tractable. For technical details, the full method is documented in Champ and Woodall (1987) and programs are available in Champ and Woodall (1990).

Example 3. Discrete CUSUM

Assume that Y_n, the i.i.d. observations for a quality control scheme, are integer-valued and that a one-sided CUSUM (Van Dobben de Bruny, 1968) is under consideration. Define $S_0 = 0$ and

$$S_n = \max\{0, Y_n + S_{n-1}\}, \qquad n = 1, 2, \ldots$$

Then the one-sided CUSUM signals an out-of-control message at stage n when $S_n \geq t$. This is a situation in which, at first sight, the decision to signal may depend on all data points up to time n, so it does not fall into the scenario described earlier for a finite Markov chain representation.

We may look at the construction somewhat differently. Since the process stops when $S_n \geq t$, obviously the important values for S_n are 0,1, ..., $t-1, t$. When $S_n \geq t$, the process stops; hence we may use a $(t+1)$-state Markov chain to describe S_n, with the last state behaving like an "absorbing state." We write the transition probability matrix as follows:

$$P = \begin{bmatrix} \mathbf{R} & (\mathbf{I}-\mathbf{R})\mathbf{1} \\ 0 & 1 \end{bmatrix} \tag{10}$$

where $\mathbf{R}$ is $t \times t$, $\mathbf{0}$ is a $1 \times t$ vector of 0's, and $\mathbf{1}$ is a $t \times 1$ vector of 1's. A typical entry of $\mathbf{R}$ is

$$r_{ij} = P[S_n = j | S_{n-1} = i], \qquad 0 \leq i, j \leq t-1$$

Expression (10) is typical for control charts represented by a finite Markov chain. Here the ARL is the average time for the process S_n to enter the absorbing state t. In symbols,

$$\text{ARL} = E(N)$$
$$N = \inf\{n \geq 1 : S_n \geq t\}$$

Example 4. Two-Sided CUSUM

The CUSUM in Example 3 is one-sided, since it detects the upward shift of the process mean only. For a two-sided (discrete) CUSUM, suppose that integer-valued random variables Y_n and Z_n are observed. Define $S_H(0) = S_L(0) = 0$, where

$$S_H(n) = \max\{0, Y_n + S_H(n-1)\}$$
$$S_L(n) = \min\{0, Z_n + S_L(n-1)\}$$

and

$$N = \inf\{n \geq 1 : S_H(n) \geq t_1 \text{ or } S_L(n) \leq -t_2\}$$

Normally, we would have $Y_i = X_i - k_1$, $Z_i = X_i + k_2$ for some known integers k_1, k_2. The X's are the basic sequence of the quality characteristic measured for control. The bivariate process $(S_H(n), S_L(n))$ takes values in $\{0,1, \ldots, t_1\} \times \{0,1, \ldots, t_2\}$, and it is possible to write a finite Markov chain with $s = (t_1 + 1)(t_2 + 1)$ states (see Lucas and Crosier, 1982). For a two-sided CUSUM, the number of states of the underlying Markov chain can be reduced to about $t_1 t_2/2$ by careful arrangement of states (Woodall, 1984). However, it is not known whether we can always reduce the Markov chain of the two-sided CUSUM to a linear function of $t_1 + t_2$.

4. GENERAL RESULTS

I have demonstrated with examples that it is often possible to represent a control chart with a Markov chain with a transition probability matrix of the form (10); i.e., states 0, 1, ..., $t-1$ are transient states, and one state, state t, is absorbing. Let N_i be the number of stages, starting from state $i \in \{0, 1, \ldots, t-1\}$, to reach the absorbing state for the first time. Then it follows from the standard Markov chain theory that the ℓth factorial moment of N_i, i.e.,

$$\mu_i^{(\ell)} = E\{N_i(N_i - 1) \ldots (N_i - \ell + 1)\} \tag{11}$$

can be found via the matrix equation

$$\begin{aligned} \underset{\sim}{\boldsymbol{\mu}}^{(\ell)} &= (\mu_0^{(\ell)}, \mu_1^{(\ell)}, \ldots, \mu_{t-1}^{(\ell)}) \\ &= \ell! \mathbf{R}^{\ell-1}(\mathbf{I} - \mathbf{R})^{-\ell} \underset{\sim}{\mathbf{1}} \end{aligned} \tag{12}$$

where $\underset{\sim}{\mathbf{1}}$ is a $t \times 1$ vector of 1's. Furthermore, the run length distributions of $N_0, N_1, \ldots, N_{t-1}$ are given by

$$(P[N_0 = r], P[N_1 = r], \ldots, P[N_{t-1} = r])' = \mathbf{R}^{r-1}(\mathbf{I} - \mathbf{R})\underset{\sim}{\mathbf{1}} \tag{13}$$

$r = 1, 2, \ldots$.

What we have described can be roughly summarized as follows. If we can find a finite Markov chain to describe the behavior of a control scheme in the form of Eq. (10), then all problems concerning the ARLs of the control chart are solved. The only technical concern is that the size of the transition matrix should be manageable.

5. APPROXIMATIONS: THE CONTINUOUS CASE

When the underlying quality characteristic is continuous, a situation may rise for which we cannot embed the control scheme into a finite Markov chain.

Example 5. One-Sided CUSUM with Continuous Observations

Let us consider a setup identical to that of Example 3 but with the Y's replaced by i.i.d. $N(k, 1)$, where k is a known positive constant. The run length is N, defined by

$$N = \inf\{n \geq 1 : S_n \geq t\}, \qquad t>0$$

We proceed to find the distribution of N. It is easy to see that N takes values 1,2, ... only, and it is sufficient to find

$$P[N>r] = P[S_1<t, S_2<t, \ldots, S_r<t]$$

Since $t>0$, it is easy to find $P[N>0]=1$ and $P[N>1] = P[Y_1<t] = \Phi(t-k)$, where Φ is the cumulative distribution function of the standard normal distribution.

The case for $P[N>2]$ is more complicated. By definition,

$$\begin{aligned} P[N>2] &= P[S_1<t, S_2<t] \\ &= \int_0^t P[S_2<t|S_1 = x]dF_1(x) \end{aligned} \tag{14}$$

where F_1 is the cumulative distribution of S_1, i.e.,

$$F_1(x) = \begin{cases} \Phi(x-k) & \text{if } x \geq 0 \\ 0 & \text{otherwise} \end{cases} \tag{15}$$

The complication in (15) results from the fact that

$$P[S_1 = 0] = P[Y_1 \leq 0] = \Phi(-k)>0$$

hence the random variable S_1 is neither continuous nor discrete. It has a jump at $S_1 = 0$ and is continuous in $(0, \infty)$. Substituting (15) into (14), omitting the algebra, we have

$$P[N > 2] = \Phi(t-k)\Phi(-k) + \int_0^\infty \Phi(t-x-k)\phi(x-k)\,dx$$

where ϕ is the probability density function for the standard normal.

But since the last integral has no simple closed form, this is about as far as we can go analytically. (We can find $P[T > 3]$ in more complicated forms, but the situation quickly runs out of our control when we try to find $P[N > r]$ for $r = 3, 4, \ldots$.) This basically shows that there is no easy way to calculate the exact ARL for the one-sided CUSUM chart if the observations are i.i.d. normal. Also, the above example demonstrated that it is necessary to use approximate methods to find an approximation for the ARL for the standard one-sided CUSUM chart.

The basic idea of how to find approximate ARLs is due to Brook and Evans (1972). Since we can find the exact ARL of CUSUM when the Y's are discrete, then when these observations are continuous it is natural to discretize the Y's first. The exact ARL for the discrete version of CUSUM serves as an approximation of the exact ARL for the continuous case.

Specifically, for the situation described in Example 5, define

$$X_n = j \qquad \text{if } (j - 1/2)w < Y_n - k < (j + 1/2)w \tag{16}$$

Then

$$\begin{aligned} P[X_n = j] &= P[(j - 1/2) < Y_n - k < w(j + 1/2)w] \\ &= \Phi((j + 1/2)w) - \Phi((j = 1/2)w) \\ &\doteq w\phi(jw) \end{aligned}$$

if w, the threshold size for our "roundoff" procedure, is small. Since $|X_n - Y_n| \le w$ for all n, we would intuitively expect $Y_n \doteq X_n$, and ARLs based on the X's, which we may find exactly via the Markov chain method, can be used to find a reasonable approximation of the ARLs for the original CUSUM based on continuous distributions.

How small should w be in order to induce a reasonable approximation? Very little is known mathematically although we believe it is workable. However, it is reported (Brook and Evans, 1972) that it is possible to obtain agreement to within 5% of the limiting value when $t = 5$ and to within 1% when $t = 10$.

The basic idea of Brook and Evans can be applied to various CUSUMs. Since the basic concept is the same, we shall only list these cases. Successful attempts have been reported for the two-sided CUSUM (Woodall, 1984) and multivariate CUSUM (Woodall and Ncube, 1985). In these cases, however, the sizes of the transition probability matrices increase exponentially with the dimension of the problem, and so far no efficient way to drastically reduce the matrix size is known. The Brook and Evans technique also applies to weighted CUSUMs (Yashchin, 1989), CUSUMs with variable sampling intervals (Reynolds et al., 1990), and the exponentially weighted moving average schemes (Saccucci and Lucas, 1990). In all these examples, the control scheme can be described in the form

$$S_{in} = g_i(X_n, S_{i,n-1}), \qquad \geq 1; i = 1, 2, ..., m$$

where g_i are fixed functions and the X's are i.i.d. continuous or discrete. For example, for the two-sided CUSUM, we have $m = 2$ and

$$g_1(x, y) = \max\{0, x - k_1 + y\}$$
$$g_2(x, y) = \min\{0, x + k_2 + y\}$$

If the S_i's are discretized to t different values, then the control scheme can be approximately described by an s-state Markov chain, $s = t^m$.

Example 6. Another Two-Sided CUSUM

For the standard two-sided CUSUM, a careful arrangement can reduce the need of t^2 states, where we assume, for simplicity, that $t_1 = t_2 = t$. If the situation is extremely lucky, it can be reduced to $2t - 1$ states; but in general, $(t^2 + t)/2$ is about the best we can do (Woodall, 1984). Hence even for the two-sided CUSUM, the Brook and Evans technique has its limitations. Another way to look at the problem is to consider a slightly different two-sided CUSUM. The version below is suggested by Crosier (1986).

Let $S_0 = 0$ and define C_n, S_n recursively by

$$C_n = |S_{n-1} + X_n - a|$$

$$S_n = \begin{cases} 0 & \text{if } C_n \leq ks \\ (S_{n-1} + X_n - a)\left(1 - \dfrac{ks}{C_n}\right) & \text{if } C_n > ks \end{cases}$$

This is clearly Markovian. Since there is essentially one equation to describe the control scheme, there is no difficulty in using a t-state Markov chain.

6. OTHER APPLICATIONS

So far we have limited our discussion to control charts. However, the Markov chain technique is so versatile that it can be applied to many quality-related topics.

A main area of application concerns various continuous sampling plans with attribute-type observations. All these plans are based on a sequence of i.i.d. discrete observations, and the decision related to these plans is normally based on at most a finite number of observations counted backward. This fits into our general scenario of Markov chains, and the only technical problem left is to find a Markov chain of reasonable size.

Most continuous sampling plans (three versions of CSP-1 and CSP-k, $k = 2, 3, 4, 5$) can be embedded into a proper Markov chain (see Blackwell, 1977). The ANSI/ASQC Z1.4 plan falls into this category (Grinde et al., 1987; Brugger, 1989). Other examples include the two-stage chain sampling plan (Stephens and Dodge, 1976), the skip-lot procedure (Brugger, 1975), process control based on within-group ranking (Bakir and Reynolds, 1979), startup demonstration test (Hahn and Gage, 1983), and precontrol sampling plans (Salvia, 1987).

A more important application of a Markov chain is to study the behavior of the quality scheme, be it discrete or continous, when the basic observations are correlated. Very little is known in this respect when the observations are continuous. But if they are discrete, we may model the dependence by assuming that the basic observations follow a finite Markov chain also. In the expression shared by many quality systems,

$$S_n = g(X_n, S_{n-1})$$

we see that S_n follows a Markov chain if X_n follows a Markov chain. Hence the general idea described in Section 4 still applies. However, studies in this respect, although workable, are rare in the literature. The only related work seems to be Chao (1989).

The Markov chain method also finds its application in various linearly connected reliability systems. A general treatment can be found in Chao and Fu (1991). Readers are referred to the review article by Chao et al. (1995).

7. CONCLUSION

In this chapter I have demonstrated, with examples and general scenario descriptions, that it is often possible to define a Markov chain such that the underlying quality control scheme can be completely described by this Markov chain.

To evaluate the system performance of a control chart, or other quality-related schemes, perhaps the most difficult quantity to calculate is the ARL and its associated sampling distributions. The Markov chain technique provides a general means for accomplishing this task.

ACKNOWLEDGEMENT

Research was partially supported by grant NSC-86-2115-M-015 from the National Science Council of ROC.

REFERENCES

Bakir ST, Reynolds MR Jr. (1979). A nonparametric procedure for process control based on within-group ranking. Technometrics 21:175–183.

Blackwell MTR. (1977). The effect of short production runs on CSP-1. Technometrics 19:259–263.

Brook D, Evans DA. (1972). An approach to the probability distribution of CUSUM run length. Cumulative sum charts; Markov chain. Biometrika 59: 539–549.

Brugger RM. (1975). A simplification of skip-lot procedure formulation. J Qual Technol. 7:165–167.

Brugger RM (1989). A simplified Markov chain analysis of ANSI/ASQC Z1.4 used without limit numbers. J Qual Technol. 21:97–102.

Champ CW, Woodall WH. (1987). Exact results for Shewhart control charts with supplementary runs rules. Quality control, Markov chain, average run length. Technometrics 29:393–399.

Champ CW, Woodall WH. (1990). A program to evaluate the run length distribution of a Shewhart control chart with supplementary runs rules. J Qual Technol 22: 68–73.

Chao MT. (1989). The finite time behavior of CSP when defects are dependent. Proceedings of the National Science Council, ROC, Part A 13:18–22.

Chao MT, Fu JC. (1991). The reliability of large series systems under Markov structure. Adv Appl Prob 23:894–908.

Chao MT, Fu JC, Koutras MV. (1995). Survey of reliability studies of consecutive-*k* out-of-*n:F* and related systems. IEEE Trans Reliab 44:120–127.

Crosier RB. (1986). A new two-sided cumulative sum quality control scheme. Technometrics 28:187–194.

Grinde R, McDowell ED, Randhawa SU. (1987). ANSI/ASQC Z1.4 performance without limit numbers. J Qual Technol 19:204–215.

Hahn GJ, Gage JB. (1983). Evaluation of a start-up demonstration test. J Qual Technol 15:103–106.

Lucas JM, Crosier RB. (1982). Fast initial response for CUSUM control schemes. Technometrics 24:199–205.

Neuman CP, Bonhomme NM. (1975). Evaluation of maintenance policies using Markov chains and fault tree analysis. IEEE Trans Reliab 24:37–45.

Reynolds MR Jr, Amin RW, Arnold JC. (1990). CUSUM charts with variable sampling intervals. Technometrics 32:371–384.

Saccucci MS, Lucas JM. (1990). Average run lengths for exponentially weighted moving average control schemes using the Markov chain approach. J Qual Technol 22:154–162.

Salvia AA. (1987). Performance of pre-control sampling plans. J Qual Technol 19: 85–89.

Stephens KS, Dodge HF. (1976). Two-stage chain sampling inspection plans with different sample sizes in the two stages. J Qual Technol 8:209–224.

Van Dobben de Bruny CS. (1968). Cumulative Sum Tests: Theory and Practice. Statistical Monographs and Courses No. 24, Griffin.

Western Electric Company. (1965). Statistical Quality Control Handbook. Western Electric Co., Indianapolis.

Woodall WH. (1984). On the Markov chain approach to the two-sided CUSUM procedure run length distribution. Technometrics 26:41–46.

Woodall WH, Ncube MM. (1985). Multivariate CUSUM quality-control procedures. Cumulative sum; Markov chain; Hotelling's T^2. Technometrics 27: 285–292.

Yashchin E. (1989). Weighted cumulative sum technique. Technometrics 31:321–338.

11

Joint Monitoring of Process Mean and Variance Based on the Exponentially Weighted Moving Averages

Fah Fatt Gan
National University of Singapore, Singapore, Republic of Singapore

1. INTRODUCTION

The Shewhart chart based on the sample mean $\bar{X}$ was first developed to monitor a process mean. The chart was then modified to plot the sample range R to monitor a process variance. Each chart was developed assuming that the other process characteristic is in control. The more advanced charting procedures such as the cumulative sum (CUSUM) and exponentially weighted moving average (EWMA) charts were later developed based on the same basic assumption. This has led to the design and evaluation of performance of the mean and variance charts separately. This kind of analysis might mislead quality control engineers into making inferences concerning the mean or the variance chart without making reference to the other. Experience with real manufacturing processes has shown that the process variance tends to increase with the process mean. A decrease in the variance when the mean is in control is highly desirable, but if a decrease in the variance is accompanied by a decrease in the mean, then it is highly undesirable. Gan (1995) gave an example of a process with a decrease in the variance coupled with a change in the mean and showed that this process state is difficult to detect. The mean chart becomes insensitive to the change in the mean because the variance of the sample mean has become smaller. Any detection of a decrease in the variance with the mean appearing to be in control could lead to the false conclusion that the process has improved. In short, the problem of monitoring the mean and variance is a bivariate one, and both the mean and variance charts need to be looked at jointly in order to make meaningful inferences.

The use of combined schemes involving simultaneous mean and variance charts based on the EWMAs of sample mean and variance is discussed in Section 2. The average run length (ARL) performance of the various schemes is assessed in Section 3. A simple design procedure of a combined EWMA scheme with an elliptical "acceptance" region is given in Section 4. A real data set from the semiconductor industry is used to illustrate the design and implementation in Section 5.

2. JOINT MONITORING OF PROCESS MEAN AND VARIANCE

Consider the simulated data set given in Gan (1995). The data set comprises 80 samples, each of sample size $n = 5$. The first 40 samples were generated from the normal distribution $N(\mu_0, \sigma_0^2)$, where $\mu_0 = 1$ and $\sigma_0^2 = 1$, and the rest were from $N(\mu_0 + 0.4\sigma_0/\sqrt{n}, (0.9\sigma_0)^2)$. Thus, the process was simulated to be in control for the first 40 samples, and between the 40th and 41st samples the mean shifted upward to $\mu_0 + 0.4\sigma_0/\sqrt{n}$ and the variance decreased to $(0.9\sigma_0)^2$. A EWMA chart for monitoring the mean is obtained by plotting $Q_0 = \mu_0$ and $Q_t = (1 - \lambda_M)Q_{t-1} + \lambda_M \bar{X}_t$ against the sample number t, where $\bar{X}_t$ is the sample mean at sample number t. A signal is issued if $Q_t > h_M$ or $Q_t < -h_M$. Similarly, a EWMA chart for monitoring the variance is obtained by plotting $q_0 = E[\log(S_t^2)] = -0.270$ (when $\sigma^2 = \sigma_0^2$) and $q_t = (1 - \lambda_V)q_{t-1} + \lambda_V \log(S_t^2)$, where S_t^2 is the sample variance at sample number t. A signal is issued if $q_t > H_V$ or $q_t < -h_V$. More details on the EWMA charts can be found in Crowder (1987, 1989), Crowder and Hamilton (1992), Lucas and Saccucci (1990), and Chang (1993). The EWMA mean and variance charts based on the parameters given in Gan (1995, Table 2, p. 448, scheme *EE*) are constructed for the data and displayed in Figure 1.

A quality control engineer has to constantly combine the information in the two charts (which might not be easily done in practice) to make meaningful inferences. To ensure that the charts are interpreted correctly, the two charts could be combined into one, and this can be done by plotting the EWMA of $\log(S^2)$ against the EWMA of $\bar{X}$ as shown in Figure 2. The chart limits of the two charts form the four sides of a rectangular "acceptance" region. Any point that falls within the region is considered an in-control point (for example, points *A* and *B*), and any point that falls outside the region is considered an out-of-control point (for example, point *C*). The thick bar on the plot is not an out-of-control region but represents the most

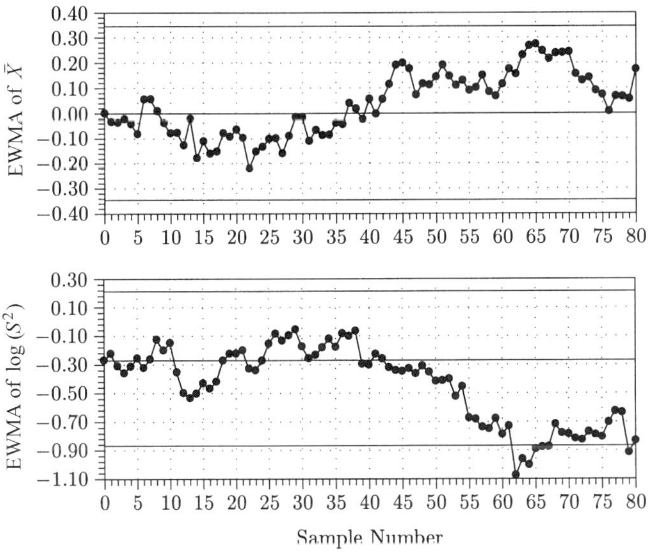

Figure 1 EWMA charts based on $\bar{X}$ and $\log(S^2)$ for a simulated data set where the first 40 samples were generated from the normal distribution $N(\mu_0, \sigma_0^2)$, where $\mu_0 = 1$ and $\sigma_0^2 = 1$, and the rest were from $N\ (\mu_0 + 0.4\sigma_0/\sqrt{n}, (0.9\sigma_0)^2)$.

desirable state, where the mean is on target and the variance has decreased substantially.

The advantage of this charting procedure is immediate: Any inference made can be based on both the EWMAs jointly. The interpretation of an out-of-control signal is easier because the position of the point gives an indication of both the magnitude and direction of the process shift. However, the order of the points is lost if they are plotted on the same

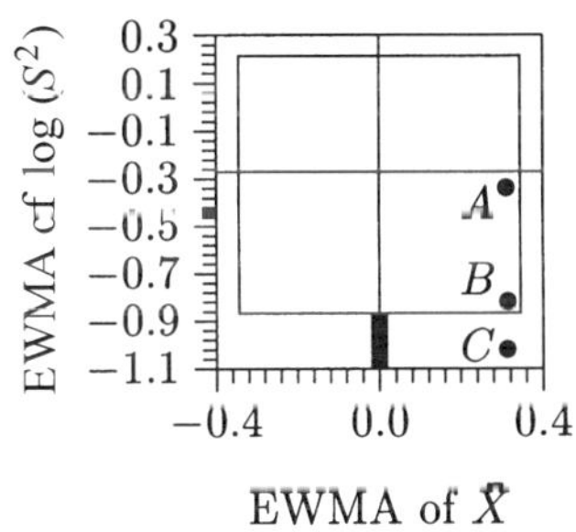

Figure 2 A combined EWMA scheme with a rectangular acceptance region.

plot. To get around this problem, each point can be plotted on a new plot in a sequence as shown later in Figure 13. The disadvantage is that it is not as compact as the traditional procedure illustrated in Figure 1.

The traditional way of plotting the mean and variance charts separately [see, for example, Gan (1995)] amounts to plotting the EWMA of $\log(S^2)$ against the EWMA of $\bar{X}$ and using a rectangular acceptance region for making decisions. The main problem with a rectangular acceptance region is that both points A and B (see Fig. 2) are considered in control, although it is fairly obvious that point B represents a far more undesirable state than that of point A. An acceptance region that is more reasonable would be an elliptical region as shown in Figure 3. Takahashi (1989) investigated an elliptical type of acceptance region for a combined Shewhart scheme based on $\bar{X}$ and S or R. An economic statistical design for the $\bar{X}$ and R charts was given by Saniga (1989). A point is considered out of control if it is outside the elliptical acceptance region. For example, point B is an out-of-control point, but point A is an in-control point, for the elliptical region given in Figure 3. This chart is called a bull's-eye chart, as any hit on the bull's-eye will provide evidence of the process being on target.

For the same smoothing constants λ_M and λ_V, in order for a EWMA scheme with an elliptical region to have the same ARL as the EWMA scheme with a rectangular region, the chart limits of the mean and variance charts have to be slightly larger, as shown in Figure 4. The idea of an elliptical region comes from the Hotelling's statistic to be discussed later. Point A is an in-control point for the rectangular region, but it is an out-of-control point for the elliptical region. Similarly, point B is an out-of-control point for the rectangular region but an in-control point for the elliptical region. Thus, an elliptical region would be expected to be more sensitive in detecting large changes in both the mean and variance and less sensitive in

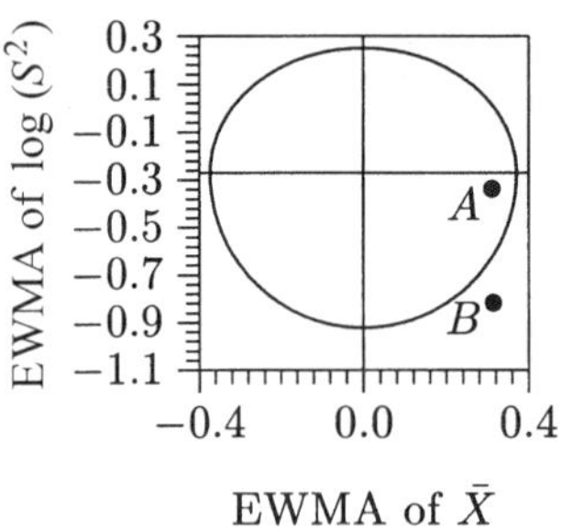

Figure 3 A combined EWMA scheme with an elliptical acceptance region.

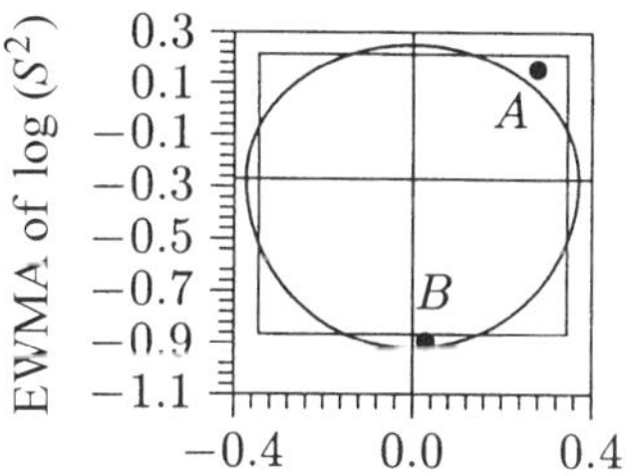

Figure 4 A combined EWMA scheme with both rectangular and elliptical acceptance regions.

detecting a large shift in one process characteristic when there is little or no change in the other characteristic.

A Shewhart bull's-eye chart and a EWMA bull's-eye chart are displayed in Figure 5. The Shewhart bull's-eye chart displays 10,000 random points ($\bar{X}$, $\log(S^2)$) when the process is in control. The EWMA bull's-eye chart displays the EWMA of the points ($\bar{X}$, $\log(S^2)$). Both the charts show that the points are roughly distributed within elliptical regions; hence an elliptical region is a natural and more appropriate decision region for a Shewhart or EWMA bull's-eye chart.

An equivalent decision procedure for the EWMA bull's-eye chart is to check the distance of the point (Q_t, q_t) from the center (μ_0, $E[\log(S^2)]$) and declare the point out of control if

$$\frac{(Q_t - \mu_0)^2}{h_M^2} + \frac{\{q_t - E[\log(S^2)]\}^2}{\{H_V - E[\log(S^2)]\}^2} > 1$$

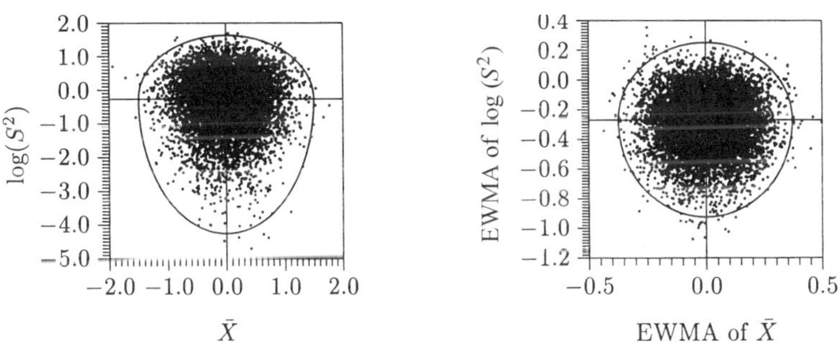

Figure 5 Shewhart bull's-eye chart and a EWMA bull's-eye chart based on 10,000 random points ($\bar{X}$, $\log(S^2)$) from an in-control normal distribution.

for a point (Q_t, q_t) located above the horizontal line $q_t = E[\log(S^2)]$ or if

$$\frac{(Q_t - \mu_0)^2}{h_M^2} + \frac{\{q_t - E[\log(S^2)]\}^2}{\{-h_V - E[\log(S^2)]\}^2} > 1$$

for a point (Q_t, q_t) located below the horizontal line. For a point above the horizontal line,

$$T^2 = \frac{(Q_t - \mu_0)^2}{h_M^2} + \frac{\{q_t - E[\log(S^2)]\}^2}{\{H_V - E[\log(S^2)]\}^2}$$

$$= (Q_t - \mu_0 \quad q_t - E[\log(S^2)]) \begin{pmatrix} 1/h_M^2 & 0 \\ 0 & 1/\{H_V - E[\log(S^2)]\}^2 \end{pmatrix} \begin{pmatrix} Q_t - \mu_0 \\ q_t - E[\log(S^2)] \end{pmatrix}$$

which is a Hotelling type of statistic. This statistic is similar to the one proposed by Lowry et al. (1992). Thus, another way of implementing the bulls's-eye chart is to plot the Hotelling type statistic T^2 against the sample number t as shown in Figure 6, which I shall refer to as a multivariate EWMA T^2 chart.

The main problem with this charting procedure is that when a signal is issued, the chart does not indicate which process characteristic gives rise to the signal. The omnibus EWMA chart proposed by Domangue and Patch

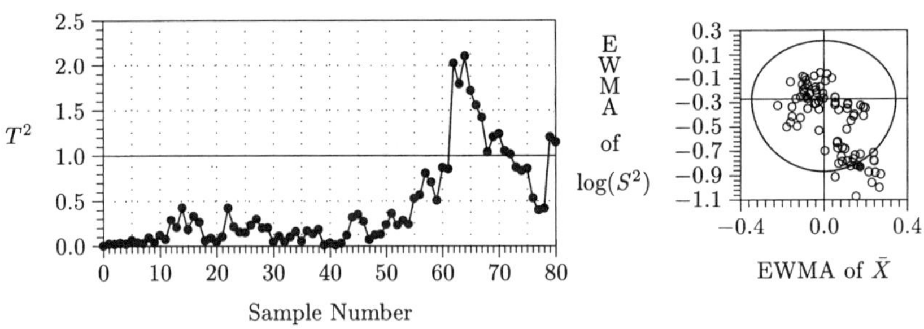

Figure 6 A multivariate EWMA T^2 chart for a simulated data set where the first 40 samples were generated from the normal distribution $N(\mu_0, \sigma_0^2)$, where $\mu_0 = 1$ and $\sigma_0^2 = 1$, and the rest were from $N(\mu_0 + 0.4\sigma_0/\sqrt{n}, (0.9\sigma_0)^2$.

(1991) has a similar problem of interpretation. The T^2 statistic indicates only the magnitude and not the direction of the shift. In process monitoring, the direction of a shift is at least as important as the magnitude of the shift. An improvement to the T^2 chart is to include a bull's-eye chart of the points with the most recent point plotted as a solid black dot. Although all the points are shown in this bull's-eye chart, in order to avoid overcrowding of points only the most recent 40 points, for example, are plotted each time. The bull's-eye chart will provide the information on both the magnitude and direction of any process shift. The interpretation of the T^2 statistic is simple and easy to understand with the bull's-eye chart. Mason et al. (1995) proposed a certain decomposition of the Hotelling statistic for interpretation of the state of a process. Their method is mathematically more complicated and hence harder for a quality control engineer to understand and appreciate.

3. COMPARISON OF SCHEMES BASED ON ARL

For a comparison of schemes based on the ARL, the in-control mean and variance are assumed to be $\mu_0 = 0$ and $\sigma_0^2 = 1$, respectively. Each sample comprises $n = 5$ normally distributed observations. The mean and variance investigated are given by $\mu = \mu_0 + \Delta\sigma_0/\sqrt{n}$ and $\sigma = \delta\sigma_0$, where $\Delta = 0.0$, 0.2, 0.4, 1.0, and 3.0 and $\delta = 0.50$, 0.75, 0.95, 1.00, 1.05, 1.25, and 3.00. Combined schemes with rectangular and elliptical acceptance regions are compared in this sectiotn. All the schemes have an approximate ARL of 250. The ARL values of the schemes EE_r and SS_r (combined EWMA and Shewhart schemes with rectangular acceptance regions) were computed exactly using the integral equation approach given in Gan (1995). The rest were simulated. Alternatively, the ARL of the EWMA bull's-eye chart EE_e could be computed by using the Markov chain approach of Brook and Evans (1972) or the integral equation approach. Waldmann's method (Waldmann, 1986a, 1986b) could be used here for approximating the run length distribution of a bull's-eye chart. Let the starting values of the EWMA mean and variance charts be u and v, respectively; then the ARL function $L_c(u, v)$ of the combined scheme with an elliptical acceptance region B can be derived as

$$L_C(u, v) = 1 + \frac{1}{\lambda_M \lambda_V} \int_B \int L_C(x, y) f_{\bar{X}}\left(\frac{x - (1 - \lambda_M)u}{\lambda_M}\right) f_{\log(S^2)}\left(\frac{y - (1 - \lambda_V)v}{\lambda_V}\right) dx\, dy$$

where $f_{\bar{x}}$ and $f_{\log(S^2)}$ are the probability density functions of $\bar{X}$ and log (S^2), respectively.

The schemes CC (combined CUSUM scheme with a rectangular acceptance region) and EE_r are the same as those given in Gan (1995). The combined scheme CC consists of a two-sided CUSUM mean chart and a two-sided CUSUM variance chart. This scheme is obtained by plotting $S_0 = T_0 = 0.0$, $S_t = \max[0, S_{t-1}+\bar{x}_t - k_M]$, and $T_t = \min\ [0, T_{t-1} + \bar{x}_t + k_M]$ against the sample number t for the mean chart and by plotting $S_0 = T_0 = 0.0$, $S_t = \max[0, S_{t-1} + \log(s_t^2) - k_{VU}]$, and $T_t = \min\ [0, T_{t-1} + \log(s_t^2) + k_{VL}]$ against t for the variance chart. The chart parameters of the various schemes are given in Table 1. More details on the CUSUM charts can be found in Gan (1991) and Chang (1993). The ARL comparisons are summarized in Table 2.

The ARL values of combined schemes CC, EE_e, and SS_e were simulated such that an ARL that is less than 10 has a standard error of 0.01; an ARL that is at least 10 but less than 50 has a standard error of 0.1; an ARL that is at least 50 but less than 100 has a standard error of about 0.2; and an ARL that is at least 100 has a standard error of about 1.0.

EE_r versus CC. The performances of these two schemes are similar except that when there is a small shift in the mean and a small decrease in the variance, the EE_r scheme is much more sensitive. When there is a large increase in the variance, the EE_r scheme is marginally less sensitive.

EE_r versus EE_e. The performances of these two schemes are similar. The EE_e scheme is generally more sensitive than the EE_r scheme in detecting increases in the variance and less sensitive in detecting decreases in the variance for the various means investigated.

Table 1 Control Chart Parameters of Combined Schemes

Scheme	Acceptance region	Control chart parameters
CC	Rectangular	$k_M = 0.224$, $h_M = 2.268$, $S_0 = T_0 = 0.0$
		$k_{VU} = 0.055$, $H_V = 4.006$, $S_0 = 0.0$
		$k_{VL} = 0.666$, $h_V = -5.054$, $T_0 = 0.0$
EE_r	Rectangular	$\lambda_M = 0.135$, $h_M = -0.345$, $H_M = 0.345$, $Q_0 = 0.000$
		$\lambda_V = 0.106$, $h_V = -0.867$, $H_V = 0.215$, $Q_0 = -0.270$
EE_e	Elliptical	$\lambda_M = 0.134$, $h_M = -0.372$, $H_M = 0.372$, $Q_0 = 0.000$
		$\lambda_V = 0.106$, $h_V = -0.922$, $H_V = 0.250$, $Q_0 = -0.270$
SS_r	Rectangular	$h_M = -1.383$, $H_M = 1.383$
		$h_V = -3.789$, $H_V = 1.531$
SS_e	Elliptical	$h_M = -1.501$, $H_M = 1.501$
		$h_V = -4.257$, $H_V = 1.635$

Table 2 Average Run Lengths of Combined Schemes with Respect to the Process Mean ($\mu_0 + \Delta \times \sigma_0/\sqrt{n}$) and Standard Deviation ($\delta\sigma_0$)

Δ	δ	CC	EE_r	EE_e	SS_r	SS_e
0.00	0.50	5.9	5.8	6.4	68.9	153.0
0.00	0.75	24.8	21.9	24.7	322.4	612.9
0.00	0.95	284.4	236.7	254.8	364.2	426.0
0.00	1.00	253.6	252.3	252.7	252.2	252.7
0.00	1.05	138.3	137.1	129.3	161.8	145.4
0.00	1.25	19.2	18.9	18.1	31.1	24.8
0.00	3.00	2.5	2.6	2.5	1.2	1.2
0.20	0.50	5.9	5.8	6.3	68.9	150.1
0.20	0.75	24.7	21.8	22.8	319.8	592.5
0.20	0.95	167.8	135.3	136.6	328.8	377.6
0.20	1.00	145.8	129.7	127.4	228.0	227.9
0.20	1.05	96.2	88.7	82.3	148.2	133.3
0.20	1.25	18.4	18.0	17.0	30.0	23.9
0.20	3.00	2.5	2.6	2.5	1.2	1.2
0.40	0.50	5.9	5.8	6.1	68.9	142.0
0.40	0.75	23.2	20.5	18.5	309.5	530.6
0.40	0.95	62.3	51.8	52.5	248.4	276.6
0.40	1.00	56.1	48.8	49.0	173.6	170.8
0.40	1.05	46.9	41.7	39.8	116.9	105.7
0.40	1.25	16.2	15.8	14.5	27.1	21.5
0.40	3.00	2.5	2.6	2.5	1.2	1.2
1.00	0.50	5.8	5.7	5.0	68.8	97.6
1.00	0.75	10.0	9.7	8.6	175.7	217.9
1.00	0.95	10.5	10.2	10.6	64.9	71.7
1.00	1.00	10.4	10.2	10.5	49.6	50.6
1.00	1.05	10.3	10.1	10.2	38.3	36.3
1.00	1.25	8.9	8.8	8.0	15.2	12.2
1.00	3.00	2.4	2.5	2.4	1.2	1.2
3.00	0.50	2.5	2.6	2.5	2.3	2.3
3.00	0.75	2.6	2.6	2.7	2.2	2.6
3.00	0.95	2.6	2.6	2.8	2.2	2.4
3.00	1.00	2.6	2.6	2.8	2.2	2.3
3.00	1.05	2.8	2.6	2.8	2.1	2.3
3.00	1.25	2.7	2.7	2.7	2.1	2.0
3.00	3.00	2.1	2.2	2.0	1.1	1.1

SS_r versus SS_e. The difference in performance is more substantial. The SS_e scheme is more sensitive than the SS_r scheme in detecting increases in the variance but substantially less sensitive in detecting decreases in the variance, especially for a small change or no change in the mean. For larger changes in the mean, the difference is smaller.

EWMA Schemes versus Shewhart Schemes. The EWMA schemes are substantially more sensitive than the Shewhart schemes except for the case when there is a big change in the variance.

In order to have a better understanding of the performance of these schemes, 10,000 random samples were simulated for four different sets of process characteristics: $\Delta = 0.0$ and $\delta = 0.75$, $\Delta = 0.4$ and $\delta = 1.00$, $\Delta = 0.4$ and $\delta = 0.75$, and $\Delta = 0.4$ and $\delta = 1.05$. These are plotted as Figures 7a, 7b, 7c, and 7d, respectively, for the combined Shewhart schemes. The EWMA of the points ($\bar{X}$, log (S^2)) are plotted as Figures 7e, 7f, 7g, and 7h, respectively. For $\Delta = 0.0$ and 0.4 and $\delta = 0.75$, the SS_r scheme is more sensitive than the SS_e scheme, and this is indicated by Figures 7a and 7c, which show that there are more points outside the rectangular acceptance region than there are outside the elliptical region.

For $\Delta = 0.4$ and $\delta = 1.05$, SS_e is slightly more sensitive, as indicated by Figure 7d, which shows that there are more points outside the elliptical region than outside the rectangular region. For $\Delta = 0.0$, $\delta = 0.75$ and $\Delta = 0.4$, $\delta = 0.75$, for example, Figures 7a and 7c show that there are very few points outside the acceptance regions. In sharp contrast, there are a substantial number of points outside the acceptance region in Figures 7e and 7g. This explains the substantial difference in the ARL's of the EWMA and Shewhart schemes. Plots 7a and 7e correspond to the case when a process improvement has taken place, and this is reflected much more clearly in a EWMA scheme than in a Shewhart scheme. This means that the EWMA scheme would be a more effective tool for quality improvement. These plots also suggest that if sufficient points are collected for a process and the points are plotted on a bull's-eye chart, then the plot will provide valuable information regarding the overall state of the process characteristics. The central location and spread of the points could also be used to estimate graphically the process characteristics.

4. DESIGN OF A EWMA BULL'S-EYE CHART AND MULTIVARIATE EWMA T^2 CHART

A simple design procedure is provided here for the design of a EWMA bull's-eye chart. Table 3 contains the chart parameters of EWMA bull's-eye charts with an in-control ARL of 300 based on a sample size $n = 5$.

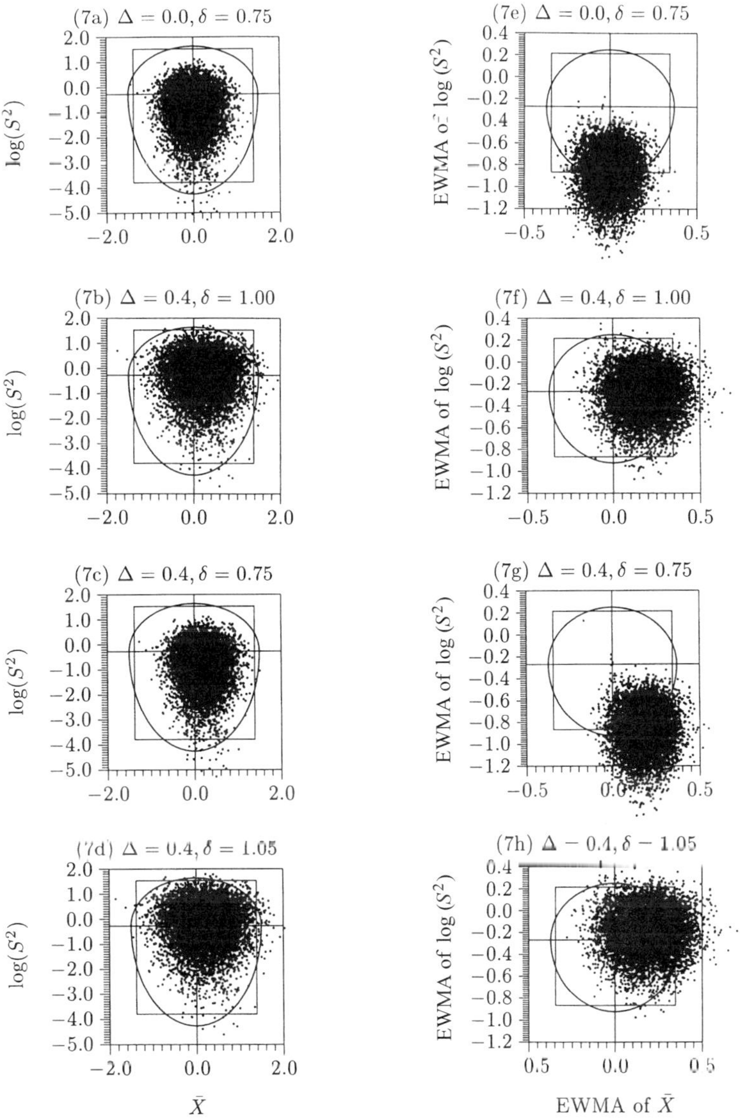

Figure 7 Combined Shewhart schemes and combined EWMA schemes based on 10,000 random points from out-of-control normal distributions.

Table 3 Control Chart Parameters of Combined EWMA Schemes with Elliptical Acceptance Region, In-Control Average Run Length of 300, and Sample Size 5

		λ_V													
λ_M		0.08	0.10	0.11	0.12	0.13	0.14	0.15	0.16	0.17	0.18	0.19	0.20	0.22	0.24
0.08	h_M	0.620	0.621	0.621	0.622	0.622	0.623	0.623	0.623	0.623	0.624	0.624	0.624	0.625	0.625
	h_{VU}	0.182	0.241	0.268	0.294	0.319	0.343	0.366	0.389	0.410	0.431	0.451	0.471	0.509	0.546
	h_{VL}	0.820	0.910	0.953	0.995	1.036	1.077	1.116	1.155	1.194	1.232	1.270	1.308	1.383	1.456
0.10	h_M	0.710	0.710	0.711	0.711	0.712	0.712	0.782	0.713	0.713	0.713	0.713	0.713	0.714	0.715
	h_{VU}	0.183	0.242	0.269	0.295	0.320	0.344	0.367	0.389	0.411	0.432	0.452	0.472	0.510	0.547
	h_{VL}	0.822	0.911	0.955	0.996	1.037	1.078	1.117	1.157	1.196	1.234	1.271	1.309	1.384	1.458
0.11	h_M	0.752	0.752	0.753	0.753	0.753	0.754	0.754	0.755	0.755	0.755	0.756	0.756	0.756	0.757
	h_{VU}	0.183	0.242	0.269	0.295	0.320	0.344	0.367	0.390	0.411	0.432	0.453	0.472	0.510	0.547
	h_{VL}	0.823	0.912	0.955	0.997	1.038	1.078	1.118	1.157	1.196	1.234	1.273	1.310	1.384	1.458
0.12	h_M	0.792	0.793	0.793	0.793	0.794	0.794	0.794	0.795	0.795	0.796	0.796	0.796	0.797	0.797
	h_{VU}	0.183	0.243	0.270	0.296	0.320	0.344	0.367	0.390	0.411	0.432	0.453	0.472	0.510	0.547
	h_{VL}	0.823	0.913	0.955	0.997	1.038	1.079	1.118	1.158	1.196	1.235	1.273	1.310	1.385	1.458
0.13	h_M	0.831	0.832	0.832	0.833	0.833	0.833	0.834	0.834	0.834	0.835	0.835	0.835	0.835	0.836
	h_{VU}	0.184	0.243	0.270	0.296	0.321	0.345	0.368	0.390	0.411	0.432	0.453	0.472	0.510	0.547
	h_{VL}	0.823	0.913	0.955	0.998	1.039	1.079	1.119	1.158	1.196	1.235	1.273	1.311	1.385	1.459
0.14	h_M	0.869	0.870	0.870	0.870	0.871	0.871	0.871	0.872	0.872	0.873	0.873	0.873	0.874	0.874
	h_{VU}	0.184	0.243	0.270	0.296	0.321	0.345	0.368	0.390	0.412	0.433	0.453	0.473	0.511	0.547
	h_{VL}	0.824	0.913	0.956	0.998	1.040	1.080	1.119	1.158	1.197	1.236	1.274	1.311	1.386	1.460
0.15	h_M	0.907	0.907	0.908	0.908	0.908	0.908	0.909	0.909	0.910	0.909	0.909	0.911	0.911	0.911
	h_{VU}	0.185	0.243	0.271	0.296	0.321	0.345	0.368	0.390	0.412	0.433	0.453	0.473	0.511	0.548
	h_{VL}	0.825	0.914	0.957	0.999	1.040	1.080	1.120	1.159	1.198	1.236	1.274	1.313	1.387	1.460
0.16	h_M	0.943	0.943	0.943	0.944	0.944	0.944	0.945	0.945	0.945	0.945	0.946	0.946	0.946	0.947
	h_{VU}	0.185	0.244	0.271	0.297	0.322	0.345	0.368	0.390	0.412	0.433	0.453	0.473	0.511	0.548
	h_{VL}	0.825	0.914	0.957	0.999	1.040	1.080	1.120	1.159	1.198	1.236	1.274	1.313	1.386	1.460
0.17	h_M	0.977	0.978	0.979	0.979	0.979	0.979	0.980	0.980	0.980	0.980	0.981	0.981	0.981	0.981
	h_{VU}	0.185	0.244	0.271	0.297	0.322	0.346	0.369	0.391	0.412	0.433	0.454	0.473	0.511	0.548
	h_{VL}	0.825	0.914	0.958	0.999	1.040	1.081	1.121	1.160	1.198	1.236	1.275	1.312	1.387	1.460
0.18	h_M	1.012	1.013	1.013	1.013	1.013	1.014	1.014	1.015	1.015	1.015	1.016	1.016	1.016	1.016
	h_{VU}	0.185	0.244	0.271	0.297	0.322	0.346	0.369	0.391	0.413	0.433	0.454	0.474	0.511	0.548
	h_{VL}	0.825	0.915	0.958	1.000	1.041	1.082	1.121	1.161	1.199	1.237	1.276	1.313	1.387	1.461
0.19	h_M	1.045	1.046	1.047	1.047	1.047	1.047	1.048	1.048	1.048	1.049	1.049	1.050	1.049	1.050
	h_{VU}	0.185	0.244	0.272	0.297	0.322	0.346	0.369	0.391	0.413	0.434	0.454	0.474	0.512	0.548
	h_{VL}	0.826	0.915	0.959	1.000	1.041	1.081	1.121	1.160	1.200	1.238	1.276	1.314	1.387	1.462
0.20	h_M	1.079	1.079	1.079	1.079	1.080	1.081	1.081	1.081	1.082	1.082	1.082	1.082	1.083	1.083
	h_{VU}	0.185	0.244	0.272	0.297	0.322	0.346	0.369	0.391	0.413	0.434	0.454	0.474	0.512	0.548
	h_{VL}	0.826	0.916	0.958	1.000	1.042	1.082	1.122	1.161	1.200	1.238	1.276	1.314	1.389	1.462

Table 3 *(continued)*

λ_M		λ_V 0.08	0.10	0.11	0.12	0.13	0.14	0.15	0.16	0.17	0.18	0.19	0.20	0.22	0.24
0.22	h_M	1.143	1.144	1.144	1.145	1.145	1.145	1.145	1.145	1.145	1.146	1.146	1.147	1.147	1.147
	h_{VU}	0.186	0.245	0.272	0.298	0.323	0.347	0.370	0.392	0.413	0.434	0.454	0.474	0.512	0.548
	h_{VL}	0.827	0.916	0.959	1.001	1.042	1.083	1.123	1.162	1.200	1.239	1.277	1.315	1.389	1.462
0.24	h_M	1.206	1.206	1.207	1.207	1.207	1.207	1.208	1.208	1.208	1.209	1.209	1.209	1.210	1.210
	h_{VU}	0.186	0.245	0.273	0.298	0.323	0.347	0.370	0.392	0.414	0.435	0.455	0.474	0.513	0.549
	h_{VL}	0.827	0.917	0.960	1.002	1.043	1.083	1.123	1.162	1.201	1.240	1.278	1.315	1.390	1.463

Similar tables covering other in-control ARLs and sample sizes are available from the author. These are obtained by using simulation such that the simulated in-control ARL has an error of 1.0. The starting value of the mean chart is given by the in-control mean μ_0, and the starting value of the variance chart is given by

$$E[\log(s_t^2)] \approx \log(\sigma_0^2) - \frac{1}{n-1} - \frac{1}{3(n-1)^2} + \frac{2}{15(n-1)^4}$$

Suppose a combined scheme with $\lambda_M = 0.14$ and $\lambda_V = 0.16$ is desired. Then the chart parameters of the combined scheme can be obtained from Table 3 easily as follows:

Mean chart:

$$\lambda_M = 0.14, \qquad h_M = \mu_0 + 0.872 \times \sigma_0/\sqrt{3},$$
$$Q_0 = \mu_0$$

Variance chart:

$$\lambda_V = 0.16, \qquad H_V = 0.390 + \log(\sigma_0^2),$$
$$h_V = -1.158 + \log(\sigma_0^2), \qquad Q_0 = E[\log(s_t^2)]$$

The elliptical acceptance region can then be constructed using

$$\frac{(Q_t - \mu_0)^2}{h_M^2} + \frac{\{q_t - E[\log(S^2)]\}^2}{\{H_V - E[\log(S^2)]\}^2} = 1$$

for the elliptical curve above the horizontal line $q_t = E[\log(S^2)]$ and using

$$\frac{(Q_t - \mu_0)^2}{h_M^2} + \frac{\{q_t - E[\log(S^2)]\}^2}{\{-h_V - E[\log(S^2)]\}^2} = 1$$

for the elliptical curve below the horizontal line.

5. A REAL EXAMPLE

Quality control engineers would like to monitor the mean ball shear strength of a connection on a microchip. From past process data, the in-control mean is estimated to be around 72 g, and the standard deviation is estimated to be around 10 g. A sample of size 5 is taken at regular intervals, and the ball shear strength of each chip is measured. The chart limits of the schemes discussed here are chosen such that a combined mean and variance scheme has an in-control ARL of about 300. The smoothing constants of the EWMA charts are chosen to be $\lambda_M = 0.14$ and $\lambda_V = 0.16$.

The individual Shewhart charts of $\bar{X}$ and $\log(S^2)$ are displayed in Figure 8. The individual EWMA charts of $\bar{X}$ and $\log(S^2)$ are displayed in Figure 9. Both variance charts suggest evidence of a decrease in the process variance. The two mean charts show that the process mean is rather unstable even though the variance has somewhat stabilized at later samples. This is an example where the process mean is unstable while the process

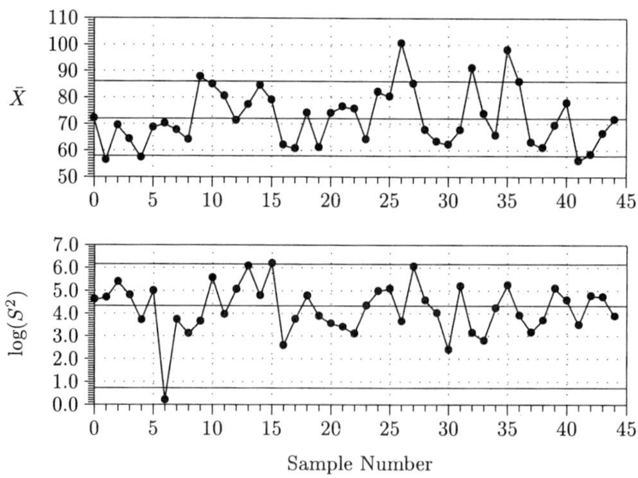

Figure 8 Shewhart charts based on $\bar{X}$ and $\log(S^2)$ for the ball shear strength data.

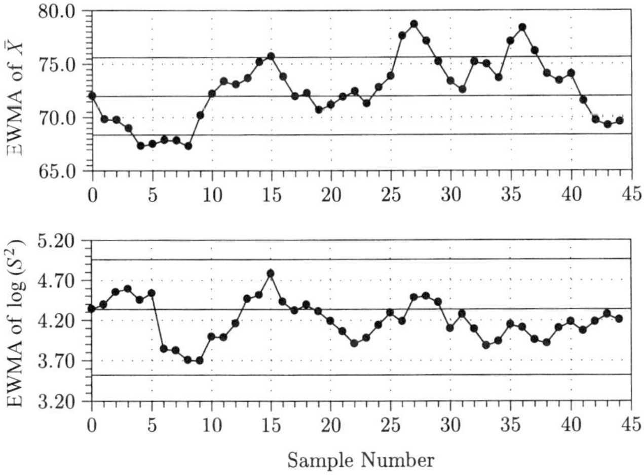

Figure 9 EWMA charts based on $\bar{X}$ and log (S^2) for the ball shear strength data.

variance is somewhat stable. This could be due to the production of chips with different mean ball shear strengths for different batches but with the variance within a batch being more stable from batch to batch. This points to the need to search for ways to ensure a more stabilized mean.

A multivariate Shewhart T^2 chart and a EWMA T^2 chart for the ball shear strength data are displayed in Figures 10 and 11, respectively. Both charts show bigger bursts of activity after the 25th sample. However, the reasons for these bursts of activity are not clear from the charts. A bull's-eye chart would help a quality control engineer to have a better understanding of the process characteristic when an out-of control signal is issued.

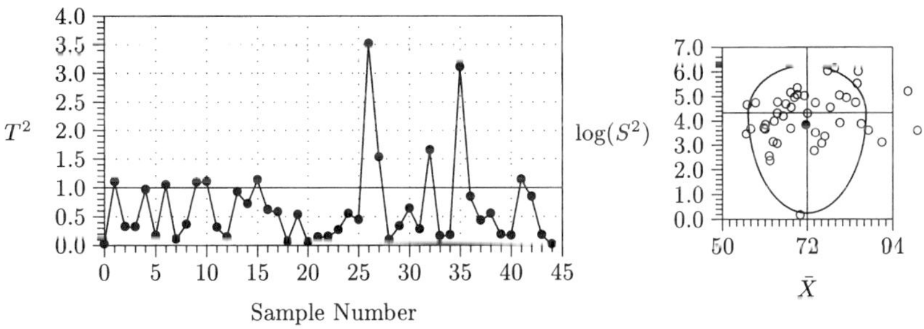

Figure 10 A multivariate Shewhart T^2 chart for the ball shear strength data.

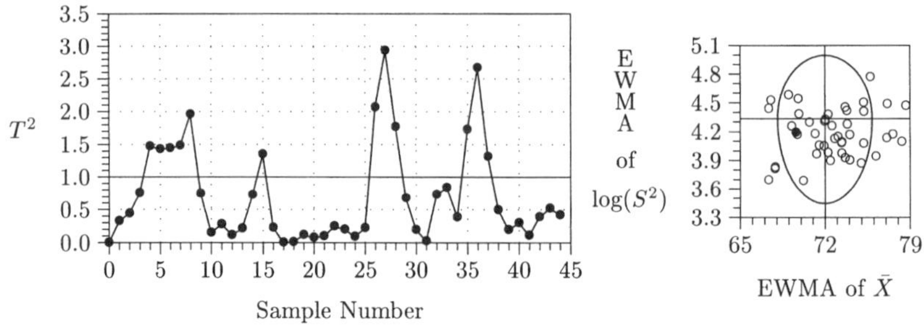

Figure 11 A multivariate EWMA T^2 chart for the ball shear strength data.

The Shewhart and EWMA bull's-eye charts are displayed in Figures 12 and 13, respectively. These types of charts should ideally be constructed using computer programs. The charts continuously provide valuable information regarding the process characeristics in a manner that is easily understood by quality control engineers. The EWMA bull's-eye chart shows that the out-of-control points for samples 6–8 are probably due to decreases in the mean and variance. Figure 13 also shows that the out-of-control points at samples 26–28 are probably due to an increase in the mean alone. Similar conclusions can be drawn from the Shewhart bull's-eye chart. If the process is in control, then the points on a Shewhart bull's-eye chart will be randomly scattered. If a sequence of plotted points are all in a particular quadrant, then the quality control engineer should be on the alert and take samples more frequently than usual (see Stoumbos and Reynolds, 1996, 1997). Alternatively, supplementary run rules could be applied to a Shewhart bull's-eye chart.

6. CONCLUSIONS

Three ways of charting $\bar{X}$ and log (S^2) for the purpose of joint monitoring of both mean and variance were discussed with respect to ease of implementation and ease of interpretation. The traditional way of plotting the mean and variance charts separately amounts to plotting log (S^2) against $\bar{X}$ based on a rectangular "acceptance" region. Using the justification of a Hotelling-type statistic, it was shown that an elliptical acceptance region is more natural and appropriate. This led to the EWMA bull's-eye chart and the multivariate EWMA chart based on a Hotelling-type T^2 statistic. A EWMA bull's-eye chart provides valuable information on both the magnitude and direction of a shift in the process characteristics. The multivariate EWMA

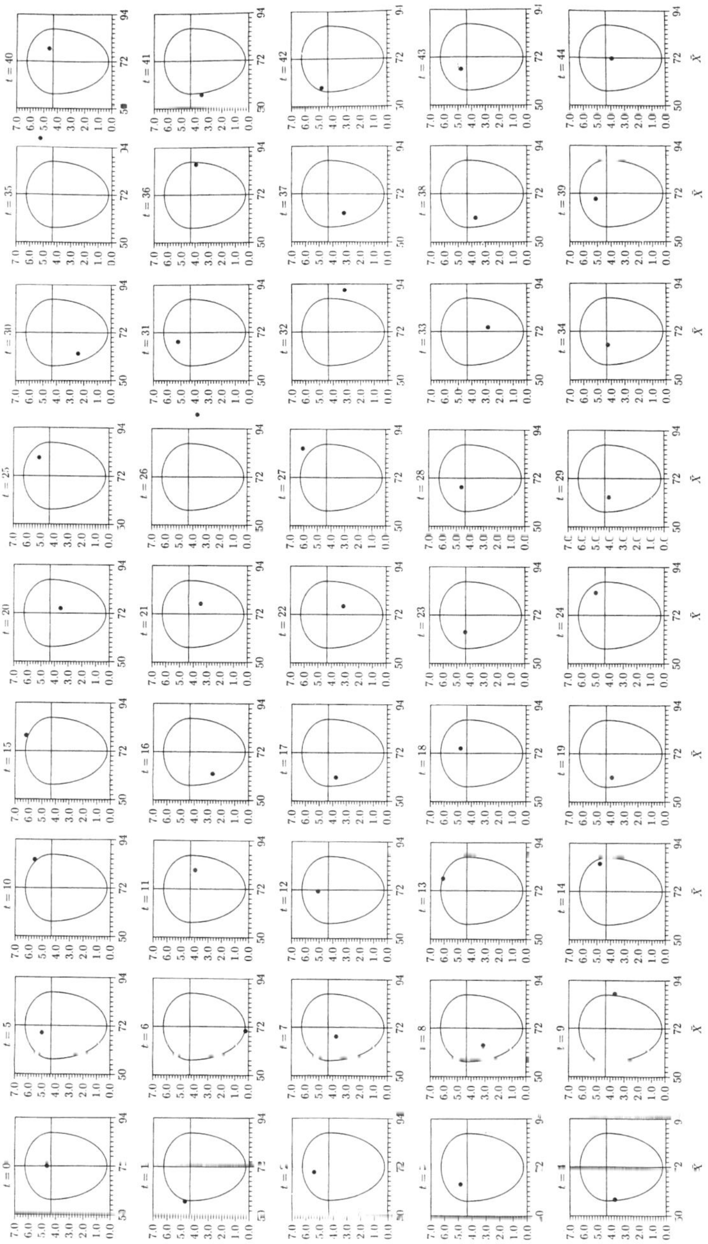

Figure 12 Shewhart bull's-eye charts for the ball shear strength data.

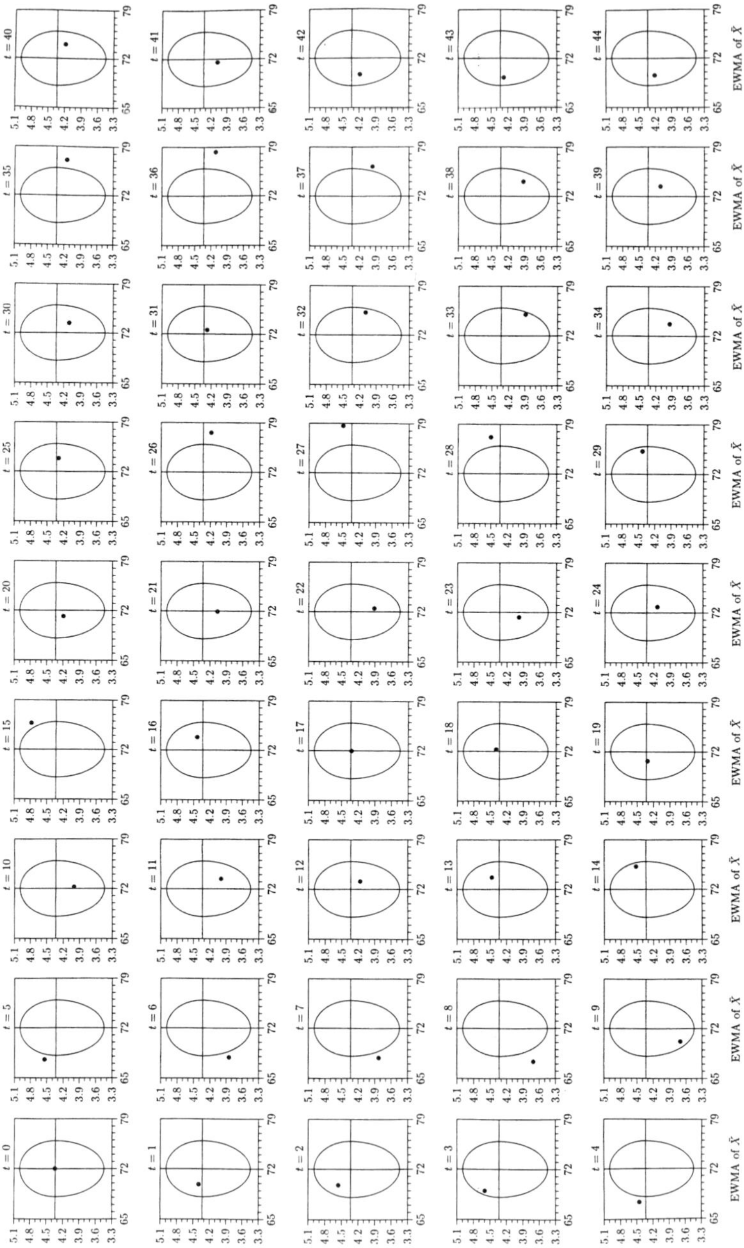

Figure 13 EWMA bull's-eye charts for the ball shear strength data.

chart provides only the magnitude and not the direction of a shift. It is recommended that a EWMA bull's-eye chart be plotted beside a multivariate T^2 chart to help quality control engineers gain a better understanding of the process characteristics. Average run length comparisons show that the performances of schemes CC and EE_r are similar except that when there is a small shift in the mean and a small decrease in the variance, the EE_r scheme is much more sensitive. When there is a large increase in the variance, the EE_r scheme is marginally less sensitive. The performances of the EE_r and EE_e schemes are also found to be similar. The EE_e scheme is generally more sensitive than the EE_r scheme in detecting increases in the variance and less sensitive in detecting decreases in the variance. The difference between SS_r and SS_e is more substantial. The SS_e scheme is more sensitive than the SS_r scheme in detecting increases in the variance but substantially less sensitive in detecting decreases in the variance, especially when there is little or no change in the mean. The EWMA schemes were found to be substantially more sensitive than the Shewhart schemes except for the case when there is a big change in the variance. Finally, a simple design procedure for an EWMA bull's-eye chart was provided.

REFERENCES

Brook D, Evans DA. (1972). An approach to the probability distribution of CUSUM run length. Biometrika 59:539–549.

Chang TC. (1993). CUSUM and EWMA charts for monitoring a process variance. Unpublished M.Sc. thesis, Department of Mathematics, National University of Singapore, Singapore.

Crowder SV. (1987). A simple method for studying run-length distributions of exponentially weighted moving average charts. Technometrics 29:401–407.

Crowder SV. (1989). Design of exponentially weighted moving average schemes. J Qual Technol, 21:155–162.

Crowder SV, Hamilton MD. (1992). An EWMA for monitoring a process standard deviation. J Qual Technol 24:12–21.

Domangue R, Patch SC. (1991). Some omnibus exponentially weighted moving average statistical process monitoring schemes. Technometrics 33:299–314.

Gan FF. (1991). An optimal design of CUSUM quality control charts. J Qual Technol 23:279–286.

Gan FF. (1995). Joint monitoring of process mean and variance using exponentially weighted moving average control charts. Technometrics 37:446–453.

Lowry CA, Woodall WH, Champ CW, Rigdon SE. (1992). A multivariate exponentially weighted moving average control chart. Technometrics 34:46–53.

Lucas JM, Saccucci MS. (1990). Exponentially weighted moving average control schemes: Properties and enhancements (with discussion). Technometrics 32: 1–29.

Mason RL, Tracy ND, Young JC. (1995). Decomposition of T^2 for multivariate control chart interpretation. J Qual Technol, 27:99–108.

Saniga EM. (1989). Economic statistical control chart designs with an application to $\bar{X}$ and R charts. Technometrics 31:313–320.

Stoumbos ZG, Reynolds MR Jr. (1996). Variable sampling rate control charts with sampling at fixed intervals. Proceedings of the (International) Industrial Engineering Research Conference, Minneapolis, MN, May 18–20, 1996 (invited paper).

Stoumbos ZG, Reynolds MR Jr. (1997). Variable sampling rate control charts for one and two-sided process changes with sampling at fixed times. Proceedings of the Second World Congress of Nonlinear Analysts, Athens, Greece, July 10–17, 1996 (invited paper).

Takahashi T. (1989). Simultaneous control charts based on $(\bar{x}, s)$ and $(\bar{x}, R)$ for multiple decision on process change. Rep Stat Appl Res, Union Jpn Sci Eng 36:1–20.

Waldmann K-H. (1986a). Bounds to the distribution of the run length in general quality-control schemes. Stat Hefte 27:37–56.

Waldmann K-H. (1986b). Bounds for the distribution of the run length of geometric moving average charts. Appl Stat 35:151–158.

12
Multivariate Quality Control Procedures

A. J. Hayter
Georgia Institute of Technology, Atlanta, Georgia

1. INTRODUCTION

In many quality control settings the product under examination may have two or more related quality characteristics, and the objective of the supervision is to investigate whether all of these characteristics are simultaneously behaving appropriately. In particular, a standard multivariate quality control problem is to consider whether an observed vector of measurements $x = (x_1, ..., x_k)'$ from a particular sample exhibits any evidence of a location shift from a set of "satisfactory" or "standard" mean values $\mu^0 = (\mu_1^0, ..., \mu_k^0)'$. The individual measurements will usually be correlated due to the nature of the problem, so that their covariance matrix Σ will not be diagonal. In practice, the mean vector μ^0 and covariance matrix Σ may be estimated from an initial large pool of observations.

$$x^1, ..., x^p$$

and the problem is then to monitor further observations x in order to identify any location shifts in any of the mean values.

If the assumption is made that the data are normally distributed, then the distribution of an observation x is $N_k(\mu, \Sigma)$, and the problem is to assess the evidence that $\mu \neq \mu^0$. In the univariate setting ($k = 1$) this problem can be handled with a Shewhart control chart with control limits set to guarantee a specified error rate α. One might consider handling the multivariate problem by constructing individual α-level control charts for each of the k variables under consideration. However, it has long been realized that such an approach is unsatisfactory since it ignores the correlation between the variables and allows the overall error rate to be much larger than α. On the other hand, if individual error rates of α/k are used, then the Bonferroni inequality ensures that the overall error rate is less than the nominal level α. However, this procedure is not sensitive enough since the actual overall error rate tends to be much smaller than α because of the correlation between the variables.

A multivariate quality control procedure that can be successfully implemented in manufacturing processes should meet the goals of

1. Controlling the error rate of false alarms
2. Providing a straightforward identification of the aberrant variables
3. Indicating the amount of deviation of the aberrant variables from their required values

In addition, for certain problems it is desirable that the multivariate quality control procedure

4. Be valid without requiring any distributional assumptions.

An overview of the multivariate quality control problem can be found in Alt (1985). In this chapter some more recent work on the problem is discussed. Specifically, Section 2 considers the situation where the normality assumption is made, and the Hayter and Tsui (1994) paper is discussed together with work by Kuriki (1997). Section 3 considers the work on nonparametric multivariate quality control procedures by Liu (1995) and Bush (1996).

2. PROCEDURES BASED ON A NORMALITY ASSUMPTION

It is clear that a basic property of a good procedure for this multivariate problem is that an overall error rate of the specified level α should be maintained exactly, so that the probability of incorrectly deciding that the process is out of control (when it is, in fact, still in control) should be equal to the specified value α. Hotelling (1947) provided the first solution to this problem by suggesting the use of the statistic

$$T^2 = (x - \mu^0)'\hat{\Sigma}^{-1}(x - \mu^0)$$

where $\hat{\Sigma}$ is an estimate of the population covariance matrix Σ. However, another prolem is that of deciding what conclusions can be drawn once the experimenter has evidence via the T^2 statistic that the process is no longer in control. Specifically, how is it determined which location parameters have moved away from their control values μ_i^0?

2.1. Confidence Intervals Procedure

Hayter and Tsui (1994) proposed a procedure that provides a solution to this identification problem and to the related problem of estimating the magnitudes of any differences in the location parameters from their standard values μ_i^0. The procedure operates by calculating a set of simultaneous confidence intervals for the variable means μ_i with an exact simultaneous coverage probability of $1 - \alpha$. The process is deemed to be out of control whenever any of these confidence intervals does not contain its respective control value μ_i^0, and the identification of the errant variable or variables is immediate. Furthermore, this procedure continually provides confidence intervals for the "current" mean values μ_i regardless of whether the process is in control or not or whether a particular variable is in control or not.

Let $X \sim N_k(\mathbf{0}, R)$, where R is a general correlation matrix with diagonal elements equal to 1 and off-diagonal elements given by ρ_{ij}, say, and define the critical point $C_{R,\alpha}$ by

$$P(|X_i| \leq C_{R,\alpha};\ 1 \leq i \leq k)$$

In the more general case when $X \sim N_k(\mu, \Sigma)$ for any general covariance matrix Σ, let the diagonal elements of Σ be given by σ_i^2, $1 \leq i \leq k$, and the off-diagonal elements by σ_{ij}. Then if R is the correlation matrix generated from Σ, so that $\rho_{ij} = \sigma_{ij}/\sigma_i\sigma_j$, it follows that

$$P(|X_i - \mu_i|/\sigma_i \leq C_{R,\alpha};\ 1 \leq i \leq k)$$

However, this equation can be inverted to produce the following exact $1 - \alpha$ confidence level simultaneous confidence intervals for the μ_i, $1 \leq i \leq k$:

$$P(\mu_i \in [X_i - \sigma_i C_{R,\alpha}, X_i + \sigma_i C_{R,\alpha}];\ 1 \leq i \leq k)$$

Notice that the correlation structure among the random variables X affects the simultaneous confidence intervals through the critical point $C_{R,\alpha}$.

The multivariate quality control procedure operates as follows. For a known covariance structure Σ and a chosen error rate α, the experimenter first evaluates the critical point $C_{R,\alpha}$. Then, following any observation $x = (x_1, \ldots x_k)'$, the experimenter constructs confidence intervals.

$$\mu_i \in [x_i - \sigma_i C_{R,\alpha}, x_i + \sigma_i C_{R,\alpha}]$$

for each of the k variables. The process is considered to be in control as long as *each* of these confidence intervals contains the respective standard value μ_i^0. However, when an observation x is obtained for which one or more of the confidence intervals does not contain its respective standard value μ_i^0, then the process is stated to be out of control, and the variable or variables whose confidence intervals do not contain μ_i^0 are identified as those responsible for the aberrant behavior.

This simple procedure clearly meets the goals set in the introduction for a good solution to the multivariate quality control problem. An overall error rate of α is achieved, since when $\mu = \mu^0$ there is an overall probability of $1 - \alpha$ that each of the confidence intervals contains the respective value μ_i^0. Also, the identification of the errant variables is immediate and simple, and furthermore, the confidence intervals allow the experimenter to assess the new mean values of the out-of-control variables. This is particularly useful when the experimenter can judge the process to be still "good enough" and hence allow it to continue.

2.2. Example

Consider first the basic multivariate quality control problem with $k = 2$ so that there are just two variables under consideration. In this case, the required critical point $C_{R,\alpha}$ depends only on the error size α and the one correlation term $\rho_{12} = \rho$, say. In tables B.1–B.4 of Bechhofer and Dunnett (1988), values of the critical point are given for $\alpha = 0.20$, 0.10, 0.05, and 0.01 and for $\rho = 0(0.1)0.9$ (the required values for $C_{R,\alpha}$ correspond to the entries for $p = 2$ and $\nu = \infty$). More complete tables are given by Odeh (1982), who tabulates the required critical points for additional values of α and ρ (the values $C_{R,\alpha}$ at $k = 2$ correspond to the entries at $N = 2$). Interpolation within these tables can be used to provide critical values for other cases not given. An alternative method is to use a computer program to evaluate the bivariate normal cumulative distribution function.

As an example of the implementation of the procedure with $k = 2$, consider the problem outlined in Alt (1985) of a lumber manufacturing plant that obtains measurements on both the *stiffness* and the *bending strength* of a particular grade of lumber. Samples of size 10 are averaged to produce an observation $x = (x_1, x_2)'$, and standard values for these averaged observations are taken to be $\mu^0 = (265, 470)'$ with a covariance matrix of

$$\Sigma = \begin{pmatrix} 10 & 6.6 \\ 6.6 & 12.1 \end{pmatrix}$$

In this case the correlation is $\rho = 0.6$, so that with an error rate of $\alpha = 0.05$, the tables referenced above give the critical point as $C_{R,\alpha} = 2.199$.

Following an observation $x = (x_1, x_2)'$, the simultaneous confidence intervals for the current mean values $\mu = (\mu_1, \mu_2)'$ are given by

$$\mu_1 \in [x_1 - 2.199\sqrt{10}, x_1 + 2.199\sqrt{10}] = [x_1 - 6.95, x_1 + 6.95]$$
$$\mu_2 \in [x_2 - 2.199\sqrt{12.1}, x_2 + 2.199\sqrt{12.1}] = [x_2 \quad 7.65, x_2 + 7.65]$$

These confidence intervals have a *joint* confidence level of 0.95. The process is considered to be in control as long as both of these confidence intervals contain their respective control values $\mu_0 = (265, 470)'$, that is, as long as $258.05 \leq x_i \leq 271.95$ and $462.35 \leq x_2 \leq 477.65$. However, following an observation $x = (255, 465)'$, say, the process would be declared to be out of control, and the first variable stiffness would be identified as the culprit. Furthermore, the confidence interval for the mean stiffness level would be $\mu_1 \in (248.05, 261.95)$ so that the experimenter would have an immediate quantification of the amount of change in the mean stiffness level. An additional example with $k = 4$ variables is given in Hayter and Tsui (1994).

2.3. Independence Assumption

A general assumption of the multivariate quality control procedures is that observations obtained from the process under consideration can be taken to be independent of each other. Specifically, if a control chart based on Hotelling's T^2 statistic is employed, then it is assumed that the two statistics

$$T_1^2 = (x^1 - \mu^0)'\hat{\Sigma}^{-1}(x^1 - \mu^0)$$

and

$$T_2^2 = (x^2 - \mu^0)'\hat{\Sigma}^{-1}(x^2 - \mu^0)$$

obtained from two observations x^1 and x^2 of the process are independent of each other. Individually, these two statistics each have a scaled F-distribution, but any lack of independence between them may seriously affect the interpretation of the control chart.

Kuriki (1997) shows how the effect of a dependence between the variables can be investigated. In general, the joint cumulative distribution function of the statistics T_1^2 and T_2^2 is

$$P(T_1^2 \leq z_1, T_2^2 \leq z_2) = P(y_1'S^{-1}y_1 \leq z_1, y_2'S^{-1}y_2 \leq z_2)$$

where $S = \hat{\Sigma}$ has a Wishart distribution and (y_1, y_2) has a $2k$-dimensional normal distribution. The random variables y_1 and y_2 may not be independent of each other due perhaps to a correlation between subsequent observations taken from the process or through μ^0, which may be an average of observations in an initial pool. This general bivariate F-distribution can be used to assess the effects of a lack of independence between observations from a process if Hotelling's control chart is employed, and Kuriki (1997) shows how it can be easily evaluated as a two-dimensional integral expression.

3. NONPARAMETRIC PROCEDURES

The flow diagram in Figure 1 illustrates how distribution-free multivariate quality control procedures can be developed. The left side of the diagram corresponds to a traditional procedure. An initial pool of "in-control" data observations is often used to determine the control values $\mu^0 = \bar{x}$, and the assumption that the data have a multivariate normal distribution is required. The dotted lines correspond to distribution-free procedures that can be employed.

The middle procedure is based on the consideration of a nonparametric test of the hypothesis

$$H_0 : \mu = \mu^0$$

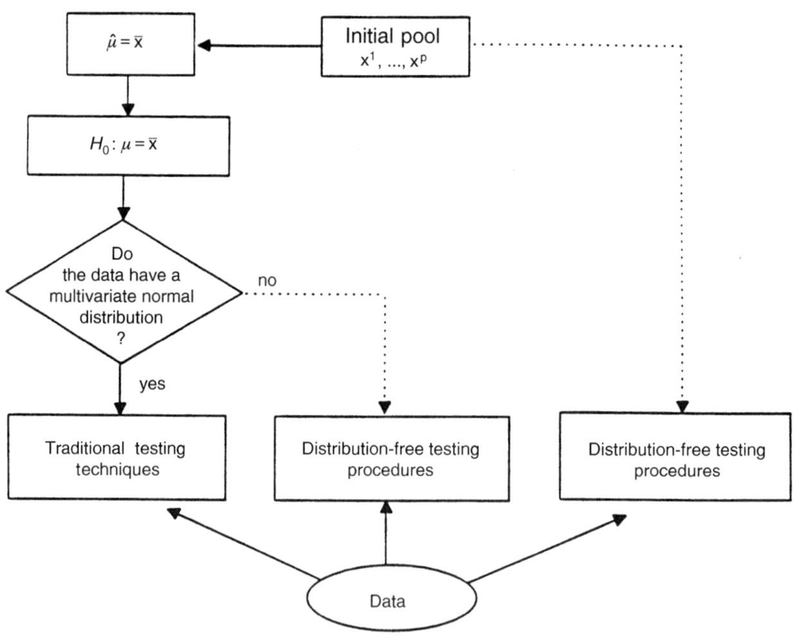

Figure 1 Flow diagram of multivariate quality control testing procedures.

with $\mu^0 = \bar{x}$. This procedure could be implemented even if there is no initial pool of data observations and μ^0 is simply some specified target value. However, in general it seems more sensible to make full use of the initial pool of observations and to develop a procedure indicated on the far right of the flow diagram in which the current data are compared with the initial pool of observations. In this case, the question of interest is whether it is plausible that the two data sets, the initial pool of observations and the current data observations, are actually observations from a common distribution. A discussion of such procedures that are developed in Liu (1995) and Bush (1996) is provided in this section.

3.1. Nonparametric Multivariate Control Charts

Liu (1995) provides some nonparametric multivariate quality control procedures that follow the right-hand dotted line of Figure 1 in that they compare current observations with an initial pool of "in-control" observations. The main idea is to reduce the current multivariate observation to a univariate index that can be plotted on a control chart. Three types of control charts are suggested that are truly nonparametric in nature and can be used to detect simultaneously any location change or variability change in the process. Liu's procedures are motivated by the "depth" of current measurements within the initial pool of observations and are conceptually equivalent to the procedures described in Bush (1996) employing functional algorithms to calculate the scores that are described in detail in the following sections.

3.2. Overview of Nonparametric Procedures

Assume that the initial pool consists of the observations

$$x^1, \ldots, x^p$$

where each x^i is a k-dimensional vector that is an observation from an unknown distribution with mean $\mu^0 = (\mu_1^0, \ldots, \mu_k^0)$ and covariance matrix Σ. Note that the observations x^i may in fact be defined to be averages of several measurements. The purpose of the quality control procedure is to determine whether or not a new observation can be considered to be an observation from this same distribution. The nonparametric procedure tests the following hypotheses:

> H_0: The new observation and the initial pool can be considered to be $p+1$ observations from the same unknown distribution.

H_A: The new observation cannot be considered to be from the same distribution as the initial pool.

If the null hypothesis is rejected, then the process is declared to be out of control.

As in other quality control procedures, the initial pool of supposedly identically distributed observations is employed to define the standards against which the new observations are measured. Traditionally, the initial pool is used to calculate control limits, but the nonparametric methods described below require a different and more direct use of the initial pool. Consider the two-dimensional case. Suppose a plot of x_1 versus x_2 reveals an elliptical shape. A new observation, x^0, is taken, and the point (x_1^0, x_2^0) is added to the graph. There is no need for concern if x^0 plots well within the borders of the ellipse. However, a point outside the ellipse or on the fringes may signal that the process is out of control. Thus the nonparametric quality control procedure operates by considering the location of the new observation with respect to the initial pool. A useful procedure will indicate whether a new observation is near the center of the initial pool, on the fringes, or outside.

3.3. Variable Transformation

It is convenient to define testing procedures in terms of a set of transformed observations. If the initial pool and the new observation are combined to form a set of $p+1$ observations, then let the sample average vector be $\bar{x} = (\bar{x}_1, ..., \bar{x}_k)$ and the sample covariance matrix be S_x. The quality control methods require calculating a distance measure between various points, and a sensible way to do this is with the Mahalanobis distance, where the distance from x^i to x^j is defined to be

$$D_{ij} = (x^i - x^j)' S_x^{-1} (x^i - x^j)$$

It can be shown that the Mahalanobis distance is equivalent to the squared Euclidean distance between "standardized" observations y^i, where

$$y^i = (x^i - \bar{x})A$$

and $AA' = S_x^{-1}$. Thus,

$$D_{ij} = (y^i - y^j)'(y^i - y^j)$$

The matrix A is easily calculated from the eigenvalues and eigenvectors of S_x^{-1}, but in practice the matrix A need never be calculated, since the testing procedures can be implemented in terms of the original observations x^i. In other words, while it is convenient to define quality control procedures in terms of the transformed observations y^i, the actual implementation may be performed in terms of the original observations x^i.

3.4. Calculation of a p-Value

The nonparametric procedure produces a set of scores $S_0, S_1, \ldots, S_p$ associated with each observation in the initial pool (S_i, $1 \leq i \leq p$) and the new observation (S_0). The score S_i reflects the "position" of observation y^i with respect to all $p+1$ observations. In general, the lower the score, the closer an observation is to the "center" of the set of observations. Let R_i, $0 \leq i \leq p$, be the rank of S_i among $S_0, \ldots, S_p$, where average ranks can be used if there are ties among the S_i in the usual manner.

The value of R_0 corresponding to the new observation is of particular interest. If the new observation and all p observations in the initial pool are observations from the same distribution (so that the process is still in control), the R_0 is equally likely to take any value from 1 to $p+1$ (supposing that there are no ties in the scores S_i). Moreover, large values of R_0 indicate that the new observation is on the fringe of the initial pool of data points, an event that has an increased probability if the process has moved out of control, and so a p-value for the null hypothesis that the process is in control can sensibly be calculated as

$$\text{p-value} = \frac{p+2-R_0}{p+1}$$

This p-value reflects the proportion of the $p+1$ observations that have scores S_i no smaller than S_0.

3.5. Decision Rules

The decision rules under which a process is declared to be out of control can be chosen by the engineers implementing the procedure. Notice that the p-value is limited by the number of observations in the pool. For example, if there are $p+1=100$ observations and $R_0=100$, then the p-value for the procedure is 0.01, and the process can be declared to be out of control if the specified probability of a false alarm, α, is greater than or equal to 0.01. Traditionally, the specified error rate for a quality control procedure is often taken to be smaller than $\alpha = 0.01$, which implies that for this nonparametric procedure a larger initial pool would be needed.

In addition to the consideration of individual p-values, "runs rules" may also be employed. In univariate control charts, several successive points on the same side of the centerline are often allowed to trigger a stopping rule suggesting that there has been a change in the mean of the distribution. Similar runs rules may be adopted for these nonparametric procedures. For example, suppose that the p-values for a series of successive observations are each less than 0.20 but that none of the individual p-values is less than the specified α level. One might declare the process to be out of control on the basis that these new observations are all near the fringes of the initial pool of observations.

Runs rules can be designed to locate changes in either the mean or the variance of the distribution. Any appearance that a set of new observations are not "well mixed" within the initial pool suggests that the distribution may have changed. Changes in the mean imply changes in the location of the distribution and may be identified by a locational shift in the new observations. Changes in the covariance structure Σ should be indicated by changes in the shape of the distribution. Specifically, increases in the variance of a variable should be characterized by frequent observations outside or on the fringes of the distribution.

In conclusion, the consideration of the individual p-values of new observations together with an awareness of the location of the new observations relative to the initial pool of observations should allow an effective determination of out-of-control signals.

3.6. Calculation of the Scores

There are two basic types of algorithms that can be used to construct the scores $S_0, S_1, ..., S_p$. These are **functional** algorithms and **linkage** algorithms.

Functional Algorithms

With functional algorithms the scores are calculated from a series of comparisons of the observations y^i with each other. Specifically, the score S_i is a function of $y^i = (y_1^i, ..., y_k^i)$ and every other point in the pool and can be written as

$$S_i = f(y^i; y^0, ..., y^p)$$

The function is defined so that observations that are far from the center of the set of observations receive high scores while observations near the center receive low scores. Three possible choices for the function are described below.

1. The easiest method to consider is

$$S_i = \sum_{r=1}^{k} \left| \sum_{j=0}^{p} (I_{\{y_r^i > y_r^j\}} - I_{\{y_r^i < y_r^j\}}) \right|$$

where $I_{\{\}}$ is the indicator function. In this case the score function can be thought of as simply being calculated from a count of how many points are on either side of a particular observation and as being similar to a multivariate sign test. The score S_i will be close to zero for points in the center of the distribution, because at the center there are roughly an equal number of observations in every direction. At the perimeter other observations tend to be to one side, and thus the score will be large. For these scores the magnitude of the difference between two points y^i and y^j is ignored, and only the direction of the difference is important. Note that there is a large potential for ties in the scores to occur with this method.

2. A second procedure is similar to the first except that the actual distances between points are used to calculate the scores. The score S_i is calculated as the sum of the Euclidean distances from y^i to every other point y^i, $0 \leq j \leq p$, so that

$$S_i = \sum_{j=0}^{p} \sqrt{D_{ij}} = \sum_{j=0}^{p} [(y^i - y^j)'(y^i - y^j)]^{1/2}$$

Thus S_i is the sum of the p distances from y^i to all points in the combined pool. It is clear that the scores of the observations at the center of the group will tend to be lower than the scores for perimeter observations.

3. The scores obtained from the third method are calculated by comparing an observation y^i with a statistic based on the combined pool. This statistic,

$$M = (M_1, \ldots, M_k)$$

is chosen to be a "middle value" of the combined pool of observations such as the mean vector or the median vector. Typically the scores can be calculated as the distances of the observations from this middle value so that

$$S_i = (y^i - M)'(y^i - M)$$

Again, note that observations near the center of the pool will have small scores while observations on the perimeter will have larger scores. Note also that this method requires far fewer calculations than method (2), although this difference should not be important with present computing facilities.

Linkage Algorithms

Linkage algorithms resemble a linking clustering algorithm in that the $p+1$ observations are linked together one point at a time. The cluster begins at the center of the distribution and branches to all of the observations in the combined pool. Points are added to the cluster in succession until all $p+1$ points are part of the cluster. The criterion for choosing the next point to be added to the cluster is that it should be the "closest" observation to the cluster. The distance to the cluster can be measured in several different ways, which are discussed below. The score S_i is defined to be equal to j when y^i is the jth point added to the cluster (note that in this case $R_i = S_i$). The first point to be added to the cluster can generally be taken to be the point closest to $\bar{y}$. Observations closest to the center will tend to be added first, and those on the perimeter will be added last. Also, observations in heavily concentrated areas will tend to be added to the cluster before observations in sparsely concentrated areas, since in dense regions observations are closer together, and therefore observations will tend to be linked in succession once the first observation in that region has been added to the cluster.

When these linkage algorithms are applied it can be useful to construct a "center value" M, which is considered to be the first point in the cluster (although it may be removed from the cluster later). Three possible ways to decide the order in which observations are added to the cluster are described below.

1. If observation y^i is not already in the cluster, then it is added to the cluster if it is the closest (among all observations not already in the cluster) observation to any observation already in the cluster. In other words, for each observation y^i not already in the cluster, the minimum distance

   $$D_{ij} = (y^i - y^j)'(y^i - y^j)$$

   is calculated over all points y^j already in the cluster. The point y^i with the smallest minimum distance is then added to the cluster.
2. Method 1 can be generalized by calculating the sum of the a smallest distances from an observation not in the cluster to observations already in the cluster, for a fixed value of a. While method (1) has $a = 1$, it may be sensible to take $a = 2$, say, whereby the sum of the two smallest distances from an observation not in the cluster to observations already in the cluster are used to determine which observation should be added to the cluster next.

3. An additional extension would be to calculate the sum of all of the distances from an observation not in the cluster to each of the observations already in the cluster. This method is different from method (2), since in this case the value of a changes and is equal to the number of observations currently in the cluster.

4. SUMMARY

A single product can be described by several correlated variables that are to be monitored by quality control procedures. The correlation structure between the variables should be taken into account when designing a quality control scheme for the product. A good multivariate quality control procedure is one that, at a specified error rate α, triggers the out-of-control alarm only with probability α when the process is still in control and triggers the alarm as quickly as possible when the process is out of control. In addition, it should provide a simple and easily implementable mechanism for deciding which of the variables are responsible when the process is determined to be out of control. Finally, it should allow easy quantification of the amount by which the out-of-control variables have changed in mean value. Recent advances in this area provide more tools for the practitioner to meet these goals.

REFERENCES

Alt FB. (1985). Multivariate quality control. In Kotz and Johnson, eds. Encyclopedia of Statistical Sciences, Vol. 6. New York: Wiley.

Bechhofer RE, Dunnett CW. (1988). Percentage points of multivariate Student t distributions. In: Selected Tables in Mathematical Statistics, Vol. 11. American Mathematical Society.

Bush HM. (1996). Nonparametric multivariate quality control. PhD Thesis, Georgia Institute of Technology, School of Industrial and Systems Engineering.

Hayter AJ, Tsui K. (1994). Identification and quantification in multivariate quality control problems. J Qual Technol 26(3):197–208.

Hotelling H. (1947) "Multivariate Quality Control" in Techniques of Statistical Analysis. Eisenhart, Rastay and Wallis, eds. New York: McGraw-Hill.

Kuriki S. (1997). A note on a bivariate F-distribution. Personal communication.

Liu RY. (1995). Control charts for multivariate processes. J Am Stat Assoc 90(432): 1380–1387.

Odeh (1982) "Tables of Percentage Points of the Distribution of The Maximum Absolute Value Equally Correlated Normal Random Variables." Communications in Statistics-Simulation and Computation 11:65–87.

13

Autocorrelation in Multivariate Processes

Robert L. Mason
Southwest Research Institute, San Antonio, Texas

John C. Young
McNeese State University, Lake Charles, Louisiana

1. INTRODUCTION

A basic assumption in most multivariate control procedures is that the observation vectors are uncorrelated over time. When this assumption is true, the graph of any process variable against time should show only random fluctuations. When the assumption is false, the patterns in such time plots are systematic and often indicate the existence of linear or quadratic trends. In these latter situations, incorrect signals can occur in the corresponding multivariate control chart, and the effectiveness of the overall control procedure may be weakened [e.g., see Alt et al. (1977) or Montgomery and Mastrangelo (1991)].

Numerous industrial processes produce data that change over time. This may occur because of such factors as the continuous wear on equipment, the degenerative effects of environmental and chemical contamination, and the depletion of the catalyst in a chemical process. Autocorrelated observations resulting because a process continuously decays over time may be detectable if one samples the process on a regular time interval. However, process decay that occurs in stages may appear to be insignificant and undetectable across short time intervals but highly significant and detectable when the process is monitored over extended time intervals. Mason et al.

(1996) present an excellent example of a situation where the autocorrelation behaves as a step function.

If autocorrelation goes undetected or ignored, it can create serious problems in multivariate control procedures. This often occurs when the effects of the autocorrelated variable are confounded with the time effects. An adjustment would be needed in such situations in order to obtain a true reading on process performance at a given point in time. Control procedures for autocorrelated data in a univariate setting make adjustments by modeling the time dependency and examining the residuals of the resultant autoregressive models. Under proper assumptions, these residuals, or adjusted values (effect of the time dependency removed), can be shown to be independent and normally distributed and are thus used as the charting statistic in the control procedure [see, e.g., Montgomery (1991)].

The problem with autocorrelated data from a multivariate process is more complicated. We have to be concerned not only with how these variables relate to the other process variables but also with how some of the process variables relate to time changes. Our procedure for analyzing such autocorrelated data centers on the use of Hotelling's T^2 as the control statistic. Many of the desirable properties of this statistic for independent observations are shown to apply to this situation.

2. DETECTION OF AUTOCORRELATION IN MULTIVARIATE PROCESSES

Why do certain types of processes have a tendency to generate observations with a time dependency? Autocorrelation may be due to a cause-and-effect relationship between a process variable and time. If this occurs, the observation on the process variable is proportional to the value of the variable at some prior time. In contrast, if the time relationship is only an empirical correlation and not a cause-and-effect one, the current value of the variable, although associated with the past value, is not determined by it. If this is the case, the association is usually due to an unobservable "lurking" variable.

Consider two process variables that are highly negatively correlated so that one variable increases as the other decreases. Suppose one of the variables, the "lurking" one, cannot be observed but is known to increase with time. Without knowledge of the relationship between the two variables, one would conclude that the second variable has a time dependency in its observations, as its values would tend to decrease as time increases. For example, if one considers the cyclical nature over time of the variable depicted in Figure 1, one might suspect that some form of time effect is

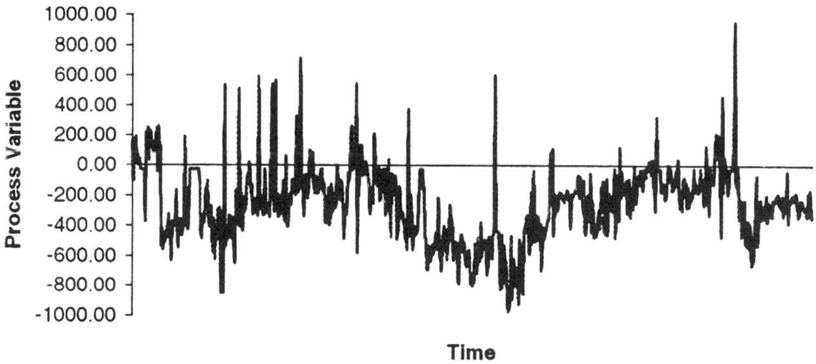

Figure 1 Process variable with cycle.

present. However, the noted trend is due to a "lurking" variable that has a seasonal component. Since the effects of such "lurking" variables, when they are known to exist, can be accounted for by making adjustments to the associated observable variable, the detection of these situations can be a great aid in the development of a proper control procedure for the process.

Detecting autocorrelation in univariate processes is accomplished by plotting the process variable against time. Depending on the nature of the autocorrelation, the plotted points will either move up or down or oscillate back and forth over time. Subsequent data analysis can be used to verify the time trend, determine lag times, and fit appropriate autoregressive models. The simple and straightforward method of graphing individual components against time can be inefficient when there are a large number of variables, and the interpretations can become confounded when these components are correlated. Despite these disadvantages, we have found that graphing each individual variable over time is still useful in multivariate processes. In addition to studying autocorrelation, it can lead to the discovery of other influential variables.

To augment the above graphical method and reduce the number of individual graphs that need to be produced, we additionally suggest that a time-sequence variable be added to the data set. If any of the other variables correlates with the time-sequence variable, it is highly probable that it correlates with itself over time. Using this approach, one can locate potential variables that are autocorrelated. Detailed analysis, including the graphing of the individual variable over time, will either confirm or deny the assertion for individual variables. Other techniques, such as that given in Tracy et al. (1993), also should be explored.

3. VARIOUS FORMS OF AUTOCORRELATION

We examine two different forms of autocorrelation: uniform decay and stage decay. It is important to recognize each type, as both play an important role in the development and implementation of a multivariate control procedure for autocorrelated data. Uniform, or continuous, decay occurs when the observed value of the process variable is dependent on some immediate past value. For example, heat transfer coefficient data behave in this fashion. During the life-cycle of a production unit, the transfer of heat is inhibited owing to equipment contamination or for other reasons that cannot be observed or measured. A new life cycle is created when the unit is shut down and cleaned. During the cycle, the process is constantly monitored to ensure maximum efficiency. Figure 2 contains the graph of a heat transfer coefficient over a number of life cycles of a production unit. The uniform decay of the unit is evident from the declining trend in the plotted curve prior to each new life cycle.

Stage decay occurs when the time change in a process variable is inconsistent on a daily basis but occurs in a stepwise fashion over extended periods of time. This form of autocorrelation is present in processes where change with time occurs very slowly. The time relationship results when the process performance in one stage is dependent on the process performance in the previous stage or stages. The graph of a stage decay process variable is presented in Figure 3. Notice that there is a distinctive shift in the process variable somewhere near the middle of the curve but that the fluctuations are around similar levels below the shift and at higher but similar levels above the shift.

4. A CONTROL PROCEDURE FOR A UNIFORM DECAY PROCESS

Our approach for obtaining a multivariate control procedure for uniform decay processes is to use Hotelling's T^2 statistic and its associated orthogo-

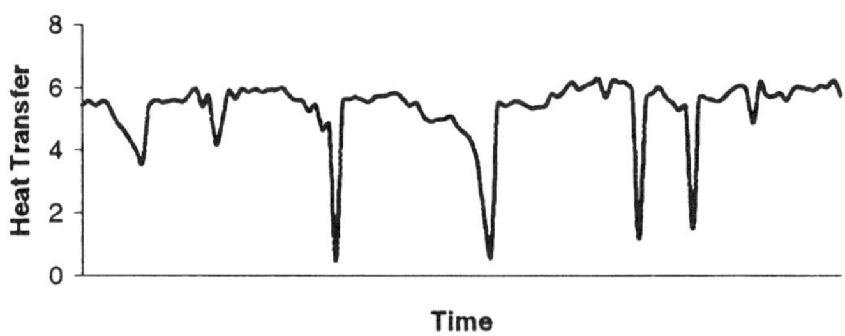

Figure 2 Life cycles over time.

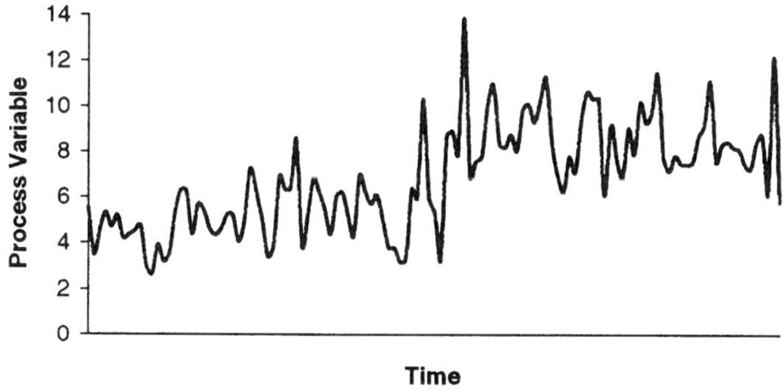

Figure 3 Step change of process variable.

nal decomposition. Mason and Young (1999) show that correct modeling of existing functional relationships between process variables increases the sensitivity of the T^2 value in signal detection. An overview of pertinent points of their work and how it relates to a multivariate control procedure for autocorrelated processes with uniform decay is discussed below. Mathematical details and data examples are provided in the original paper.

One example of an orthogonal decomposition of the T^2 value associated with a p-dimensional data vector, $X' = (x_1, ..., x_p)$, is given as

$$\begin{aligned} T^2 &= (X - \bar{X})' S^{-1} (X - \bar{X}) \\ &= T_1^2 + T_{2.1}^2 + \cdots + T_{p.12...(p-1)}^2 \end{aligned}$$

where $\bar{X}$ and S are the usual estimates of the population mean vector and covariance matrix obtained by using an in-control historical data set. In this procedure [see Mason et al. (1995) for a complete description], the first component of a particular decomposition, termed the unconditional term, is used to determine whether the observation on the jth variable of a signaling data vector is within the operational range of the process. The general form of the jth unconditional T^2 is given by

$$T_j^2 = \frac{(x_j - \bar{x}_j)^2}{s_j^2} \tag{1}$$

where x_j is the jth component of X, and $\bar{x}_j$ and s_j^2 are the corresponding mean and variance estimates as determined using the in-control data set. The remaining components, termed conditional terms of the decomposition, are used in detecting deviations in relationships among the variables that produced the signal. The general form of a conditional T^2 term is given by

$$T^2_{j.12...(j-1)} = \frac{(x_j - \bar{x}_{j.12...(j-1)})^2}{s^2_{j.12...(j-1)}} \tag{2}$$

This is the square of the jth variable adjusted by the estimates of the mean and variance of the conditional distribution of x_j given $x_1, x_2, ..., x_{j-1}$.

The ordering of the components in the data vector determines the representation of each term of the decomposition. As pointed out by Mason et al. (1995), there are $p!$ different arrangements of the p components of a data vector, and these lead to $p!$ decompositions, each consisting of p terms. Mason and Young (1997) show that the unique terms of all such decompositions will contain all possible regressions of an individual variable on all possible subgroups of the remaining p–1 variables. For example, the first component, x_1, of a three-dimensional data vector would be regressed against all possible subgroups of the other two variables. These regressions and the corresponding conditional T^2 terms are presented in Table 1. Using the tabulated results, a control procedure based on the T^2 statistic can be developed for a set of process variables that exhibit uniform time decay in the observations and, at the same time, are correlated with other process variables. As an example, consider a bivariate vector (x,y) where the variable y exhibits a first-order autoregressive relationship [i.e., AR(1)]. Note that the observations are actually of the form (X_t, Y_t, Y_{t-1}), where t represents the time sequence of the data. The AR(1) relationship for y can be represented in model form as

$$y_t = b_0 + b_1 y_{t-1} + \text{error} \tag{3}$$

where b_0 and b_1 are unknown constants. If y were being monitored while its relationship with x was ignored, a signal would be produced when the observed value of y was not where it should be as predicted by the estimate of the model in Eq. (3). However, if one chooses to examine the value of y

Table 1 List of Possible Regressions for x_1 When $p = 3$

Regression of	Conditional T^2
x_1 on x_2	$T^2_{1.2}$
x_1 on x_3	$T^2_{1.3}$
x_1 on x_2, x_3	$T^2_{1.23}$

adjusted for the effect of x and the time dependency, a model of the form

$$y_t = b_0 + b_1 y_{t-1} + b_2 x_t + \text{error} \tag{4}$$

would be more appropriate.

The modeling of time relationships existing among the process variables requires adding additional lag variables to the historical data. For example, a historical data set for a bivariate process is a data matrix consisting of observations on the vector (x_t, y_t), where $t = 1, ..., n$. Assuming autocorrelation exists among the observations on y and is of the AR(1) form given in (4), the data set will have to be reconstructed to have the form (x_t, y_t, y_{t-1}), $t = 2, ..., n$, in order to estimate the model. The ordering of the vector components is arbitrary but is important to the notation scheme for the T^2 terms. Interpretation of a signal for this situation is achieved by examining appropriate terms from all possible decompositions of the signaling T^2 value. Details are provided in Table 2.

Higher order autoregressive relationships can be examined by adding other lag variables to the historical data set. For example, suppose the variable y has an AR(2) time dependency so that

$$y_t = b_0 + b_1 y_{t-1} + b_2 y_{t-1}^2 + \text{error}. \tag{5}$$

The reconstructed data vector would be of the form $(x_t, y_t, y_{t-1}, y_{t-1}^2)$. The use of such time-dependent models requires process knowledge and an extensive investigation of the historical data.

Table 2 Interpretation of Useful T^2 Components in AR(1) Model

T^2 component	Interpretation
T_1^2	Checks if x component of data vector is in operational range of x.
T_2^2	Checks if y component of data vector is in operational range of y.
$T_{2.3}^2$	Determines if current value of y is in agreement with the value predicted using previous y value, or examines the value of y with the effect of y_{t-1} removed.
$T_{1.2}^2$	Checks if x and y are countercorrelated. Effect of time is not removed.
$T_{2.1}^2$	Checks if y and x are countercorrelated. Not symmetrical with $T_{1.2}^2$. Effect of time is not removed.
$T_{2.13}^2$	Determines if present value of y is in agreement with the value predicted using x and previous value of y.

5. EXAMPLE OF A UNIFORM DECAY PROCESS

Consider a chemical process where observations are taken on a reactor used to convert ethylene (C_2H_4) to ethylene dichloride (EDC), the basic building block for much of the vinyl products industry. Feedstock for the reactor is hydrochloric acid gas (HCl) along with ethylene and oxygen (O_2). Conversion of the feedstock to EDC takes place in a reactor under high temperature, and the process is referred to as oxyhydrochlorination (OHC). There are many different types of OHC reactors available to convert ethylene and HCl to EDC. One type, a fixed-life or fixed-bed reactor, must have critical components replaced at the end of each run cycle, as the components are slowly depleted during operation. Performance of the reactor follows the depletion of the critical components; i.e., the best performance of the reactor is at the beginning of the run cycle, and it gradually becomes less efficient during the remainder of the cycle. This inherent uniform decay in the performance of the reactor produces a time dependency in many of the resulting process and quality variables.

Consider a steady-state process where the reactor efficiency is at 98%. The efficiency variable will contain very little variation (due to the steady-state conditions), and its operation range will be small. Any significant deviation from this range should be detected by the process control procedure. However, over the life cycle of a uniformly decaying reactor, the unit efficiency might have a very large operational range. For instance, it might range from 98% at the beginning of a cycle to 85% at the end of the cycle and would thus contain more variation than a steady-state variable. If we failed to consider the decay in the process, any efficiency value between 85% and 98% would be acceptable, even 85% at the beginning of a cycle.

A deviation beyond the operational range (established using in-control historical data) for a process variable can be detected by using its unconditional T^2 term. In addition, incorrect movement of the variable within its range (occurring because of improper relationships with other process variables) can be detected by using the conditional T^2 terms. However, this approach does not account for the effects of movement due to time dependencies. Adjusting for a time effect will provide additional monitoring of the movement of an individual variable within its operational range when the effect of its previous value(s) has been removed. Including time-lag variables in the computation of the T^2 statistic adds corresponding terms to the T^2 decompositions that can be used to monitor movement of the variables through time. This enhances the signal detection performance of the overall T^2 statistic.

Although the above reactor process is controlled by many variables, we will use only four of them in this example in order to demonstrate the

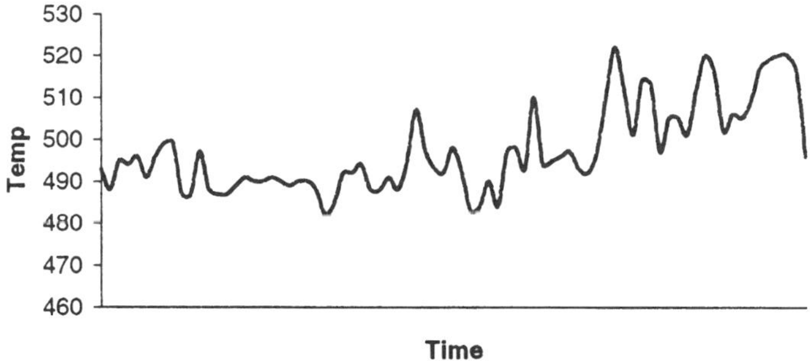

Figure 4 Reactor temperature versus time.

proposed control chart procedure. These include three process variables, labeled TEMP, L3, and L1, and a measure of feed rate, labeled RP1. All, with the exception of feed rate, show some type of time dependency.

Temperature measurements are available from many different locations on a reactor, and together these play an important role in the performance and control of the reactor. To demonstrate the time decay in all of the measured temperatures, we present in Figure 4 a graph of their average over a good production run. The plot indicates that the average temperature of the reactor gradually increases over the life cycle of the unit.

Figures 5 and 6 contain graphs of the other two process variables, L3 and L1, over time. The decay effect for L3 in Figure 5 has the appearance of an AR(1) relationship, while that for L1 in Figure 6 has the appearance of some type of quadratic (perhaps a second-order quadratic) or an exponential autoregressive relationship.

Feed flow (RP1) to a reactor consists of three components: the flows of O_2, HCl gas, and C_2H_4. However, since these components must be fed in at a constant ratio, one graph is sufficient to illustrate the feed. During a run

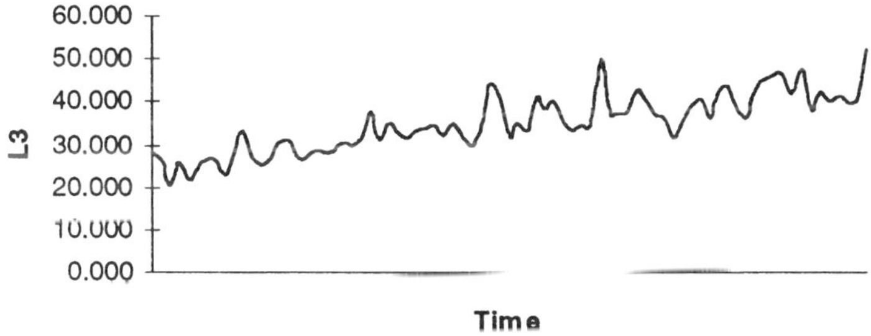

Figure 5 L3 versus time.

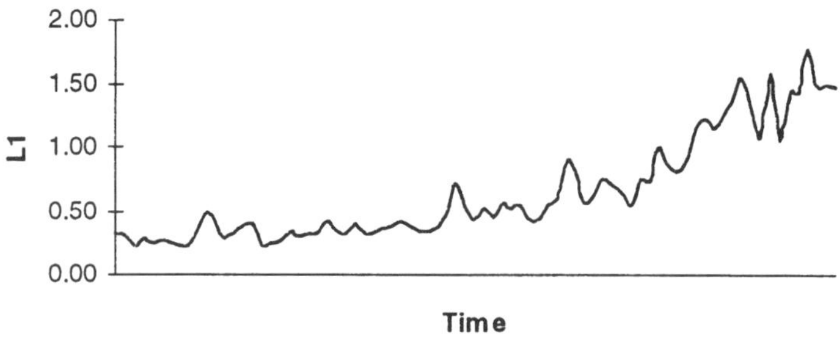

Figure 6 L1 versus time.

cycle, the feed to the reactor is somewhat consistent and does not systematically vary with time. This is illustrated in Figure 7.

The correlation matrix for the four variables RP1, L1, L3, and TEMP, including the first-order lag variables for L1, L3, and temperature (LL1, LL3, and LTEMP), is presented in Table 3. Note the very strong lag correlation for the three process variables. For example, L1 has a correlation of 0.93 with its lag value, indicating that over 80% of the variation on this variable can be explained by the relationship with its lag value. This strong correlation implies that an AR(1) model is a good approximation to the true time dependency. Also, note the strong relationship between L1 and the lag of the temperature. The correlation of 0.80 implies that over 64% of the variation in the present value of L1 can be explained by the temperature of the unit during the last sampling period.

To see the effect of these time-lag variables on a T^2 control procedure, we will compare the T^2 values obtained with and without the lag variables. For comparison purposes, we denote the T^2 based on the chosen four variables RP1, L1, L3, and TEMP by T_4^2 and the T^2 based on all seven variables, including the three lag variables LL1, LL3, and LTEMP, by T_7^2. Assume that each observation vector is repre-

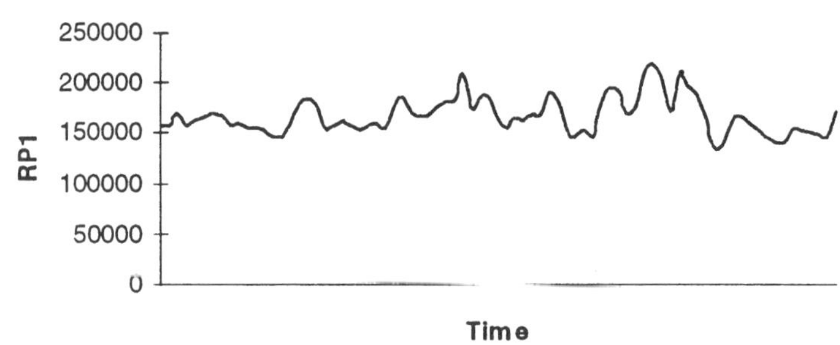

Figure 7 RP1 versus time.

Table 3 Correlation Matrix for Reactor Data

	RP1	L1	L3	TEMP	LL1	LL3	LTEMP
RP1	1.00						
L1	–0.23	1.00					
L3	–0.02	0.79	1.00				
TEMP	0.12	0.74	0.39	1.00			
LL1	–0.22	0.93	0.72	0.72	1.00		
LL3	–0.03	0.70	0.75	0.49	0.78	1.00	
LTEMP	–0.02	0.80	0.53	0.76	0.76	0.42	1.00

sented as (RP1, L1, L3, TEMP, LL1, LL3, LTEMP). Since the statistic T_4^2 is based on the first four components of this vector, it is contained in the overall vector T_7^2. Also, all of the terms in the possible decompositions of T_4^2 are contained in the various decompositions of T_7^2. Since T_7^2 contains information on the time-lag variables, it will be more sensitive to any change in the process.

The inclusion of lag variables in the historical data will produce new conditional terms in the decomposition of the T_7^2 statistic. For example, the unconditional term T_{L3}^2, which is contained in both T_4^2 and T_7^2, is used to determine if L3 is in its operational range. However, including the lag variable LL3 adds the new conditional term, $T_{\text{L3.LL3}}^2$, to T_7^2 and allows one to monitor the location of L3 based on its previous value. For lag values of one sampling period, this term contains the AR(1) model

$$\text{L3} = b_0 + b_1\text{LL3} + \text{error}.$$

To compare the performance of T_7^2 to T_4^2, consider a sequence of 14 observations (in time order) on the above four reactor variables and the corresponding three time-lag variables. The data are presented in Table 4. Our interest lies in the process variables L1 and TEMP. The values of L1 are relatively high for the first two observations, drop dramatically for the next two observations, and then gradually increase in value to near the end of the data set. In contrast, the TEMP values start relatively low, gradually rise until the middle observations, and then stabilize near the end.

Table 5 contains the T_7^2 and T_4^2 values for the 14 sample points. The α level for both statistics is 0.0001. Note that a signal is detected by T_7^2 at observations 4 and 6, but no signal is detected by T_4^2 at any of the observations.

Interpretation of T^2 signals for autocorrelated dats is no different than for data without time dependencies. When a signal is detected, the T^2 statistic is decomposed to determine the variable or set of variables that caused the signal. When T_7^2 for observation 4 is decomposed, using the procedure

Table 4 Reactor Data

Obs. No.	RP1	L1	L3	TEMP	LL1	LL3	LTEMP
1	188,300	0.98	44.13	510	1.40	50.47	498
2	189,600	0.81	33.92	521	0.98	44.13	510
3	198,500	0.46	28.96	524	0.81	33.92	521
4	194,700	0.42	29.61	521	0.46	28.96	524
5	206,800	0.58	29.31	530	0.42	29.61	521
6	198,600	0.63	28.28	529	0.58	29.31	530
7	205,800	0.79	29.08	534	0.63	28.28	529
8	194,600	0.84	30.12	526	0.79	29.08	534
9	148,000	0.99	39.77	506	0.84	30.12	526
10	186,000	1.19	34.13	528	0.99	39.77	506
11	200,200	1.33	32.61	532	1.19	34.13	528
12	189,500	1.43	35.52	526	1.33	32.61	532
13	186,500	1.10	34.42	524	1.43	35.52	526
14	180,100	0.88	37.88	509	1.10	34.42	524

Table 5 T^2 Values for Reactor Data

Observation No	T_7^2 (Critical value = 39.19)	T_4^2 (Critical value = 28.73)
1	16.98	4.75
2	14.46	9.95
3	37.28	24.27
4	41.88	22.78
5	39.10	27.82
6	42.71	23.79
7	37.18	25.27
8	31.74	13.51
9	23.58	3.65
10	18.07	10.99
11	20.76	19.49
12	20.43	15.50
13	22.02	8.39
14	18.73	1.67

described in Mason et al. (1997), several large conditional T^2 components are produced, and each includes some subset of the variables L1, lag L1, TEMP, and lag TEMP. For example, $T^2_{\text{L1.LTEMP}}$ has a value of 18.40. Such a large conditional T^2 term implies that something is wrong with the relationship between L1 and temperature. The predicted value of L1 using LTEMP as a predictor is not within the range of the error of the model as determined from the in-control historical data set. On closer examination, the data in Table 4 for observation 4 suggest that the value of L1 is too small for the temperature value. With the removal of these two components from the signaling observation vector, the subvector containing the remaining five variables produces no signal. The T^2 value for the subvector is 15.31, which is insignificant compared to the critical value of 32.21 ($\alpha = 0.0001$).

Given the dependency of L1 on time, as illustrated in Figure 6, it may be surprising that we did not find a problem with the relationship between L1 and its lag value. However, in examining the values in Table 4, it is clear that the trend in L1 from observation 3 to observation 4 is not unusual, as there is a downward trend in L1 from observation 1 to observation 4. However, at observation 4, the downward movement in L1 is not in agreement with the upward movement in the temperature, particularly when one considers the positive correlation between these two variables noted in Table 3 for the historical data set. Thus, a process problem is created, and the T^2 statistic signals.

Analysis of the signaling observation 6 produces similar results. The conditional terms involving subsets of L1, lag L1, TEMP, and lag TEMP are generally large in value. For example, $T^2_{\text{L1.LTEMP}}$ has a value of 17.72, $T^2_{\text{TEMP.L1}}$ has a value of 21.98, and $T^2_{\text{L1.TEMP,LTEMP}}$ has a value of 21.53. All these values indicate that there is a problem in the relationship between L1 and TEMP relative to that seen in the historical data.

Note that T^2_4, which did not include the effects of the time dependencies between the process variables, failed to detect the above two data problems. However, this is not due to a failure of the T^2 statistic, as its performance is based solely on the provided process information. Clearly, T^2_7 is more sensitive than T^2_4, since it has included information on the autocorrelation that is present in three of the four variables. Thus, one would expect its performance in signal detection to be superior.

6. A CONTROL PROCEDURE FOR STAGE DECAY PROCESSES

Process decay that occurs in stages was illustrated in Figure 2. As a general rule, this type of decay occurs over many months or years, and the time dependency is between different stages in the process. For example, process performance in the second stage might be dependent on performance in the first stage, and performance in the third stage might be dependent on performance in the previous stages. A process-monitoring procedure at any given stage must adjust the process for its performance in the previous stages. Thus, control procedures are initiated to detect when significant deviation occurs from the expected adjusted performance as determined by the historical database. An overview of how this is done is briefly discussed in this section, and more extensive details and examples can be found in Mason et al. (1996). Consider a situation where a three-stage life has been determined for a production facility consisting of n units. Observations are homogeneous within each stage but heterogeneous between stages. An in-control historical data set, composed of observations on p variables for each unit during each stage of operation, is available. This is represented symbolically in Table 6, where each X_{ij} is a p-dimensional vector that represents an observation on p process variables; i.e.,

$$X'_{ij} = (x_{ij1}, x_{ij2}, \dots, x_{ijp})$$

where $i = 1, \dots, n$, and $j = 1, 2, 3$.

The proposed solution for the T^2 control procedure for use with such stage-decay process data is to use a $3p$-dimensional observation vector given by $X'_k = (X_{k1}, X_{k2}, X_{k3})$, $k = 1, 2, \dots, n$. The vector X_k represents all the observations taken on the p variables from a given processing unit across the three stages of its life. For a given production unit, the observations across the three stages are time-related and thus dependent. However, within a given stage, observations are independent between production units. Since X_k has three components corresponding to the three life cycles of the unit, it will be possible to adjust the p process variables in the T^2 statistic for the corresponding stage dependencies.

Suppose X_k can be described by a multivariate normal distribution with a mean vector represented as $\mu' = (\mu_1, \mu_2, \mu_3)$, where the μ_i, $i = 1, 2, 3$, are the p-dimensional mean vectors of the process variables at the ith stage. The covariance structure for X_k is given as

Table 6 Three-Stage Life History

Unit	Stage 1	Stage 2	Stage 3
1	X_{11}	X_{12}	X_{13}
2	X_{21}	X_{22}	X_{23}
$\vdots$	$\vdots$	$\vdots$	$\vdots$
n	X_{n1}	X_{n2}	X_{n3}

$$\Sigma = \begin{bmatrix} \Sigma_{11} & \Sigma_{12} & \Sigma_{13} \\ \Sigma_{21} & \Sigma_{22} & \Sigma_{23} \\ \Sigma_{31} & \Sigma_{32} & \Sigma_{33} \end{bmatrix}$$

where Σ_{ii} represents the covariance structure of the observations for the ith stage, $i = 1, 2, 3$; and Σ_{ij}, $i \neq j$, denotes the covariance structure of the observations between stages. Using a historical data set, standard estimates $(\bar{X}, S)$, of the unknown population parameters (μ, Σ) can be obtained, and a control procedure based on an overall T^2 can be developed. Note that the estimates are partitioned in the same fashion as the parameters.

As an example of the proposed control procedure, suppose a new observation, X, is taken on a given unit in its third stage. The overall T^2 for this observation is given by

$$T^2 = (X - \bar{X})' S^{-1} (X - \bar{X})$$

and will be used as the charting statistic. Interpretation of a signaling vector is keyed to the partitioned parts of X (i.e., the subvectors representing observations on the unit at the various stages). Significant components of the T^2 decomposition and how they pertain to the observation vector X taken in stage 3, assuming satisfactory performance in stages 1 and 2, are presented in Table 7.

When a signalling T^2 component is identified, it can be decomposed to locate the signaling variable or group of variables. Suppose a problem is located in the conditional $T^2_{3.2}$ term. This implies, from Table 7, that the observation vector taken at stage 3, adjusted for the process performance at stage 2, is out of control. With this result, however, we will not know if the process performance is better or worse than that indicated by the historical situation unless we further examine the source of the problem in terms of the

Table 7 Interpretation of Components in Stage Decay, $p = 3$

Component	Interpretation of component
T^2_3	Checks if the p components of the observation vector X_3 are within tolerance.
$T^2_{3.1}$	Checks process performance on stage 3, i.e., X_3, adjusting for performance in stage 1 as given by X_1.
$T^2_{3.2}$	Checks process performance in stage 3, adjusting for performance in stage 2.
$T^2_{3.12}$	Checks process performance in stage 3, adjusting for performance in stages 1 and 2.

individual variables. To do this, we will need to perform a second decomposition, but this one will involve decomposing the signaling conditional T^2 component.

For $p = 3$, one possible decomposition of $T^2_{3.2}$ is given by

$$T^2_{3.2} = (T^2_1)_{3.2}(T^2_{2.1})_{3.2} + (T^2_{3.21})_{3.2}$$

Interpretation of these doubly decomposed terms is the same as for any T^2 with variable components. For example, $(T^2_1)_{3.2}$ represents an unconditional T^2 term and can be used to check the tolerance of the first component of the observation vector.

In general, incoming observations on a new unit are monitored in a sequential fashion. When a unit is in stage 1, only the observation X_1 is available, and monitoring is based on use of the statistic

$$T^2 = (X_1 - \bar{X}_1)' S_{11}^{-1} (X_1 - \bar{X}_1)$$

If a signal is observed, the T^2 is decomposed and the signaling variables are determined. For signaling observations in the remaining stages, the procedure is the same as that outlined above for an observation in stage 3.

7. SUMMARY

The charting of autocorrelated multivariate data in a control procedure presents a number of serious challenges. A user must not only examine the relationships existing between the process variables to determine if any are unusual but must also adjust the control procedure for the effects of the time dependencies existing among these variables. This chapter presents one possible solution to problems associated with constructing multivariate control procedures for processes experiencing either uniform decay or stage decay.

The proposed procedure is based on exploiting certain properties of Hotelling's T^2 statistic. The first useful property is the inherent dependency of this statistic on the relationships that exist between and among the process variables. If time dependencies exist, they can be identified by including time variables in the observation vector and then examining their relationships with the process variables. A second important property of T^2 is that its signaling values can be decomposed into components that lead to clearer interpretation of signals. The resulting decomposition terms can be used to monitor relationships with the other variables and to determine if they are in agreement with those found in the historical data set. This property is particularly helpful in examining stage-decay processes, as the decay occurs sequentially and thus lends itself to analysis by repeated decompositions of the T^2 statistic obtained at each stage.

REFERENCES

Alt FB, Deutch SJ, Walker JW. (1977). Control charts for multivariate, correlated observations. ASQC Technical Conference Transactions. Milwaukee, WI: American Society for Quality Control, pp 360–369.

Mason, RL, Young, JC. (1999). improving the sensitivity of the T^2 statistic in multivariate process control. J Qual Technol 31. In press.

Mason RL, Tracy ND, Young JC. (1995). Decomposition of T^2 for multivariate control chart interpretation. J Qual Technol 27:99–108.

Mason RL, Tracy ND, Young JC. (1996). Monitoring a multivariate step process. J Qual Technol 28:39–50.

Mason RL, Tracy ND, Young JC. (1997). A practical approach for interpreting multivariate T^2 control chart signals. J Qual Technol 29:396–406.

Montgomery DC. (1991). Introduction to Statistical Quality Control. New York: Wiley.

Montgomery DC, Mastrangelo CM. (1991). Some statistical process control methods for autocorrelated data. J Qual Technol 23:179–193.

Tracy ND, Mason RL, Young JC. (1993). Use of the covariance matrix to explore autocorrelation in process data. In: Proceedings of the ASA Section on Quality and Productivity. Boston, MA: American Statistical Association, pp 133–135.

14
Capability Indices for Multiresponse Processes

Alan Veevers
Commonwealth Scientific and Industrial Research Organization, Clayton, Victoria, Australia

1. INTRODUCTION

Production processes can be characterized by the simple fact that something is produced as a result of a number of deliberate actions. The product may be an item such as a glass bottle, a brake drum, a tennis ball, or a block of cheese. Alternatively, it may be a polymer produced in a batch chemical process or a shipment of a mineral ore blended from stockpiles that are being continuously replenished. Whatever the case, there will usually be several measurable quality characteristics of the product for which specifications exist. These are often a pair of limits between which the appropriate measurement is required to lie. Sometimes a specification is a one-sided limit such as an upper limit on the amount of an impurity in the product of a chemical reaction.

The extent to which a process could or does produce product within specifications for all its measured quality characteristics is an indication of the *capability* of the process. Capability can be measured both with and without reference to targeting, and it is important to distinguish between these two situations. The principal reasons why product may be produced out-of-specification, i.e., nonconforming, are either poor targeting of the process mean or excessive variation or a combination of both. In process development or improvement campaigns, the two situations relate to the following questions.

1. Are the ranges of variation in my product characteristics small enough to fit within the specification ranges?
2. How shall I choose the aim-point for my process mean in order to minimize the proportion of nonconforming product?

Capability potential is concerned with the first question. It is a comparison of a measure of process dispersion with the amount of dispersion allowed by the specifications. *Capability performance* addresses the second question and is concerned with what actually happens during a period of stable production. These concepts have been formalized for a single response by the introduction of capability indices; see, for example, Kane [1], of which C_p (for potential) and C_{pk} (for performance) are the most commonly used. These, and other, indices are discussed in the book by Kotz and Johnson [2], which, together with the references therein and other chapters of the present volume, provide a good summary of single-response capability indices. For multiresponse processes, the question arises as to whether or not suitable and useful multivariate capability indices exist. If so, they will need to provide answers to the above two questions. Several indices have been proposed for multiresponse processes, and some of them are discussed later. However, it is first necessary to deal with some important issues of clarification.

2. CAPABILITY STUDIES, PROCESS MONITORING AND CONTROL

Since capability indices were brought to the attention of mathematical and statistical researchers in the 1980s, there has been some self-perpetuating confusion in the literature. A number of authors, for example Chan, et al. [3] and Spiring [4], argue that C_p is a poor capability measure because it fails to take account of the target. What seems to be forgotten is that the p in C_p stands for potential and there was never any intention that it should take account of the target. C_p is meant as an aid to answering question 1 posed in Section 1, and concerns variation but not location. On the other hand, C_{pk} was devised to help answer question 2 and refers to the actual performance of the process when targeting has taken place. There is no need to compare C_p with C_{pk} (or with any other performance measures), because they measure different things. The fact that C_p and C_{pk} are both routinely reported during the performance phase in automotive and other manufacturing processes might cloud the issue but should not lead to them being regarded as alternative measures of the same thing. For example, if a stable process is reporting $C_p = 2.1$ and $C_{pk} = 0.9$, then the most likely explanation is that the process mean is not optimally targeted. The information provided by the C_p value tells us that the process is potentially capable without further need to reduce variation. Process performance will be improved, monitored by C_{pk}, by suitably adjusting the target for the process mean.

Similar considerations apply to multiresponse capability indices. Specifically, there is a clear justification for developing analogs of C_p for the multivariate case that, of course, take no account of targeting. Such an index will measure the potential of the process to meet specifications (addressing question 1) but will not, by intent, measure actual performance. Different measures must be devised for the latter purpose.

Another source of confusion arises when process capability and process control issues are not separated. An illustration of the point being made here is based on the following example. During the 1997 Australian Open Tennis tournament, some of the top players complained about the quality of the balls being used. International regulations specify that they shall weigh not less than 56.7 g and not more than 58.5 g and must be between 6.35 cm and 6.67 cm in diameter. The tennis ball production process must be set to achieve both these specifications simultaneously. This defines a rectangular specification region for the bivariate quality measure consisting of the weight and diameter of a tennis ball. A small sample of measurements on ordinary club tennis balls was obtained that showed a correlation of 0.7 between weight and diameter. This information was used to contrive the situation shown in Figure 1 to illustrate the difference between capability and control considerations. Suppose that a period of stable production produced data approximately following a bivariate normal distribution with a correlation coefficient of 0.7. A 99% probability ellipse for such a distribution is shown in Figure 1. Now suppose that the next two measured balls are represented by the + signs in the figure. Two conclusions can be drawn, first that the process has gone out of statistical control and second that the two new balls are perfectly capable of being used in a tournament. In fact, the two new balls are arguably better, in the sense of being nearer to the center of the specification region, than any of the balls produced in the earlier stable phase.

From the process control point of view, the out-of-control signals must be acted upon and steps taken to bring the process back into stable production. Multivariate process control techniques, such as that introduced by Sparks et al. [5] or those discussed in a previous chapter of this book, are available for this purpose. Based on multivariate normal theory, ellipsoidal control regions form the natural boundaries for in-control observations. Points falling outside the control region are usually interpreted as meaning that something has gone wrong with the process. From the process capability point of view, it is whether or not production will consistently

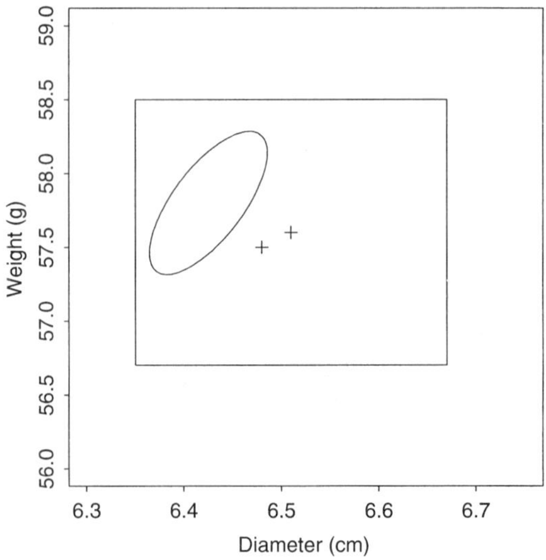

Figure 1 A 99% probability ellipse representing the bivariate distribution of the tennis ball quality characteristics lies comfortably inside the specification rectangle.

meet specifications that is of primary importance. In this case, the fact that the region bounding the swarm of data points may be ellipsoidal is of minor importance. The main concern is whether or not it fits into the specification region. Capability indices are not tools for process control and should not be thought of as measures by which out-of-control situations are detected. They are simply measures of the extent to which a process could (potential) or does (performance) meet specifications. Issues of control and capability need to be kept separate; otherwise unnecessary confusion can occur. For example, although correlation is of critical importance in control methodology, it is largely irrelevant for many capability considerations.

3. MULTIVARIATE CAPABILITY INDICES

As pointed out by Kotz and Johnson [2], most multivariate capability indices proposed so far are really univariate indices derived from the vector of quality characteristics. An exception is the three-component vector index introduced by Hubele et al. [6]. While a complete review of the subject to date is not intended, some of the significant developments are mentioned here. The indices fall broadly into two groups: those that use a hyperrectangular specification region and those that use an ellipsoidal specification region. Within those groups there are indices that measure capability potential and some that measure capability performance.

Let $\mathbf{X}_q = (X_1, X_2, ..., X_q)'$ represent the vector of q quality characteristics, and suppose that an adequate model for $\mathbf{X}_q$ under stable process conditions is multivariate normal with mean vector μ and variance–covariance matrix Σ. Taking the widely accepted value of 0.27% to be the largest acceptable proportion of nonconforming items produced, a process ellipsoid

$$(\mathbf{X} - \mu)'\Sigma^{-1}(\mathbf{X} - \mu) = c^2$$

where c^2 is the 0.9973 quantile of the χ^2 distribution on q degrees of freedom, can be defined. More generally, c^2 can be chosen to correspond to any desired quantile.

Referring to the ellipsoid as the process region, the two questions of interest can be rephrased as follows.

1. With freedom of targeting, would it be possible for the process region to fit into the specification region?
2. During stable production with the mean of the process distribution targeted at the point $\mathbf{T}$, what proportion of nonconforming product can be expected?

Attempts at direct extension of C_p set out to compare a measure of process variation with a measure of the variation allowed by the specifications. A difficulty immediately arises because the specification region is almost always a hyperrectangle. Even if it is not, it is unlikely to be ellipsoidal and even more unlikely to be ellipsoidal with the same matrix Σ^{-1} as the process region. Nonetheless, capability indices have been proposed based on ellipsoidal specification regions. Davis et al. [7] assume $\Sigma = \sigma^2\mathbf{I}$ and define a *spread ratio*, U/σ, for the special case of circular and spherical specification regions. Here, U is the radius of the circle or sphere, and the target is the center point. Thus, they are addressing questions 1 and 2 together. The focus of their article is on nonconforming parts, and they present a table giving the number of nonconforming parts per billion corresponding to any spread ratio between 3.44 and 6.85. Chan et al. [8] define an ellipsoidal specification region with the same matrix as the process region and offer an extension of C_{pm} to address question 2. Taam et al. [9] also extend C_{pm}, using an index that is the ratio of the volume of a modified specification region to the volume of a scaled process region. These last two articles (apparently an earlier version of the second one) are discussed by Kotz and Johnson [2] together with the suggestions of Pearn et al. [10], who introduce two indices based on the ratio of a generalized process length to a generalized length allowed by the specifications.

Tang and Barnett [11] introduce three indices for multiresponse processes. The first involves projecting the process ellipsoid onto its component axes and taking the minimum of the one-dimensional C_p values each scaled by a projection factor and a deviation from target factor. They note that this index does not involve the correlations between elements of $\mathbf{X}_q$. The second index is similar to the first but uses the Bonferroni inequality to determine a process hyperrectangle such that each side is a $100(1-\alpha/q)\%$ centered probability interval for the marginal distribution. A usual choice would be to take $\alpha = 0.0027$. The third index is based on a process region obtained using Sidak's probability inequality but is otherwise of a similar form to the first two. Tang and Barnett [11] show that the third index is the least conservative and is favored over the other two.

Chen [12] defines a general specification region, or tolerance zone, consisting of all values of $\mathbf{X}_q$ for which $h(\mathbf{X}_q - \mathbf{T}) \leq r_0$, where $h\ (\cdot)$ is a positive function with the same scale as $\mathbf{X}_q$ and r_0 is a positive number. The process is capable if

$$P(h(\mathbf{X}_q - \mathbf{T}) \leq r_0) \geq 0.9973$$

so, taking r to be the minimum value for which

$$P(h(\mathbf{X}_q - \mathbf{T}) \leq r) \geq 0.9973$$

a capability index can be defined as r_0/r. The formulation includes ellipsoidal and hyperrectangular specification regions as special cases. Hubele et al. [6] propose a three-component vector index for bivariate response processes. The first component is an extension of C_p, namely the ratio of the area of the specification rectangle to the area of the process rectangle. The second component is the significance level of Hotelling's T^2 statistic testing for a location shift, and the third is an indicator of whether or not the process rectangle falls entirely within the specification rectangle. This last component is necessary because the first component can give a C_p-like value suitably greater than 1 despite one of the quality characteristics being, in itself, not capable.

A completely different approach is taken by Bernardo and Irony [13], who introduce a general multivariate Bayesian capability index. They use a decision-theoretic formulation to derive the index

$$C_b(D) = \frac{1}{3}\Phi^{-1}\{P(\mathbf{X}_q \in A|D)\}$$

where A is the specification region, D represents the data, and Φ is the standard normal distribution function. The distribution of $\mathbf{X}_q$ can be of any type, and exploration of the posterior predictive distribution of C_b given D is limited only by available computing power.

Most of the above indices are not easy to use in practice and present difficult problems in the exploration of their sampling distributions. Two approaches that don't suffer from this are given by Boyles [14] and Veevers [15]. Boyles moves away from capability assessment and promotes capability improvement by using exploratory capability analysis. Further developments in this area are described by Boyles (in the present volume). Veevers' approach is based on the concept of process viability, which is discussed in the next section.

4. PROCESS VIABILITY

Veevers [15, 16] realized the difficulties associated with extensions of C_p and C_{pk} to multiresponse processes and concluded that the reasons lay in the logic underlying the structure of C_p and C_{pk}. This led to the notion of process viability as a better way of thinking about process potential than the logic underlying C_p. He introduced the *viability index* first for a single-response process and then for a multiresponse process.

Basically, viability is an alternative to capability potential, leaving the word "capability" to refer to capability performance. For a single-response process it is easy to envisage a window of opportunity for targeting the process mean. Consider the process distribution, which need not be normal and, conventionally, identify the lower 0.00135 quantile and the upper 0.99865 quantile. Place this distribution on a scale of measurement that has the lower and upper specification limits (LSL and USL, respectively) marked on it, with the lower quantile coincident with the LSL. If the USL is to the right of the upper quantile, slide the distribution along the line until the upper quantile coincides with the USL. The line segment traced out by the mean of the distribution is the window of opportunity for targeting the mean. The interpretation of the window is that if the mean is successfully targeted anywhere in it, then the proportion of nonconforming items will be no greater than 0.27%. A process for which a window of opportunity such as this exists is said to be *viable*; i.e., all that needs to be done is to target the mean in the allowable window. If, however, the USL is to the left of the upper quantile (after the first positioning), then there is clearly more variation in the response than is allowed for by the specifications, and the process is not viable. Sliding the distribution to the left until the upper quantile and the USL coincide causes the mean to trace out a line segment that, this time, can be thought of as a "negative" window of opportunity for targeting the mean. Referring to the length of the window, in both cases, as w, a viable process will have a positive w and a nonviable process a negative w. The viability index is defined as

$$V_r = \frac{w}{\text{USL} - \text{LSL}}$$

If the process is comfortably viable, then w will be a reasonable portion of USL − LSL, but if the process is only just viable, w will be zero and $V_r = 0$. Processes that are not viable will have V_r negative.

If the quality characteristic has a normal distribution with standard deviation σ, it is easy to see that $6\sigma + w = \text{USL} - \text{LSL}$ for both positive and negative w, hence $V_r = 1 - 1/C_p$. Some readers will know that an early capability ratio was $C_r = 1/C_p$ (see, e.g., Amsden et al. [17]), so $V_r = 1 - C_r$. Statistical properties of estimators of V_r are relatively straightforward to establish, as indicated in Veevers [15]. It must be remembered that the viability index is a measure of capability potential and addresses only question 1. The knowledge that a process is viable is valuable even if an unacceptable proportion of nonconforming parts is produced when the process is operating. It means that the process must be targeted better (question 2) to achieve acceptable capability performance, but there is no need, at this stage, to reduce variation. Of course, in a continuous improvement environment, steps would be taken to reduce variation in the longer term, but that is separate from the point being made here.

Extension of V_r to multiresponse processes requires the definition of a multidimensional window of opportunity for targeting the mean. Because the process is viable, the distribution of $\mathbf{X}_q$ can be located almost entirely within the hyperrectangular specification region, A. And since targeting is not at issue, the distribution can be thought of as free to move around. The shape of the distribution will not change with this movement, only its location. In particular, because the correlations are fixed, the orientation of a process ellipsoid for a multivariate normal distribution will remain constant during location shifts. The window of opportunity for targeting the mean of the distribution consists of all points μ for which the proportion of nonconforming items would be less than 0.27%. The boundary of the window can be envisaged as the locus of μ as the distribution is moved around inside A while keeping exactly 0.27% of the probability mass outside A and 99.73% inside A. Figure 2 shows the window of opportunity for a viable bivariate normally distributed process. The window is almost a rectangle, with sides parallel to the specification rectangle, except that its corners are rounded due to simultaneous breaching of the two marginal specifications.

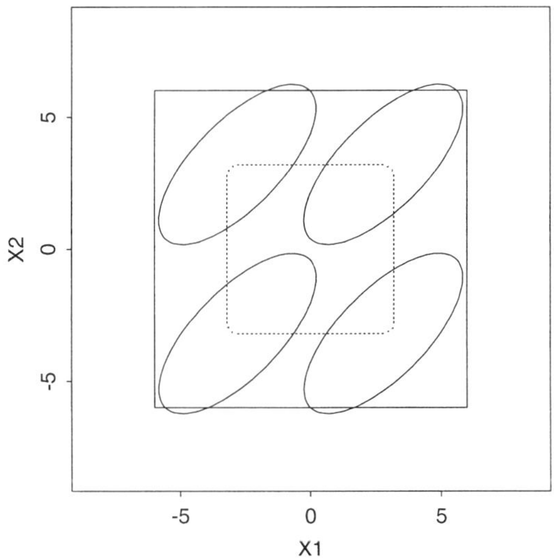

Figure 2 The window of opportunity (dotted rectangle) for targeting the mean for a viable bivariate process. The solid rectangle is the specification region.

The viability index for a q-dimensional multiresponse process is defined as

$$V_{rq} = \frac{\text{volume of } w}{\text{volume of } A}$$

A process is viable only if it is separately viable in all its individual quality characteristics. Otherwise it is not viable, and variation must be reduced, at least in the characteristics that prompted the nonviable decision. In order to produce a practically useful index, Veevers [15] represents the process distribution by a process rectangle that has as its sides the widths of the one-dimensional marginal distributions. The width of a univariate distribution is the difference between the 0.99865 quantile and the 0.00135 quantile (or as appropriate, depending on the amount of probability to be excluded).

The window of opportunity for a viable process can thus be envisaged by sliding this rectangle around just inside the specification region and ignoring the rounding at the corners. For a viable process this leads to the expression

$$V_{rq} = \prod_{i=1}^{q} V_r(X_i)$$

where $V_r(X_i)$ is the viability index for the ith quality characteristic X_i. For nonviable processes, Veevers [15] defines negative windows of opportunity in such a way as to ensure that the viability value obtained for a $(q-1)$ dimensional process is the same as would be obtained from the q-dimensional process by setting the marginal variance of the qth characteristic equal to zero. Hence, V_{rq} is defined in all nonviable cases to be

$$V_{rq} = 1 - \prod_{i=1}^{q} [1 - V_r(X_i)]^{I_i}$$

where

$$I_i = \begin{cases} 0 & \text{if } V_r(X_i) \geq 0 \\ 1 & \text{if } V_r(X_i) < 0 \end{cases}$$

As with any index for multiresponse processes, the viability index is best used in a comparative fashion. In a process improvement campaign the viabilities can be compared after each improvement cycle, thus providing a simple measure of the progress being made. V_{rq} depends only on the marginal viabilities and is therefore independent of the correlation structure of $\mathbf{X}_q$. The correlation coefficients do, however, affect the proportion of nonconforming items that would occur if the process was in production. If an upper bound of 0.27% is required, then a conservative choice of quantiles to use for the calculation of the marginal viabilities is $0.00135/q$ and $1 - 0.00135/q$. The specific choice in an improvement campaign is unimportant, since the emphasis is on changes in V_{rq} rather than the proportion nonconforming.

Having had some experience with multiresponse viability calculations, the following modification to the V_{rq} index is proposed. First, note that a viable process with, say, $q = 6$ and marginal viabilities of 0.25 each (corresponding to C_p values of 1.33) has $V_{rq} = 0.00024$. It is difficult to relate this small number to the reasonable level of viability it represents. Further, it depends on q, and for larger values of q the viability index would be very small. These difficulties can be overcome by defining a modified index

$$V_{rq}^* = V_{rq}^{1/q}$$

for viable processes. This has the benefit of being interpretable on the scale of V_r, independently of q. For nonviable processes, V_{rq} is negative, so V_{rq}^* must be defined as

$$V_{rq}^* = \text{sign}(V_{rq})|V_{rq}|^{1/q}$$

which is also valid for viable processes and provides a general definition of V_{rq}^*. A plot of V_{r2}^* for a two-response process is shown in Figure 3. If desired, V_{rq}^* can be converted to a capability potential index, C_{pq}^*, by $C_{pq}^* = 1/(1 - V_{rq}^*)$.

Viability calculations are illustrated in the following example used by Sparks et al. [5] to demonstrate the dynamic biplot for multivariate process monitoring. A flat rolled rectangular metal plate is supposed to be of uniform thickness (gauge) after its final roll. Measurements are made at four positions on the plate, giving a four-dimensional response for the process. The positions can be conveniently referred to as FL (front left), FR (front right), BL (back left), and BR (back right). The original data are subject to a confidentiality agreement, so they have been transformed before being plotted as pairwise scatter diagrams in Figure 4. Typical specification limits are superimposed, but it must be remembered that this is being done to visualize process dispersion relative to specifications and does not represent actual process performance with respect to targeting. The two-, three-, and four-dimensional specification regions are squares, cubes, and a hypercube, as appropriate.

The individual viabilities for FL, FR, BL, and BR are calculated as 0.147, 0.185, 0.111, and 0.137, respectively. This implies the existence of a positive window of opportunity for targeting the mean and gives $V_{r4} = 0.000415$ and $V_{r4}^* = 0.143$. Using the relationship between viability and cap-

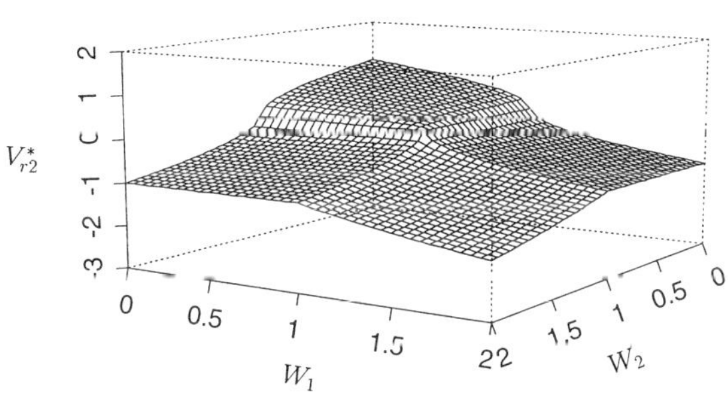

Figure 3 The viability index V_{r2}^* plotted against the widths, W_1 and W_2, of the marginal distributions for a bivariate process with unit specification ranges.

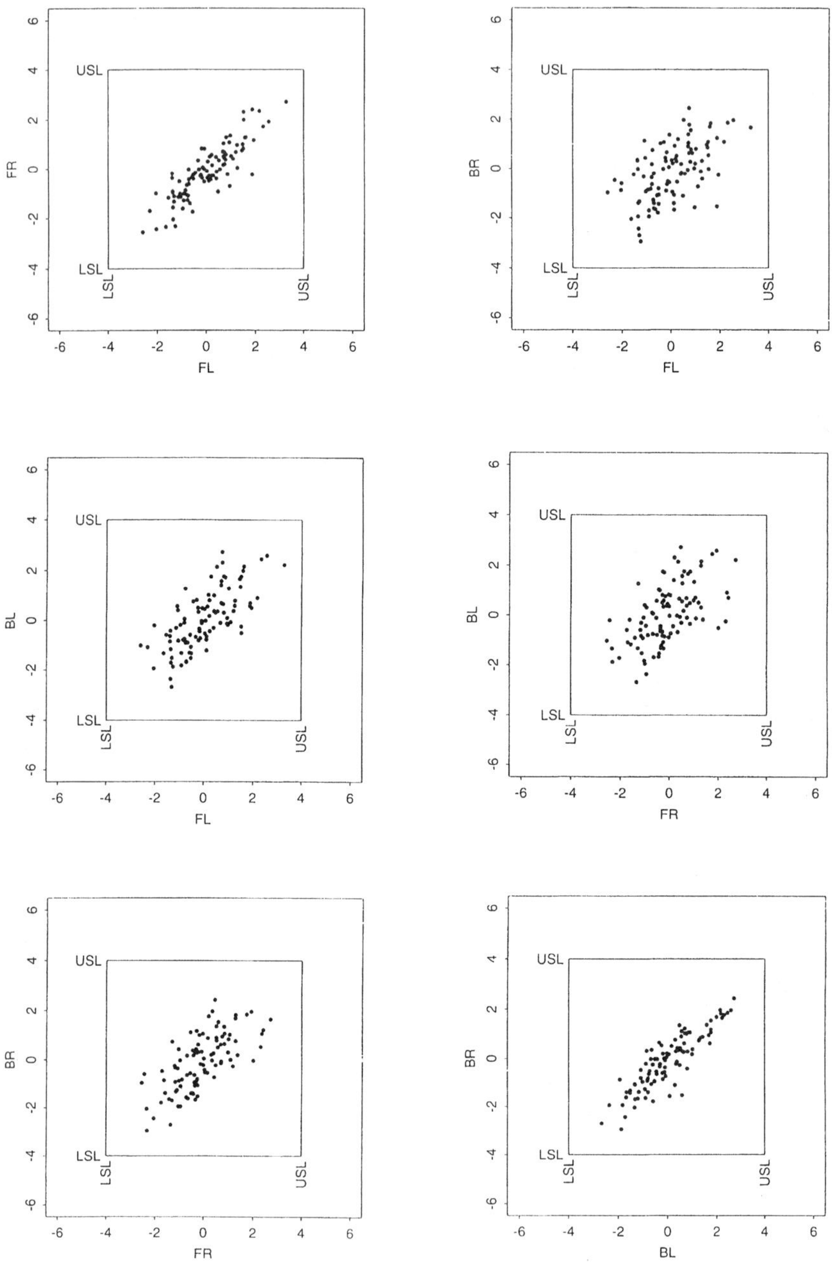

Figure 4 Pairwise scatter diagrams of thickness data at four locations—FL, FR, BL, BR—on 100 metal sheets. Specification rectangles are superimposed.

ability, this corresponds to a capability potential of $C_{p4}^* = 1.167$. Since all these values are intended for use in comparative situations, suppose some process changes gave individual viabilities for FL, FR, BL, and BR of 0.190, 0.225, 0.175, and 0.210, respectively. Then, $V_{r4} = 0.00157$ and $V_{r4}^* = 0.199$, indicating the improvement in viability. Experience in the use of viability indices is necessary in order to get a feel for the extent of the improvement. Converting to a capability potential value gives $C_{p4}^* = 1.248$. Practitioners used to working with C_p may feel more comfortable on this scale of measurement in the first instance.

5. PRINCIPAL COMPONENT CAPABILITY

Although specification regions are generally hyperrectangular, support for differently shaped regions determined by loss functions is growing. Consider a situation where the marginal specifications have ranges $2d_i$, $i = 1, 2, \ldots, q$. By transforming $\mathbf{X}_q$ to $\mathbf{Y}_q$, where the elements of $\mathbf{Y}_q$ are $Y_i = X_i/d_i$, the specification region becomes a hypercube of side 2. If, on this scale of measurement, the loss associated with an item is proportional to the distance between $\mathbf{Y}_q$ and the center of the region, then a hyperspherical tolerance region would be appropriate. The word "tolerance" is used here to distinguish the region from the specification region, which remains a hypercube.

For the purpose of developing a capability index there are several choices of centered hyperspheres that approximate the specification region. For example, there is the unit-radius inscribing hypersphere, the $\sqrt{2}$-radius outscribing hypersphere, and the hypersphere with the same volume as the specification hypercube.

If $\mathbf{Y}_q$ is adequately modeled by a multivariate normal distribution with variance–covariance matrix Σ_y, then the question of capability potential revolves around whether or not the process ellipsoid will fit inside the hypersphere. This is governed only by the "length" of the principal axis of the ellipsoid. A suitable length can be obtained by taking a multiple of the standard deviation of the first principal component, Z_1, of Σ_q, since this is along the principal axis of the ellipsoid. Denoting by λ_1 the eigenvalue associated with Z_1, it follows that the standard deviation of Z_1 is $\sqrt{\lambda_1}$. Hence, taking $6\sqrt{\lambda_1}$ to be the length of the principal axis of the ellipsoid, a capability potential index can be constructed by comparing this length with the diameter of the tolerance region. Using the unit-radius hypersphere gives

$$C'_{pc} = \frac{1}{3\sqrt{\lambda_1}}$$

and using the $\sqrt{2}$-radius hypersphere gives

$$C''_{pc} = \frac{\sqrt{2}}{3\sqrt{\lambda_1}}$$

each of which could be used in its own right as a capability index. However, it seems a sensible compromise to take the average of these two as a measure of capability potential. Hence, a principal component capability index is defined as

$$C_{pc} = \frac{1+\sqrt{2}}{6\sqrt{\lambda_1}}$$

More generally, C_{pc} could be defined as $k/\sqrt{\lambda_1}$, where k is a constant to be determined from considerations of the maximum acceptable proportion of the centered process distribution allowed to be outside the specification region. Since C_{pc} is meant to be an index of capability potential that is intended for use as a comparative measure, fine-tuning of k is unimportant and will not be further considered here.

The sampling distribution of the natural estimator of C_{pc} can be studied using the sampling distribution of the eigenvalue associated with the first principal component of the estimated variance–covariance matrix $\hat{\Sigma}_y$.

The following example shows the spirit in which C_{pc} may be used. The plastic bracket and metal fitting attached to a car's internal sun visor are manufactured to specifications relating to the torque involved in the swivel action. Four torque quality characteristics, $\mathbf{X}_4$, are measured which, in disguised units, have nominal values 2, 2.25, 2, 2.25 and specifications (1, 3), (1, 3.5), (1.3), (1, 3.5), respectively. Data on 30 items from a batch gave

$$\hat{\Sigma}_y = \begin{bmatrix} 0.0390 & 0.0306 & -0.0008 & -0.0004 \\ 0.0306 & 0.0423 & -0.0032 & -0.0018 \\ -0.0008 & -0.0032 & 0.0589 & 0.0519 \\ -0.0004 & -0.0018 & 0.0519 & 0.0579 \end{bmatrix}$$

with entries rounded to four decimal places. From this, $\hat{\lambda}_1 = 0.1105$, giving $\hat{C}_{pc} = 1.21$. As an absolute value this should be interpreted with caution, but for process improvement purposes it is useful as a comparative value. A sample of 25 items from a batch produced under slightly different conditions gave $\hat{\lambda}_1 = 0.086$ and $\hat{C}_{pc} = 1.37$, showing a marked improvement. The manufacturer's aim is to keep the process at these conditions, which show it to be potentially capable, and then concentrate on targeting at the nominal values to ensure a capable performance.

6. CONCLUSION

Capability indices for multiresponse processes have been discussed. It has been stressed that capability potential indices are useful in their own right and should not be confused or unfairly compared with capability performance indices. Most of the literature on indices for multiresponse processes concerns extensions to C_p, C_{pk}, and C_{pm}. The viability index, V_r, however, offers an alternative way of thinking about capability potential and extends naturally to multiresponse processes. A modification to the multiresponse viability index is proposed that makes it easier to interpret in practice. Calculations are illustrated on real data from a rolling mill. A new principal component capability index is presented that is based on a loss function proportional to the distance from the process mean to the target point. Another real example from the motor parts industry is used to illustrate the use of this index. In all cases it is emphasized that capability indices for multiresponse processes are best used in comparative fashion and should be treated with caution as individual values.

REFERENCES

1. Kane VE. Process capability indices. J Qual Technol 18:41–52, 1986.
2. Kotz S, Johnson NL. Process Capability Indices. London: Chapman and Hall, 1993.
3. Chan LK, Cheng SW, Spiring FA. A new measure of process capability: C_{pm}. J Qual Technol 20:162–175, 1988.
4. Spiring FA. A unifying approach to process capability indices. J Qual Technol 29:49–58, 1997.
5. Sparks RS, Adolphson AF, Phatak A. Multivariate process monitoring using the dynamic biplot. Int Stat Rev 65:325–349, 1997.
6. Hubele NF, Shahriari H, Cheng C-S. A bivariate process capability vector. In: JB Keats, DC Montgomery, eds. Statistical Process Control in Manufacturing. New York: Marcel Dekker, 1991, pp 299–310.
7. Davis RD, Kaminsky FC, Saboo S. Process capability analysis for processes with either a circular or a spherical tolerance zone. Qual Eng 5:41–51, 1992.
8. Chan LK, Cheng SW, Spiring FA. A multivariate measure of process capability. J Modeling Simulation 11:1–6, 1991.

9. Taam W, Subbaiah P, Liddy JW. A note on multivariate capability indices. J Appl Stat 20:339–351, 1993.
10. Pearn WL, Kotz S, Johnson NL. Distributional and inferential properties of process capability indices. J Qual Technol 24:216–231, 1992.
11. Tang PF, Barnett NS. Capability indices for multivariate processes. Technical Report 49EQRM14, Victorian University of Technology, Melbourne, Australia, 1994.
12. Chen H. A multivariate process capability index over a rectangular solid tolerance zone. Stat Sin 4:749–758, 1994.
13. Bernardo JM, Irony TZ. A general multivariate Bayesian process capability index. Statistician 45:487–502, 1996.
14. Boyles RA. Exploratory capability analysis. J Qual Technol 28:91–98, 1996.
15. Veevers A. Viability and capability indexes for multi-response processes. J Appl Stat 25:545–558, 1998.
16. Veevers A. A capability index for multiple responses of a process. Technical Report DMS–D95/1, CSIRO Division of Mathematics and Statistics, Sydney, Australia, 1995.
17. Amsden RT, Butler HE, Amsden DM. SPC Simplified: Practical Steps to Quality. New York: Quality Resources, 1986.

15

Pattern Recognition and Its Applications in Industry

R. Gnanadesikan
Rutgers University, New Brunswick, New Jersey

J. R. Kettenring
Telcordia Technologies, Morristown, New Jersey

1. INTRODUCTION

In a very general sense, pattern recognition is often considered to be the essence of intelligence. For example, an often heard argument for the ability of human chess masters to beat state-of-the-art computer programs is that whereas the latter may be fast in enumerating a large number of moves and consequences, the masters tend to rely on some innate "pattern recognition" abilities based on extensive experience. In a more limited sense, pattern recognition arises in many guises in industrial settings, e.g., robotics in manufacturing, detection of errors in massive software systems, and widely used image analysis applications in medicine and in such things as airport luggage scanners.

For purposes of this chapter, the phrase "pattern recognition" is used to indicate an even more specific statistical methodological area, that of classification and clustering. The term "classification" is used for situations wherein so-called training samples that can be labeled by their origin (the case of "known" groups) are available and one is interested in using these as the bases for classifying so-called test samples. Other terminology for this class of pattern recognition methods include discriminant analysis and supervised learning. In the clustering scenario, on the other hand, all one has are the data at hand, with no labels to identify sources (the case of "unknown" groups), and the analysis leads to finding groupings of the observations that are more similar within groups than across them. This setting is also known as unsupervised learning. There are, of course, many real-world situations that fall between the two scenarios, and often one needs a combination of the two approaches to find useful solutions to the problem at hand. For instance, while the early development of so-called neural networks, which basically are automatic classifiers implemented in either software or hardware, focused on supervised learning methods, the current uses of these encompass both supervised and unsupervised learning algorithms.

This chapter has three objectives. First, taking a broad view of business and industry, it seeks to identify a variety of aspects of such enterprises, as well as examples of specific problems arising in such facets, wherein classification and clustering techniques are used to find appropriate solutions. Second, using the theme of quality and productivity as a focus, it describes a sample of applications (drawn from both the literature and our experience) in which this theme is a clear objective of using such techniques. Third, it is aimed at discussing some methodological issues that cut across applications and need to be addressed by practitioners to ensure effective use of the methods as well as by researchers to improve the options available to practitioners.

More specifically, Section 2 identifies areas of business and industry, as well as some specific examples of problems in such areas, where classification and clustering techniques have been used. It also describes in a bit more detail a subset of the examples where assessment and improvement of quality, efficiency, and/or productivity are explicitly involved as a goal of the analysis. Section 3 discusses some general methodological issues that need to be considered. Section 4 consists of concluding remarks.

2. ASPECTS AND EXAMPLES OF BUSINESS AND INDUSTRIAL PROBLEMS AMENABLE TO PATTERN RECOGNITION

Perhaps the better known industrial applications of pattern recognition, including some that were mentioned in the introduction, are in manufacturing. However, one can identify a number of facets that are integral parts of business and industry as a whole and give rise to problems that are amenable to the meaningful use of pattern recognition methods. Table 1 contains a partial list of different facets of a business enterprise and some specific examples of applications of classification and clustering methods in each category. A subset of the examples (identified by asterisks) in Table 1,

Table 1 Applications of Classification and Clustering Methods Within a Business Enterprise

Finance

Use of discriminant analysis for effective development of credit ratings of individuals and firms, including bond ratings [See, e.g., Chapters IV and V of Altman et al. (1981).]

*Use of discriminant analysis and clustering for developing "comparable risk" groups of companies for the purpose of determining appropriate "rates of return" (Chen et al., 1973, 1974; Cohen et al., 1977)

Marketing

Use of cluster analysis for market segmentation on the basis of geodemographic similarity [See, e.g., Chapter 12 of Curry (1993)] and the recent development of database marketing

*Use of cluster analysis for identifying "lead users" and for product development in light of the needs of such lead users (Urban and Von Hippel, 1988)

Resource allocation

Utilization of robotics (entailing the recognition of "shapes" and "sizes" of objects to be assembled into a product) in assembly line manufacturing with gains in quality and productivity arising from decreased variability and speed as well as lower costs in the long run [See, e.g., Dagli et al. (1991).]

Niche applications of neural networks for such things as speech and writing recognition (e.g., voice-activated dialing of telephones; automatic verification of payments of bills paid by customers via checks)

Use of cluster analysis for grouping similar jobs prior to the development of regression models for aiding assessment and improvement of utilization of computing resources (Benjamin and Igbaria, 1991)

*Use of cluster analysis in the development of a curriculum that better meets job needs and is likely to enhance worker productivity (Kettenring et al., 1976)

Software engineering

Use of fuzzy clustering to improve the efficiency of a database querying system (Kamel et al., 1990)

Use of discriminant analysis for predicting which software modules are error-prone (Conte et al., 1986)

*Use of neural networks for "clone" recognition in large software systems (Carter et al., 1993; Barson et al., 1995)

Strategic planning

*Use of cluster analysis for identifying efficient system-level technologies (Mathieu, 1992; Mathieu and Gibson, 1993)

wherein assessment and improvement of quality, efficiency, or productivity was an explicit goal, is now described in a bit more detail.

2.1. Finance

As noted in Table 1, classification and clustering are used to establish categories of comparable risk so as to determine appropriate rates of return.

Historically, and particularly during the 1970s, one role of governmental regulatory bodies in the United States was to set allowed rates of return on equity for the companies they regulated. The regulated companies argued that in order to attract investors they needed higher rates of return, while the regulators felt pressured to keep them low. An accepted tenet for resolving the two conflicting aims was that the rate of return should be commensurate with the "risk" associated with the firm. For implementing this principle, one formal approach employs the capital assets pricing model espoused by Lintner (1965), Markowitz (1959), and Sharpe (1964). Chen et al. (1973) took a different and more empirical approach by using data concerning several variables that are acknowledged to be risk-related (e.g., debt ratio, price/earnings, ratio, stock price variability) and finding companies with similar risk characteristics that could then be compared in terms of their rates of return. Standard & Poor's COMPUSTAT database pertaining to over 100 utilities and over 500 industrials was the source, and a particular interest of the analysis was to compare AT&T's rate of return within the group of firms that shared its risk characteristics.

At an initial, general level of analysis, Chen et al. (1973) addressed the question of AT&T's classification as belonging to either the utility group or the industrial group through the use of discriminant analysis. They found strong evidence that AT&T belonged with the industrials. To provide a different look, one could use cluster analysis to find groups of firms with similar risk features and further investigate the particular cluster to which AT&T belongs. Since the primary interest of the authors was in the latter, and also partly because the number of firms was large, an attempt was made to find a "local" cluster near AT&T in terms of the risk measures rather than clustering all the firms [see Cohen et al. (1977) for details of the algorithm involved]. This analysis led to detecting a cluster of 100 industrial firms with risk comparable to AT&T's. In terms of the performance measure of rate of return, AT&T's value was found to lie below the median of the rates of return of this cluster, thus providing a quantitative basis for arguing a higher rate of return.

2.2. Marketing

In market research, classification and clustering can serve as aids in product development in light of the needs of lead users.

Urban and Von Hippel (1988) describe an innovative approach to product development in situations where the technology may be changing very rapidly. Efficiency in developing a product with an eye to capturing a significant share of the market is the desired goal. The efficiency arises from studying a carefully chosen subset of the potential market and yet ending up having a product that is likely to satisfy the needs of and be adopted by a much larger group of customers. The main steps of the approach proposed by Urban and Von Hippel are to use cluster analysis for identifying a set of "lead" users of the product, then seek information from such users about what features and capabilities they would like the product to have, and finally apply this information not only to develop the product but also to test its appeal and utility for a wider group of users. The specific product used to illustrate the approach is software for computer-aided design of printed circuit boards (PC-CAD). Careful choice of variables that are likely to indicate "lead" users is a key part of and reason for the success of the initial cluster analysis. Variables used included measures of in-house building of PC-CAD systems, willingness to adopt systems at early stages of development, and degree of satisfaction with commercially available systems. A total of 136 firms were clustered on the basis of such variables. Both two- and three-cluster solutions were studied, and the former was chosen as satisfactory with one of the two clusters being predominantly "lead" users. Treating the two clusters as if they were prespecified—i.e., the discriminant analysis framework, for instance—the authors report that the fraction correctly classified in the two clusters was almost 96%. More interesting, when information gathered from the lead users was used to design a new PC-CAD system and this new design was presented to the participants in the study, about 92% of the lead user group and 80% of the non-lead group rated it as their first choice! Urban and Von Hippel (1988) also discuss the advantages and disadvantages of their lead-user methodology in general contexts.

2.3. Resource Allocation

Kettenring et al. (1976) describe the role of cluster analysis to assess the current validity of course objectives in a multifaceted industrial training curriculum for workers with evolving training needs. The approach involves three major components: (1) careful preparation of an inventory of the p current elements of the job, (2) collection of data about the nature of their jobs and training needs from a sample of n workers engaged in the job at which the training is directed, and (3) cluster analysis of the resulting ($p \times n$) matrices in various ways. In one analysis, the $p = 169$ rows of a matrix indicating elements performed on the job by the sample of $n = 452$ workers yielded insights into clusters of elements of the job that fit together and might potentially be taught together as a module. These helped identify gaps in the existing curriculum where new resources were needed. In another analysis, the $n = 452$ workers were clustered into groups with common training needs. The range of needs across the clusters suggested that a training program with flexible options would be an efficient way to train the workers.

2.4. Software Engineering

Carter et al. (1993) (see also Barson et al., 1995) tackle the problem of clone detection in large telecommunications software systems. A clone is a unit of software source code that is very similar to some other unit of code in the same system. In large systems with a long history, it may happen that there are several clones of the same piece of software. These can unnecessarily inflate the size of the overall system and make it less efficient to maintain. For example, should there be a fault in one of the clones, it would probably be present and need to be corrected in the others as well.

The two papers mentioned above discuss different neural network approaches to software clone detection. In Carter et al. (1993), an unsupervised neural net is used to form clusters of software units based on a set of features or variables. The variables characterize different aspects of a unit of source code such as its physical layout. New units of code can be compared against existing clusters to see if they fall within one of these clusters. The overall approach is attractive, even though it does not yet appear to have been widely applied.

2.5. Strategic Planning

Mathieu (1992) (see also Mathieu and Gibson, 1993) discusses an interesting use of cluster analysis for prioritizing critical technologies in national policy making and guiding the choice of an efficient system-level technology. One of the prime difficulties in such situations is the interdependencies among the technologies. This work claims to be the first in the literature to provide a systematic quantitative method for explicitly identifying "high performance" technologies for aiding national policy making. As stated by Mathieu, "the purpose of using cluster analysis in technology planning is to determine natural groupings of system level technologies based upon the scientific interdependencies that link these technologies." The particular application discussed in this work concerned satellite technologies and policy making related to these in Washington, DC. Thirty (=n) system-level technologies were considered for the clustering, and 72 (=p) binary variables that measure the presence or absence of 72 element-level support technologies in each of the system-level technologies were used for the cluster analysis. The analysis led to six clusters of the system-level technologies, with the smallest of the clusters containing only two technologies and the largest group containing seven.

For aiding the identification of high perormance technologies, two variables extraneous to the cluster analysis were introduced, market share and sales growth rate, and average values of these for all U.S. companies for the system-level technologies grouped in each cluster were computed. Mathieu (1992) used an interesting graphical scheme (see Fig. 1) for a two-dimensional display of these averages for the six clusters. The six clusters are represented by circles and labeled with the names given to them by Mathieu. The circles are centered at the average values with diameter proportional to the total U.S. market size for each technology group and thickness proportional to a measure of cluster "tightness." The display thus contains information on four characteristics. Relatively large and thick circles located toward the top left corner of the display would indicate the

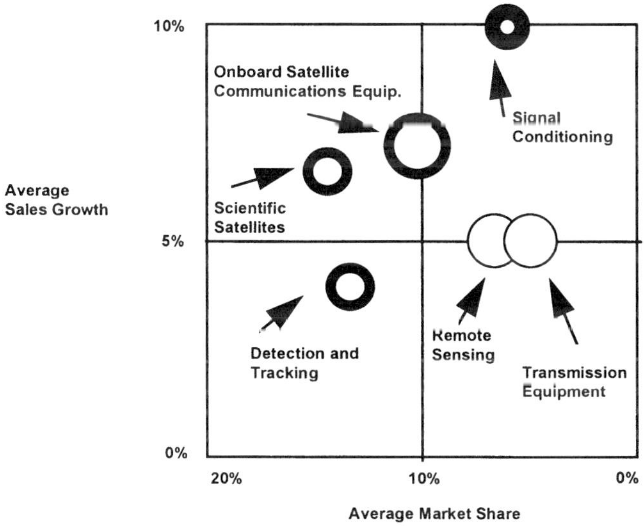

Figure 1 Mathieu's six clusters of system-level technologies. (Copyright 1991 IEEE.)

system-level technologies that were preferred. From the configuration shown here, Mathieu concluded that while no single technology group is uniformly dominant with respect to all four characteristics, the two labeled "onboard satellite communications equipment" and "scientific satellites" appear to be favorable choices, while the two labeled "remote sensing" and "transmission equipment" are clearly ruled out in terms of the desire to choose high performance technologies.

3. SOME STATISTICAL METHODOLOGICAL ISSUES

The discussion in the previous section was designed to leave the impression that methods of pattern recognition are used in many facets of business and are having considerable impact on matters of quality and productivity. Indeed, if one takes a reasonably holistic view of quality management, it is not a stretch to conclude that these methods are a potent part of the arsenal of tools for quality improvement.

At the same time, practitioners of these methods need to be aware of the care that is necessary for their successful use. The applications literature, unfortunately, is not reassuring in this regard; subtle details are seldom discussed, and canned programs appear to be heavily, even totally, relied upon.

The difficulties start at the earliest part of the analysis when a commitment is made to what data and which variables to use. The temptation is to include every variable of possible value to avoid missing out on an important one. The price one pays for this ranges from a needlessly watered down analysis to full-blown distortion of the results. In cluster analysis, the risk is particularly severe: Clear-cut clusters confined to a subspace of the variables can be completely overlooked.

The traditional methods of discriminant analysis have the nice mathematical property of being invariant under nonsingular linear transformation of the data. However, in most cluster analysis procedures, this is not the case. There is explicit or implicit commitment to a metric that at one extreme may be invariant but otherwise without rationale (as when one uses the total covariance matrix of the entire data set to form a weighting matrix for the metric) and at the other may involve no reweighting of the variables and therefore no such invariance (as in the case of Euclidean distance). An intermediate, and far too popular, example is autoscaling or weighting to equalize the total sample variances of all the variables. This works against detecting clusters by all methods that take autoscaled data, or distances derived from them, as input. Rather than putting the variables on an equal footing, according to their within-cluster variation (which is what one would prefer to do), it places variables with cluster structure on the same overall footing as those without such structure and thereby makes it more difficult to find the clusters via standard algorithms. See Gnanadesikan et al. (1995) for further discussion and mitigating alternatives.

Another worry is which method or algorithm to choose for all analyses. Neural networks? Classical discriminant analysis, or classification trees? A hierarchical method, or a partitioning method of cluster analysis? There are many choices. Users need to be sensitive to the pros and cons of them and to resist having the analysis driven by the content of the nearest software package. A very appealing strategy in pattern recognition work is, in fact, to apply a thoughtful variety of methods to the data. The hope is that major well-formed patterns will emerge from different looks at the data, and others that are less pronounced but still potentially noteworthy will reveal themselves in at least one of the alternative calculations.

The findings can also be made more credible by subjecting them to a variety of sensitivity analyses for a particular method. For example, controlled jiggling of the data or systematic deletion of variables and/or observations followed by reapplication of the method can help one to appreciate just how stable or fragile the results are (see Gnanadesikan et al., 1977; Cohen et al., 1977).

As the number of variables, p, or observations, n, grows—and this is clearly the trend in many industrial applications—a much more daunting challenge arises. Many of the standard pattern recognition methods become impractical or literally break down. The irony of this is that with massive sets of data one needs just such pattern recognition approaches to bring the data under control by dividing them into manageable chunks.

To illustrate the point, consider what is probably the most popular and widely available form of clustering, hierarchical cluster analysis. This method operates on $n(n - 1)/2$ interpoint distances to produce hierarchical trees with n leaves at the top and one trunk at the bottom. The distances present data management challenges when n is large and the trees, which ought to be studied, become so big that they cannot be readily drawn or digested.

Other popular algorithms, such as k-means, may be more suitable as n increases, but they are not a panacea. Brand new approaches are really needed. For example, "localizing" the analysis so that one is looking for patterns of a particular type in a particular region of space may be one effective way to reduce the problem to a reasonable size. See Section 2.1 for an example.

When p is too large, other complexities arise. As indicated already, masking of patterns is a serious limitation, and available methods for variable selection and dimensionality reduction, whether graphical or numerical, are unlikely to work well.

To make matters worse, even current practice for reducing the number of variables when p is only moderately large is open to criticism. Again in the context of cluster analysis, a widely advocated and practiced technique is to reduce dimensionality via principal components analysis. Although this can work well in some situations, the logic of this approach is suspect, and it is easy to give examples of when it fails.

The relative size of n and p can matter a lot for some types of pattern recognition problems. If n is small relative to p, the already suspect reduction of variables via principal components will also suffer from numerical instability problems. When both are very large, entirely new approaches to pattern recognition may be the answer. For example, one can envisage extensive distributed computations of massive data sets. Local exploration may be handled by burrowing deeply into the local detail. The global solution would be obtained by ultimately stitching the local solutions together.

In summary, there is much to worry about in terms of methodological issues if one is to take advantage of pattern recognition techniques in complex industrial problems. A "black box" or "canned program" approach will not cut it and can easily do more harm than good.

4. CONCLUDING REMARKS

Pattern recognition methods are natural ones for helping to improve quality and productivity in industrial settings. Applications are prevalent, and several rather different ones were given to illustrate this point. Nevertheless, careful attention to detail is needed to ensure that the methods, which are far from infallible, are effectively applied. When they are, they can be powerful tools in the search for total quality management.

REFERENCES

Altman EI, Avery RB, Eisenbeis RA, Sinkey JF Jr. (1981). Applications of Classification Techniques in Business, Banking and Finance. Greenwich, CT: JAI Press.

Barson P, Davey N, Frank R, Tansley, DSW. (1995). Dynamic competitive learning applied to the clone detection program. Proceedings International Workshop on Applications of Neural Networks to Telecommunications, pp 234–241.

Benjamini Y, Igbaria M. (1991). Clustering categories for better prediction of computer resources utilizations. Appl Stat 40:295–307.

Carter S, Frank RJ, Tansley DSW. (1993). Clone detection in telecommunications software systems. Proceedings International Workshop on Applications of Neural Networks to Telecommunications, pp 273–280.

Chen HJ, Gnanadesikan R, Kettenring JR. (1973). A Statistical Study of Groupings of Corporations. Bell Labs Technical Memorandum.

Chen H, Gnanadesikan R, Kettenring JR. (1974). Statistical methods for grouping corporations. Sankhya B 36:1–28.

Cohen, A, Gnanadesikan R, Kettenring JR, Landwehr JM. (1977). Methodological developments in some applications of clustering. In: Krishnaiah PR, ed. Applications of Statistics. New York: North-Holland, pp 141–162.

Conte SO, Dunsmore HE, Shen VY. (1986). Software Engineering Metrics and Models. Menlo Park, CA: Benjamin/Cummings.

Curry DJ. (1993). The New Marketing Research Systems. New York: Wiley.

Dagii CH, Kumara, SRT, Shin YC. (1991). Intelligent Engineering Systems Through Artificial Neural Networks, New York: ASME Press.

Gnanadesikan R, Kettenring JR, Landwehr JM. (1977). Interpreting and assessing the results of cluster analyses. Bull Int Stat Inst 47:451–463.

Gnanadesikan R, Kettenring, JR, Tsao SL. (1995). Weighting and selection of variables for cluster analysis. J Classif 12:113–136.

Kamel M, Hadfield B, Ismail M. (1990). Fuzzy query processing using clustering techniques. Info Process Manage 26:279–293.

Kettenring JR, Rogers WH, Smith ME, Warner JL. (1976). Cluster analysis applied to the validation of course objectives. J Educ Stat 1:39–57.

Lintner J. (1965). The valuation of risky assets and the selection of risky investments in stock portfolios and capital budgets. Rev Econ Stat 47:13–37.

Markowitz H. (1959). Portfolio Selection: Efficient Diversification of Investments. New York: Wiley.

Mathieu RG. (1992). A method based on cluster analysis for national and regional technology policy development. Proceedings of the 1991 Portland International Conference on Management of Engineering and Technology, PICMET'91. Piscataway, NJ: IEEE, pp 685–688.

Mathieu RG, Gibson JE. (1993). A methodology for large-scale R&D planning based on cluster analysis, IEEE Trans Eng Manage 40:283–292.

Sharpe W. (1964). Capital assets prices: A theory of market equilibrium under conditions of risk. J Finance 19:425–442.

Urban GL, Von Hippel E. (1988). Lead user analysis for the development of new industrial products. Manage Sci 34:569–582.

16
Assessing Process Capability with Indices

Fred A. Spiring
The University of Manitoba, Winnipeg, and Pollard Banknote Limited, Manitoba, Canada

1. GENESIS

The automotive industry has been a leading promoter of process capability indices as tools for quality improvement. It is no longer alone, as process capability indices are now embraced by a wide variety of industries interested in assessing the ability of a process to meet customers' requirements. The popularity of these indices is generally attributed to their ability to provide a single-number summary that relates process performance to process requirements. Practitioners use the single-number summary in many ways including (1) awarding supplier audit points based on the magnitude of the summary value, (2) documented evidence of process performance relative to customers' requirements, and (3) in identifying processes in need of improvement.

The use of single-number summaries to assess the overall performance of a process has been criticized; however, when used in conjunction with other quality tools, the information provided by these summaries can be invaluable. Under the assumption that meeting or exceeding customer requirements is the focus of most quality programs and considering process capability indices to be the quantification of the process's ability to meet customer requirements, the increasing use of process capability measures seems only natural. Unfortunately, users of process capability indices have developed several "bad habits," in part due to a lack of practical, statistically sound techniques.

2. PROCESS CAPABILITY INDICES

Process capability indices are used to assess a process's ability to meet a set of requirements. When used correctly these indices provide a measure of process performance that in turn can be used in the ongoing assessment of process improvement. Indices allow statistically based inferences to be used in the assessment of process capability as well as in the identification of changes in the ability of the process to meet requirements.

It is generally acknowledged that Japanese companies initiated the use of process capability indices when they began relating process variation to customer requirements in the form of a ratio. The ratio, now referred to as the *process capability index*, is defined to be

$$C_p = \frac{\text{USL} - \text{LSL}}{6\sigma}$$

where the difference between the upper specification limit (USL) and the lower specification limit (LSL) provides a measure of allowable process spread (i.e., customer requirements) and 6σ, σ^2 being the process variance, a measure of the actual process spread (see Fig. 1).

C_p uses only the customer's USL and LSL in its assessment of process capability and fails to consider a target value. The five processes depicted by

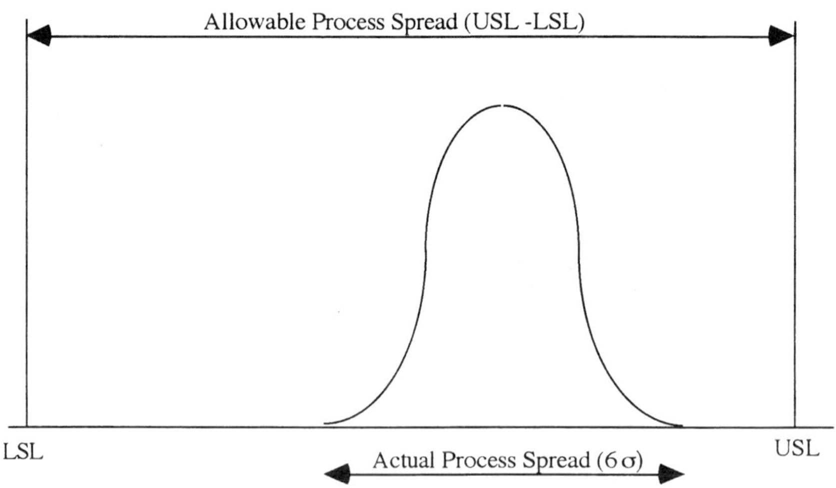

Figure 1 Allowable process spread versus actual process spread.

the numbered normal curves in Figure 2 have identical values of σ^2 and hence identical values of C_p. However, because the means of processes 2, 3, 4, and 5 all deviate from the target (T), these processes would be considered less capable of meeting customer requirements than process 1.

Processes with poor proximity to the target have sparked the derivation of several indices that attempt to incorporate a target into their assessment of process capability. The most common process capability indices assume T to be the midpoint of the specification limits and include

$$C_{pm} = \frac{\mathrm{USL} - \mathrm{LSL}}{6[\sigma^2 + (\mu - T)^2]^{1/2}}$$

$$C_{pu} = \frac{\mathrm{USL} - \mu}{3\sigma}$$

$$C_{pl} = \frac{\mu - \mathrm{LSL}}{3\sigma}$$

$$C_{pk} = \min(C_{pl}, C_{pu})$$

and

$$C^*_{\mathrm{pk}} = (1 - k)C_p$$

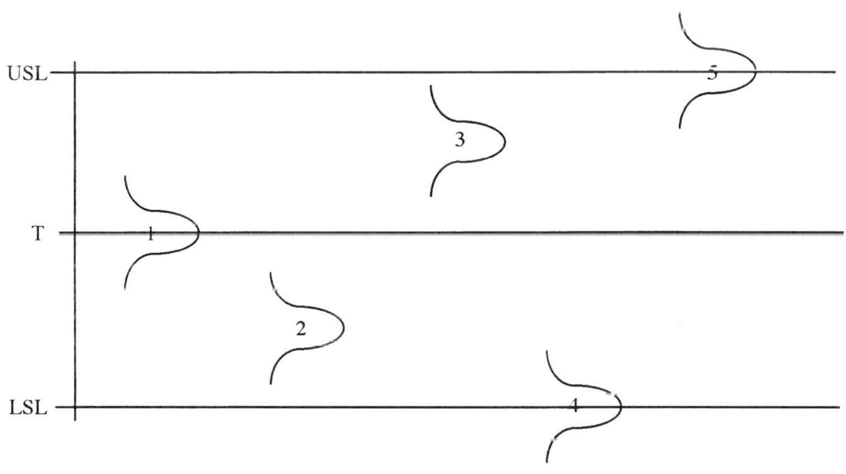

Figure 2 Five processes with identical values of C_p.

where $k = 2|T - \mu|/(\text{USL} - \text{LSL})$ and μ represents the process mean such that LSL $< \mu <$ USL. The two definitions C_{pk} and C^*_{pk} are numerically equivalent when $0 \leq k \leq 1$.

Individually, C_{pu} and C_{pl} consider only unilateral tolerances (i.e., USL or LSL, respectively) when assessing process capability. Both use 3σ as a measure of actual process spread, while the distance from where the process is centered (μ) to the USL (for C_{pu}) or to the LSL (for C_{pl}) is used as a measure of allowable process spread. Both C_{pu} and C_{pl} compare the length of one tail of the normal distribution (3σ) with the distance between the process mean and the respective specification limit (see Fig. 3). In the case of bilateral tolerances, C_{pu} and C_{pl} have an inverse relationship and individually do not provide a complete assessment of process capability. However, conservatively taking the minimum of C_{pu} and C_{pl} results in the bilateral tolerance measure defined as C_{pk}.

Similar to C_p, C_{pm} uses USL − LSL as a measure of allowable process spread but replaces the process variance in the definition of C_p with the process mean square error around the target. For all processes, C_p and C_{pm} are identical when the process is centered at the target [i.e., $\mu = T = (\text{USL} + \text{LSL})/2$]; however, as the process mean drifts from T, C_{pm} becomes smaller while C_p remains unchanged.

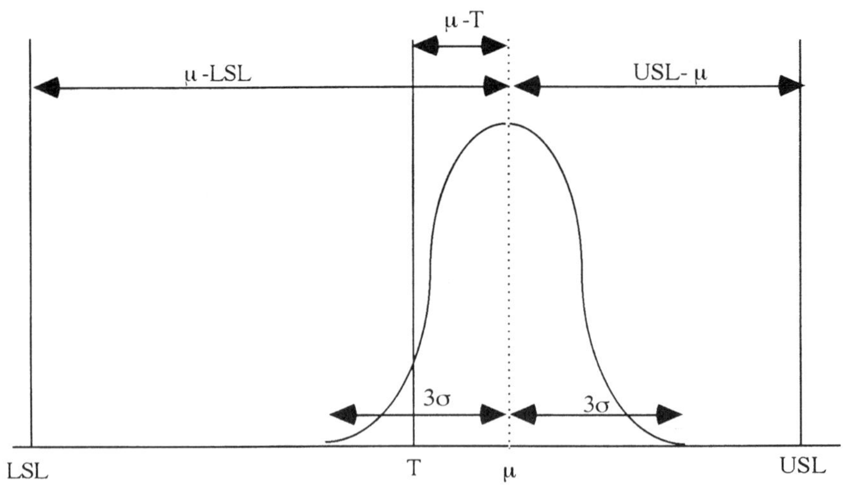

Figure 3 Target is the midpoint of the specification limits.

The generalized analogs of these measures do not assume T to be the midpoint of the specifications (see Fig. 4) and are of the form

$$C_{pm} = \frac{\min[\mathrm{USL} - T, T - \mathrm{LSL}]}{3[\sigma^2 + (\mu - T)^2]^{1/2}}$$

$$C_{pu} = \frac{\mathrm{USL} - T}{3\sigma}\left(1 - \frac{|T - \mu|}{\mathrm{USL} - T}\right)$$

$$C_{pl} = \frac{T - \mathrm{LSL}}{3\sigma}\left(1 - \frac{|T - \mu|}{T - \mathrm{LSL}}\right)$$

and

$$C_{pk} = \min(C_{pl}, C_{pu})$$

Note that the original definitions of C_{pm}, C_{pl}, C_{pu} and C_{pk} are special cases of the generalized analogs with $T = (\mathrm{USL} + \mathrm{LSL})/2$.

The process capability indices C_p, C_{pl}, C_{pu}, C_{pk}, and C_{pm} and their generalized analogs belong to the family of indices that relate customer requirements to process performance as a ratio. As process performance improves, through either reductions in variation and/or moving closer to

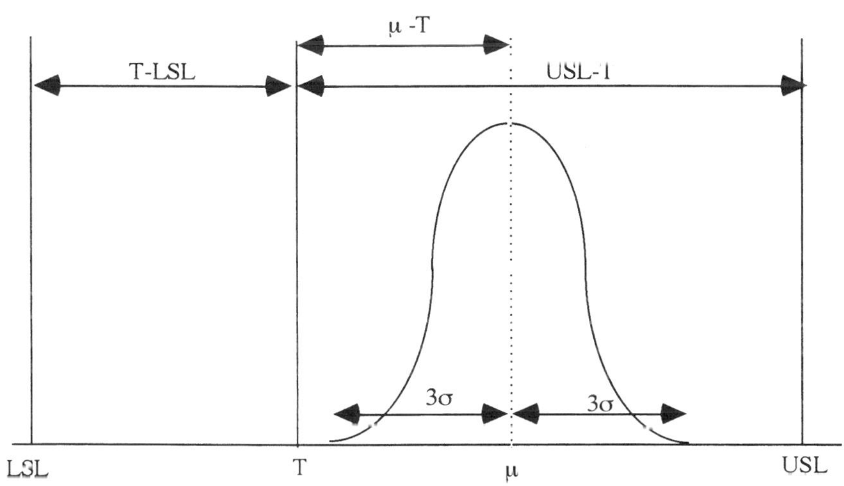

Figure 4 Target is not the midpoint of the specification limits.

the target, these indices increase in magnitude for fixed customer requirements. In each case larger index values indicate a more capable process.

Many modifications to the common indices, as well as several newly developed indices, have been proposed but are not widely used in practice. With remarkably few exceptions these recent developments can be represented using the generic process capability index

$$C_{pw} = \frac{\min[\mathrm{USL} - T, T - \mathrm{LSL}]}{3[\sigma^2 + w(\mu - T)^2]^{1/2}}$$

where w is a weight function. Allowing the weight function to take on different values permits C_{pw} to assume equivalent computational forms for a host of potential capability measures. For example, with $T = (\mathrm{USL} + \mathrm{LSL})/2$ and $w = 0$, C_{pw} is simply C_p, while for $w = 1$, C_{pw} assumes the generalized form of C_{pm}. Letting $p = |\mu - T|/\sigma$ denote a measure of "off-targetness," the weight function

$$w = \left(\frac{d^2}{(d - |a|)^2} - 1\right)\left(\frac{1}{p^2}\right)$$

for $0 < p$ where $d = (\mathrm{USL} - \mathrm{LSL})/2$ and $a = \mu - (\mathrm{USL} + \mathrm{LSL})/2$ allows C_{pw} to represent C^*_{pk}. The weight function

$$w = \frac{k(2 - k)}{(1 - k)^2 p^2}$$

for $0 < k < 1$ allows C_{pw} to represent C_{pk}, or alternatively, defining w as a function of C_p,

$$w = \frac{6C_p - p}{(3C_p - p)^2 p}$$

for $0 < p/3 < C_p$ again results in C_{pw} representing C_{pk}.

A recent refinement that combines properties of both C_{pk} and C_{pm} is defined to be

$$C_{pmk} = \frac{\min[\mathrm{USL} - \mu, \mu - \mathrm{LSL}]}{3[\sigma^2 + (\mu - T)^2]^{1/2}}$$

and can be represented by C_{pw} using the weight function

$$w = \frac{(1 - c^2) + p^2}{c^2 p^2}$$

for $c = \min[\mathrm{USL} - \mu, \mu - \mathrm{LSL}]/\min[\mathrm{USL} - T, T - \mathrm{LSL}]$ such that $0 < c < 1$.

3. INTERPRETING PROCESS CAPABILITY INDICES

Traditionally, process capability indices have been used to provide insights into the number (or proportion) of product beyond the specification limits (i.e., nonconforming). For example, practitioners cite a C_p value of 1 as representing 2700 parts per million (ppm) nonconforming, while 1.33 represents 63 ppm; 1.66 corresponds to 0.6 ppm; and 2 indicates <0.1 ppm. C_{pk} has similar connotations, with a C_{pk} of 1.33 representing a maximum of 63 ppm nonconforming.

Practitioners, in turn, use the value of the process capability index and its associated number conforming to identify capable processes. A process with $C_p \geq 1$ has traditionally been deemed capable, while $C_p < 1$ indicates that the process is producing more than 2700 ppm nonconforming and is deemed incapable of meeting customer requirements. In the case of C_{pk}, the automotive industry frequently uses 1.33 as a benchmark in assessing the capability of a process. Several difficulties arise when process capability indices are used in this manner, including (1) the robustness of the indices to departures from normality, (2) the underlying philosophy associated with converting index values to ppm nonconforming, and (3) the use of estimates as parameters.

The number of parts per million nonconforming is determined directly from the properties of the normal distribution. If the process measurements do not arise from a normal distribution, none of the indices provide a valid measure of ppm. The problem lies in the fact that the actual process spread (6σ) does not provide a consistent measure of ppm nonconforming across distributions. For example, suppose that 99.73% of the process measurements fall within the specification limits for five processes, where the statistical distributions associated with the processes are (1) uniform, (2) triangular, (3) normal, (4) logistic, and (5) double exponential (see Fig. 5). The values of C_p for the five processes are 0.5766, 0.7954, 1.0000, 1.2210, and 1.4030, respectively. Hence as long as 6σ carries a practical interpretation when assessing process capability and the focus is on ppm nonconforming, none of the indices should be considered robust to departures from normality.

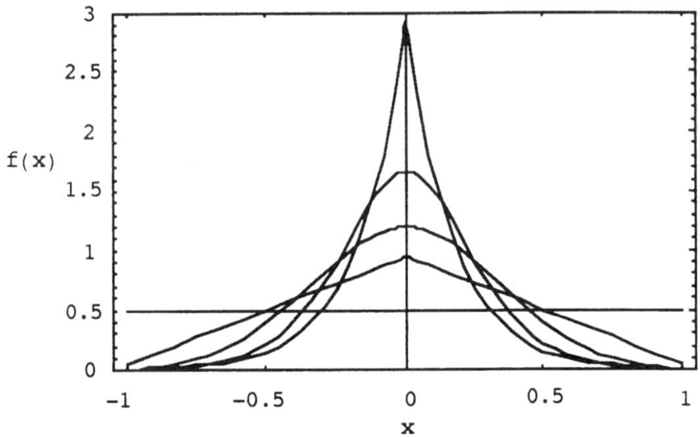

Figure 5 Five processes with equivalent nonconforming but different values of C_p.

Inherent in any discussion of the number nonconforming as a measure of process capability is the assumption that product produced just inside the specification limit is of equal quality to that produced at the target. This is equivalent to assuming a square-well loss function (see Fig. 6) for the quality variable. In practice, the magnitudes of C_p, C_{pl}, C_{pu}, and C_{pk} are interpreted as a measure of ppm nonconforming and therefore follow this square-well loss function philosophy. Any changes in the magnitude of these indices (holding the customer requirements constant) is due entirely to changes in the distance between the specification limits and the process mean. C_p, C_{pu}, C_{pl}, and C_{pk} do not consider the distance between μ and T but are used to identify changes in the amount of product beyond the specification limits

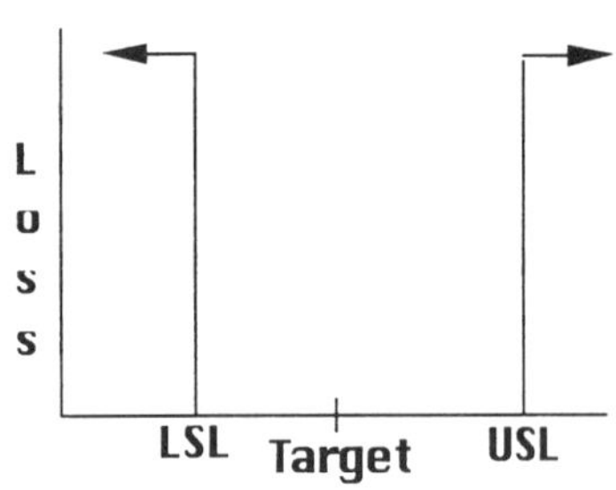

Figure 6 Square-well loss function.

(not proximity to the target) and are therefore consistent with the square-well loss function.

Taguchi uses the quadratic loss function (see Fig. 7) to motivate the idea that a product imparts "no loss" only if that product is produced at its target. He maintains that small deviations from the target result in a loss of quality and that as the product increasingly deviates from its target there are larger and larger losses in quality. This approach to quality and quality assessment is different from the traditional approach, where no loss in quality is assumed until the product deviates beyond its upper or lower specification limit (i.e., square-well loss function). Taguchi's philosophy highlights the need to have small variability around the target. Clearly in this context the most capable process will be one that produces all of its product at the target, with the next best being the process with the smallest variability around the target.

The motivation for C_{pm} does not arise from examining the number of nonconforming product in a process but from looking at the ability of the process to be in the neighborhood of the target. This motivation has little to do with the number of nonconforming, although upper bounds on the number of nonconforming can be determined for numerical values of C_{pm}. The relationship between C_{pm} and the quadratic loss function and its affinity with the philosophies that support a loss in quality for any departure from the target set C_{pm} apart from the other indices.

C_{pk} and C_{pm} are often called second generation measures of process capability whose motivations arise directly from the inability of C_p to consider the target value. The differences in their associated loss functions demarcate the two measures, while the magnitudinal relationship between

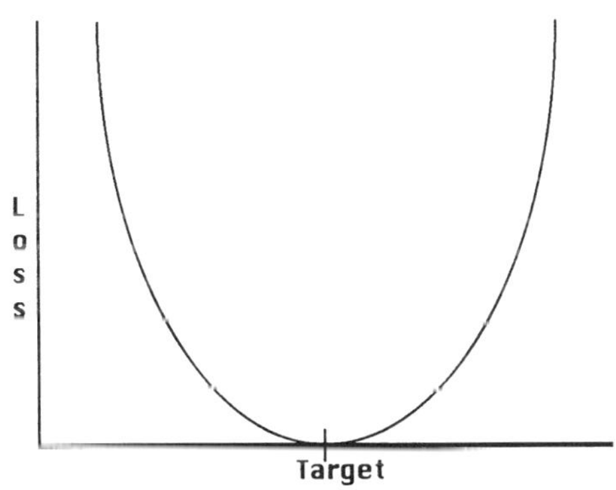

Figure 7 Quadratic loss function.

C_p and C_{pk}, C_{pm} are also different. C_{pk} and C_{pm} are functions of C_p that penalize the process for not being centered at the target. Expressing C_{pm} and C_{pk} as

$$C_{pm} = \frac{1}{\{1 + [(\mu - T)/\sigma]^2\}^{1/2}}$$

and

$$C_{pk} = \left(1 - \frac{2|\mu - T|}{\text{USL} - \text{LSL}}\right) C_p$$

illustrates the "penalizing" relationship between C_p and C_{pm}, C_{pk}, respectively. As the process mean drifts from the target (measured by $p = |\mu - T|/\sigma$), both C_{pm} and C_{pk} decline as a percentage of C_p (Fig. 8). In the case of C_{pm}, this relationship is independent of the magnitude of C_p, while C_{pk} declines as a percentage of C_p, with the rate of decline dependent on the magnitude of C_p. For example, in Figure 8, C_{pk} (5) represents the relationship between C_{pk} and C_p for $C_p = 5$, and is different from C_{pk} (1), which represents the relationship between C_{pk} and C_p for $C_p = 1$.

C_{pk} and C_{pm} have different functional forms, are represented by different loss functions, and have different relationships with C_p as the process drifts from the target. Hence although C_{pm} and C_{pk} are lumped together as second generation measures, they are very different in their development and assessment of process capability.

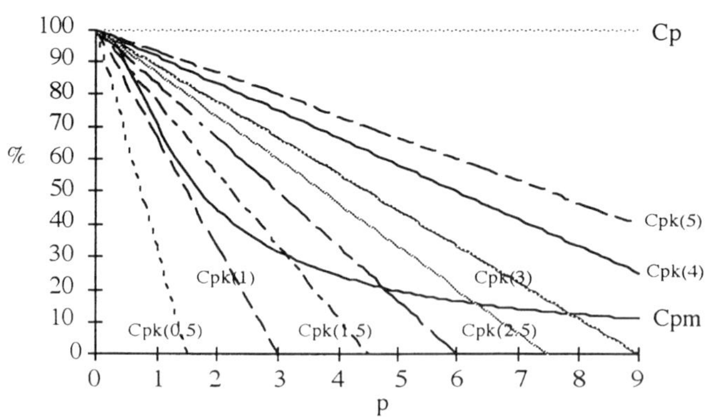

Figure 8 Relationships between C_p and C_{pk}, C_{pm}.

4. ANALYZING PROCESS CAPABILITY STUDIES

The usual estimators of the process capability indices are

$$\hat{C}_p = \frac{\text{USL} - \text{LSL}}{6s}$$
$$\hat{C}_{pm} = \frac{\text{USL} - \text{LSL}}{6s\{s^2 + [n/(n-1)](\bar{x} - T)^2\}^{1/2}}$$
$$\hat{C}_{pu} = \frac{\text{USL} - \bar{x}}{3s}$$
$$\hat{C}_{pl} = \frac{\bar{x} - \text{LSL}}{3s}$$
$$\hat{C}_{pk} = \min(\hat{C}_{pu}, \hat{C}_{pl})$$

or

$$\hat{C}_{pk^*} = (1 - \hat{k})\hat{C}_p$$

where $\hat{k} = 2|T - \bar{x}|/(\text{USL} - \text{LSL})$, s is the sample standard deviation, and $\bar{x}$ is the sample mean. The probability density functions (pdf's) of $\hat{C}_p$, $\hat{C}_{pm}$, $\hat{C}_{pl}$, and $\hat{C}_{pu}$ are easily determined, assuming the process measurements follow a normal distribution. However, the distributions of $\hat{C}_{pk}$ and $\hat{C}_{pk^*}$ raise some challenges, as their pdf's are functions of dependent noncentral t distributions for which only asymptotic solutions currently exist.

4.1. Confidence Intervals

Several inferential techniques have recently been developed, most of which have had little impact on the practice of judging a process capable. In defense of the practitioners, several notable texts promote the use of estimates as parameters with the proviso that large sample sizes (i.e., $n > 50$) are required. A general confidence interval approach for the common indices can be developed using C_{pw} and its associated estimator $\hat{C}_{pw}$. The general form of the estimator for C_{pw} is

$$\hat{C}_{pw} = \frac{\text{USL} - \text{LSL}}{6[\hat{\sigma}^2 + w(\bar{x} - T)^2]^{1/2}}$$

where $\hat{\sigma}^2 = \sum_{i=1}^{n}[(x_i - \bar{x})^2/n]$ and $\bar{x} = \sum_{i=1}^{n}(x_i/n)$ Assuming that the process measurements are normally distributed it follows that (1) $\hat{\sigma}^2 \sim (\sigma^2/n)\chi^2_{n-1}$, (2) $\bar{x} \sim N[\mu, \sigma^2/n]$, and (3) $\bar{x}$ and $\hat{\sigma}^2/n$ are independent.

Assuming w and T to be nonstochastic, it follows that $(\bar{x}-T)^2 \sim (\sigma^2/n)\chi^2_{1,\lambda}$ with noncentrality parameter $\lambda = n(\mu - T)^2/\sigma^2$ and $w(\bar{x}-T)^2 \sim (w\sigma^2/n)\chi^2_{1,\lambda}$. Defining

$$Q^2_{n,\lambda} = \frac{n}{\sigma^2}[\hat{\sigma}^2 + w(\bar{x}-T)^2]$$

$Q^2_{n,\lambda}$ is a linear combination of two independent chi-square distributions, $\chi^2_{n-1} + w\chi^2_{1,\lambda}$, whose cumulative distribution function (cdf) $Q^2_{n,\lambda}(x)$ can be expressed as a mixture of central chi-square distributions with the general form

$$Q^2_{n,\lambda}(x) = \sum_{i=0}^{\infty} d_i \chi^2_{n+2i}(x)$$

The d_i's are simply weights such that $\sum_{i=0}^{\infty} d_i = 1$ and functions of the degrees of freedom ($n-1$ and 1), the noncentrality parameter (λ), and the weight function (w) of the linear combination of chi-square distributions. The functional form of the d_i's for the general $Q^2_{n,\lambda}(x)$ are

$$d_0 = w^{-1/2}e^{-\lambda/2}$$

$$d_i = \sum_{j=0}^{i}\sum_{k=0}^{j} e^{-\lambda/2}\left(\frac{\lambda}{2}\right)^{j-k} w^{-1/2-j+k}(1-w^{-1})^{k+i-j} \left(\frac{\Gamma(i-j+0.5)}{\Gamma(0.5)\Gamma(i-j+1)}\right)\left(\frac{(j-1)!}{k!(j-k-1)!}\right)$$

for $i = 1, 2, 3, \ldots$, when λ denotes the value of the noncentrality parameter and w the value of the weight function. The value of the d_i's and $Q^2_{n,\lambda}(x)$ can be calculated using the following Mathematica code:

```
In[1]:
** To determine the di's for the number of specified **
** i's enter the values of l and w                   **
l=       ;w=       ;
Do[Print[Sum[Sum[Exp[-(l)/2](((l)/2)^(b-k))(((b-k)!)^-1)*
(w^(-.5-b+k))((1-w^(-1))^(k+g-b))Gamma[(.5+g-b)]*
Binomial [b-1,k]/(Gamma[(g-b+1)]Gamma[.5]),
{k,0,b}],{b,0,g}]],{g,1,i}]
```

```
In[2]:
 ** Approximate the value of the distribution by     **
 ** replacing an infinite sum with an finite sum of **
 ** i+1 terms using values of n, a , l and w         **
<<Statistics'ContinuousDistributions'
l=   ;w=   ;n=   ;a=   ;
Sum[Quantile[ChiSquareDistribution[n+2g],a]*
Sum[Sum[Exp[-(l)/2](((l)/2)^(b-k))(((b-k)!)^-1)*
(w^(-.5-b+k))((1-w^(-1))^(k+g-b))Gamma[(.5+g-b)]*
Binomial[b-1,k]/(Gamma[(g-b+1)]Gamma[.5]),
{k,0,b}],{b,0,g}],{g,1,i}]+
(Exp[-l/2](w^(-0.5))*Quantile[ChiSquareDistribution[n],a])
```

The pdf of $\hat{C}_{pw}$ can then be expressed as a function of $Q^2_{n,\lambda}(x)$, allowing confidence intervals and statistical criteria to be used in assessing $\hat{C}_{pw}$ while also providing small sample distribution properties for $\hat{C}_p$, $\hat{C}_{pm}$, $\hat{C}_{pk}$, and $\hat{C}_{pk^*}$. Returning to the general form of the index,

$$C_{pw} = \frac{\text{USL} - \text{LSL}}{6[\sigma^2 + w(\mu - T)^2]^{1/2}}$$

it follows that $[(1 + w\lambda/n)]^{1/2} C_{pw} = (\text{USL} - \text{LSL})/6\sigma$. By considering

$$Pr\left(Q^2_{n,\lambda}\left(\frac{\alpha}{2}\right) \le \frac{n}{\sigma^2}[\hat{\sigma}^2 + w(\bar{x} - T)^2] \le Q^2_{n,\lambda}\left(1 - \frac{\alpha}{2}\right)\right) = 1 - \alpha$$

where $Q^2_{n,\lambda}(\beta)$ represents the value of the $Q^2_{n,\lambda}(x)$ variate for n, λ and probability β. It follows that

$$Pr\left(\left[Q^2_{n,\lambda}\left(\frac{\alpha}{2}\right)\right]^{1/2} \le \left[\frac{n}{\sigma^2}[\hat{\sigma}^2 + w(\bar{x} - T)^2\right]^{1/2} \le \left[Q^2_{n,\lambda}\left(1 - \frac{\alpha}{2}\right)\right]\right)^{1/2} = 1 - \alpha$$

which implies

$$Pr\left(\left[\frac{n(1 + w\lambda/n)}{Q^2_{n,\lambda}(\alpha/2)}\right]^{1/2} C_{pw} \ge \hat{C}_{pw} \ge \left[\frac{n(1 + w\lambda/n)}{Q^2_{n,\lambda}(1 - \alpha/2)}\right]^{1/2} C_{pw}\right) - 1 - \alpha$$

resulting in a general confidence interval for C_{pw} of the form

$$Pr\left(\left[\frac{Q^2_{n,\lambda}(\alpha/2)}{n(1+w\lambda/n)}\right]^{1/2}\hat{C}_{pw} \le C_{pw} \le \left[\frac{Q^2_{n,\lambda}(1-\alpha/2)}{n(1+w\lambda/n)}\right]^{1/2}\hat{C}_{pw}\right) = 1-\alpha \tag{1}$$

For $w = 0$,

$$C_{pw} = \frac{\text{USL} - \text{LSL}}{6\sigma} = C_p$$

and the confidence interval in Eq. (1) becomes

$$Pr\left(\left[\frac{\chi^2_{n-1}(\alpha/2)}{n}\right]^{1/2}\hat{C}_p \le C_p \le \left[\frac{\chi^2_{n-1}(1-\alpha/2)}{n}\right]^{1/2}\hat{C}_p\right) = 1-\alpha$$

where $\hat{C}_p = (\text{USL} - \text{LSL})/6s$.

Similarly, for $w = 1$,

$$C_{pw} = \frac{\text{USL} - \text{LSL}}{6[\sigma^2 + (\mu - T)^2]^{1/2}} = C_{pm}$$

with confidence interval

$$Pr\left(\left[\frac{\chi^2_{n,\lambda}(\alpha/2)}{n(1+\lambda/n)}\right]^{1/2}\hat{C}_{pm} \le C_{pm} \le \left[\frac{\chi^2_{n,\lambda}(1-\alpha/2)}{n(1+\lambda/n)}\right]^{1/2}\hat{C}_{pm}\right) = 1-\alpha \tag{2}$$

for

$$\hat{C}_{pm} \frac{\text{USL} - \text{LSL}}{6\{s^2 + [n/n-1](\bar{x} - T)^2\}^{1/2}}$$

The weight function

$$w = \frac{k(2-k)}{(1-k)^2 p^2}$$

for $0 < k < 1$, and assuming p and k to be nonstochastic, results in $C_{pw} = C^*_{pk}$ with confidence interval

$$Pr\left(\left[\frac{Q^2_{n,\lambda}(\alpha/2)}{n(1+wp^2)}\right]^{1/2}\hat{C}^*_{pk} \le C^*_{pk} \le \left[\frac{Q^2_{n,\lambda}(1-\alpha/2)}{n(1+wp^2)}\right]^{1/2}\hat{C}^*_{pk}\right) = 1-\alpha$$

where $\hat{C}^*_{pk} = (1-\hat{k})\hat{C}_p$.

For

$$w = \left(\frac{d_2}{(d-|a|)^2} - 1\right)\left(\frac{1}{p^2}\right)$$

assuming that $p(0 < p)$, d, and a are known (i.e., nonstochastic), $C_{pw} = C_{pk}$ results in the confidence interval

$$Pr\left(\left[\frac{Q^2_{n,\lambda}(\alpha/2)}{nd^2/(d-|a|)^2}\right]^{1/2}\hat{C}_{pk} \le C_{pk} \le \left[\frac{Q^2_{n,\lambda}(1-\alpha/2)}{nd^2/(d-|a|)^2}\right]^{1/2}\hat{C}_{pk}\right) = 1-\alpha$$

where $\hat{C}_{pk} = \min(\hat{C}_{pl}, \hat{C}_{pu})$.

The weight function may have to be estimated on occasion. However, it is often possible to obtain good information regarding the weight functions from the data used to ensure that the process is in control. Since we require that the process be in a state of statistical control prior to determining any process capability measure, this generally requires that control charts be kept on the process. In most situations the control charts will provide very good information regarding values necessary in determining the weight function. For example, $\bar{x}$ and $\bar{s}$ from the control chart can provide information regarding μ and σ, respectively, which in turn provides an alternative method for determining the distribution function and associated confidence interval for each of the estimated indices.

4.2. Monitoring Process Capability

A criticism of the traditional process capability study is that it provides only a snapshot in time of the process's ability to meet customer requirements. Process capability studies are often conducted at startup and then again during a supplier's audit or after changes have been made to the process. As a result, practitioners have little knowledge of the process's capability over time. With the advent of small-sample properties for the various measures of process capability, it is now easier to incorporate stochastic inferences into the assessment and analysis of process capability measures and to assess capability on a continuous basis.

If all other requirements are met, it is possible to estimate process capability using the information gathered at the subgroup level of the traditional control charts. The usual control chart procedures are used to first verify the assumption that the process is in control. If the process is deemed in control, then estimates of the process capability can be calculated from the subgroup information. These estimates are then plotted, resulting in a chart that provides insights into the nature of a process's capability over its lifetime. The proposed chart is easily appended to an $\bar{X}\&R$ (or s) control chart and facilitates judgments regarding the ability of a process to meet requirements and the effect of changes to the process, while also providing visual evidence of process performance.

Letting $x_1, x_2, x_3, \ldots, x_n$ represent the observations in subgroup t of an $\bar{X}\&s$ control chart used to monitor a process, consider

$$\hat{C}_{pm} = \frac{\min[\text{USL} - T, T - \text{LSL}]}{3[s_t^2 + n(\bar{x}_t - T)^2/(n-1)]^{1/2}}$$

where s_t^2 is the subgroup sample variance and $\bar{x}_t$ the average of the observations in subgroup t. If an $\bar{X}\&R$ chart is used, consider

$$\hat{C}_{pm} = \frac{\min[\text{USL} - T, T - \text{LSL}]}{3[(R_t/d_2)^2 + n(\bar{x}_t - T)^2/(n-1)]^{1/2}}$$

where R_t denotes the range for subgroup t and d_2 the usual control chart constant. Each subgroup in the process provides a measure of location, $\bar{x}_t$, and a measure of variability (either R_t or s_t). Hence an estimate of $\hat{C}_{pm}$ can be determined for each subgroup, which results in a series of estimates for C_{pm} over the life of the process.

A mean line as well as upper and lower limits can be created for a capability chart using information gathered from the control chart. Similar to Shewhart control charts, the upper and lower limits for $\hat{C}_{pm}$ will represent the interval expected to contain 99.73% of the estimates if the process has not been changed or altered. The mean line, denoted $\overline{C_{pm}}$, will be

$$\overline{C_{pm}} = \frac{\min[\text{USL} - T, T - \text{LSL}]}{3\{(\bar{s}/c_4)^2 + [n/(n-1)](\bar{\bar{x}} - T)^2\}^{1/2}}$$

when using an $\bar{X}$&s chart. Assuming equal subgroup sizes, $\bar{\bar{x}}$ denotes the average of the subgroup averages $\bar{x}_i$,

$$\bar{\bar{x}} = \frac{\bar{x}_1 + \bar{x}_2 + \cdots + \bar{x}_k}{k}$$

$\bar{s}$, the average of the subgroup standard deviations s_i,

$$\bar{s} = \frac{s_1 + s_2 + \cdots + s_k}{k}$$

and c_4 the traditional constant.

Assuming that the process measurements are $X \sim N[\mu, \sigma^2]$ and using $\hat{\sigma}^2 = \sum_{i=1}^{n}[(x_i - \bar{x})^2/(n-1)]$, we can rewrite Eq. (2) as

$$Pr\left(\left[\frac{(n-1)(1+\lambda/n)}{\chi^2_{n-1,\lambda}(1-\alpha/2)}\right]^{1/2} C_{pm} \le \hat{C}_{pm} \le \left[\frac{(n-1)(1+\lambda/n)}{\chi^2_{n-1,\lambda}(\alpha/2)}\right]^{1/2} C_{pm}\right) = 1 - \alpha$$

Simplifying this expression we get

$$Pr(I_2 C_{pm} \le \hat{C}_{pm} \le I_3 C_{pm}) = 1 - \alpha$$

where

$$I_2 = \left[\frac{(n-1)(1+\lambda/n)}{\chi^2_{n-1,\lambda}(1-\alpha/2)}\right]^{1/2}$$

and $$I_3 = \left[\frac{(n-1)(1+\frac{\lambda}{n})}{\chi^2_{n-1,\lambda}(\frac{\alpha}{2})}\right]^{1/2}$$

The upper (U_l) and lower (L_l) limits for $\hat{C}_{pm}$ in conjunction with an $\bar{X}$&s chart depend on the subgroup size n and noncentrality parameter λ. Analogous to the use of $\bar{x}$ and s in Shewhart charts, the noncentrality parameter $\lambda = n(\mu - T)^2/\sigma^2$ can be estimated from the control chart using $[(\bar{\bar{x}} - T)/(\bar{S}/c_4)]^2$.

When using $\bar{X}$&R charts with equal subgroup sizes,

$$\overline{C_{pm}} = \frac{\min[\text{USL} - T, T - \text{LSL}]}{3[(\bar{R}/d_2)^2 + [n/(n-1)](\bar{\bar{x}} - T)^2]^{1/2}}$$

where $\bar{R}$ denotes the average of the subgroup ranges R_i,

$$\bar{R} = \frac{R_1 + R_2 + \cdots + R_k}{k}$$

and d_2 the traditional constant. The upper and lower limits for $\overline{\hat{C}_{pm}}$ in conjunction with an $\bar{X}\&R$ chart are of the form $U_l = J_3\overline{C_{pm}}$ and $L_l = J_2\overline{C_{pm}}$, where J_2, J_3 are constants that depend on the subgroup size and a noncentrality parameter λ. Again analogous to the use of $\bar{\bar{x}}$and $\bar{R}$ in Shewhart charts, the noncentrality parameter $\lambda = n(\mu - T)^2/\sigma^2$ can be estimated from the control chart using $[(\bar{\bar{x}} - T)/(\bar{R}/d_2)]^2$.

5. EXAMPLE

5.1. The Process

In this example 20 subgroups of size 10 were gathered from a process for which the customer had indicated that USL $= 1.2$, $T = 1$ and USL $= 0.8$. In this case T is the midpoint of the specification limits; however, all calculations use the general definitions in determining $\hat{C}_{pm}$ and the associated limits. From the 20 subgroups we found $\bar{\bar{x}} = 1.1206$ and $\bar{s} = 0.11$, which resulted in an upper control limit of 1.230 and a lower control limit of 1.014 for $\bar{x}$ and an upper control limit of 0.189 and a lower limit of 0.031 for s. Looking first at the s chart, the process variability does not appear unusual (i.e., no out-of-control signals), which also seems to be the case with the $\bar{x}$ chart. The control limits and centerlines for the $\bar{X}\&s$ charts are included in Figure 9.

Since the process appears to be in control, we proceed to determine $\hat{C}_{pm}$ for each subgroup. In the case of subgroup 1, $\bar{x}$ and s were found to be 1.15 and 0.136, respectively, resulting in

$$\hat{C}_{pm_1} = \frac{\min[1.2 - 1, 1 - 0.8]}{3[0.136^2 + 10(1.15 - 1)^2/9]^{1/2}} = 0.32$$

$\hat{C}_{pm_1}$ and the subsequent 19 subgroup values of $\hat{C}_{pm_t}$ are plotted in Figure 9. Using (1) the customer's requirements USL $= 1.2$, $T = 1$, and LSL $= 0.8$, (2) the process results $\bar{\bar{x}} = 1.1206$ and $\bar{s} = 0.11$, and (3) the constants $n = 10$ and $c_4 = 0.9727$, we determined that

$$\lambda = n\left(\frac{\bar{\bar{x}} - T}{\bar{s}/c_4}\right)^2 = 10\left(\frac{1.1206 - 1}{0.11/0.9727}\right)^2 = 11.4$$

and

$$\overline{C_{pm}} = \frac{\min[\mathrm{USL} - T, T - \mathrm{LSL}]}{3\left[\left(\dfrac{\bar{s}}{c_4}\right)^2 + \dfrac{n}{n-1}(\bar{\bar{x}} - T)^2\right]^{1/2}}$$

$$= \frac{0.2}{3\left[\left(\dfrac{0.11}{0.9727}\right)^2 + \dfrac{10}{9}(1.1206 - 1)^2\right]^{1/2}} = 0.3918$$

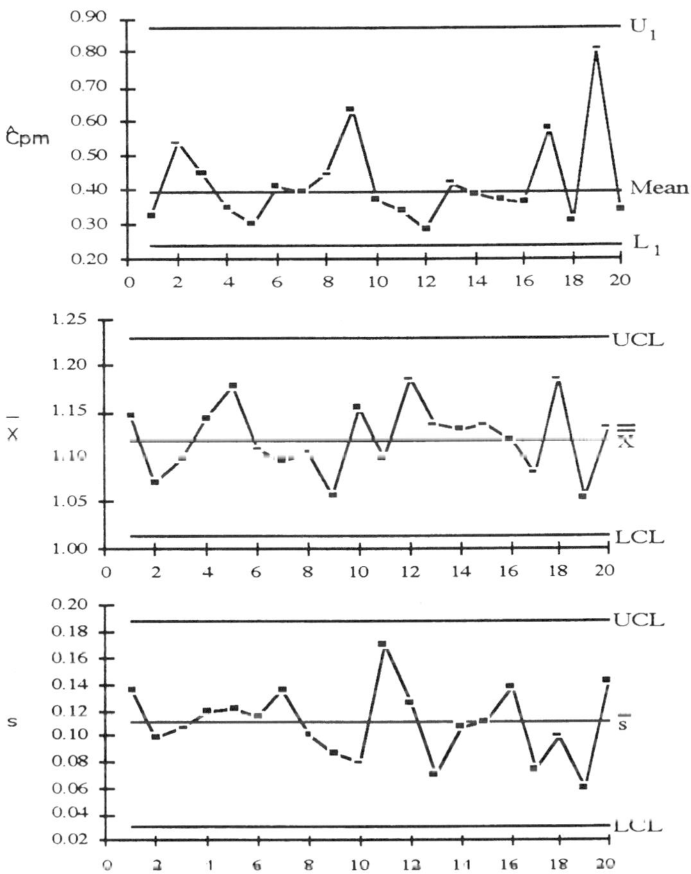

Figure 9 Capability chart appended to an $\bar{X}$&s chart.

The values of I_2 and I_3 for $n = 10$ and $\lambda = 11.4$ are

$$I_2 = \left[\frac{9(1 + 11.4/10)}{\chi^2_{9,11.4}(0.99865)}\right]^{1/2} = \left[\frac{9(1 + 11.4/10)}{51.42}\right]^{1/2} = 0.6120$$

and

$$I_3 = \left[\frac{9(1 + 11.4/10)}{\chi^2_{9,11.4}(0.00135)}\right]^{1/2} = \left[\frac{9(1 + 11.4/10)}{3.839}\right]^{1/2} = 2.23985$$

resulting in the limits

$$U_l = 2.23985(0.3918) = 0.87757 \qquad \text{and} \qquad L_l = 0.6120(0.3918) = 0.2398$$

which are sketched in Figure 9.

5.2. Observations and Insights

Several things are evident from Figure 9. Clearly, the estimates of the process's capability vary from subgroup to subgroup. Except for subgroup 19, the fluctuations in $\hat{C}_{pm}$ appear to be due to random causes. In period 19 the process capability appears to have increased significantly and warrants investigation. Practitioners would likely attempt to determine what caused the capability to rise significantly and recreate that situation in the future.

If the estimated process capability had dropped below L_l, this would signal a change in the process, and if the process capability was not at the level required by the customer, changes in the process would be required. In a continuous improvement program the process capability should be under constant influence to increase. The capability chart used in conjunction with the traditional Shewhart variables charts will provide evidence of improvement. It may also assist in ending the unfortunate practice of including specification limits on the $\bar{x}$ chart, as the additional chart will incorporate the limits and target into the calculation of process capability.

Much like the effect of first-time control charts, practitioners will see that process capability will vary over the life of the process, illustrating the idea that the estimates are not parameter values and should not be treated as such. The procedures provide evidence of the level of process capability attained over the lifetime of the process rather than at snapshots taken, for example, at the beginning of the process and not until some change in the process has been implemented. They will also provide evidence of the ongoing assessment of process capability for customers. The effect of any changes to the process will also show up on the chart, thereby providing feedback to the practitioner regarding the effect changes to the process have on process capability.

6. COMMENTS

Several ideas have been presented that address some concerns of two distinguished quality practitioners in the area of process capability, Vic Kane (Kane, 1986) and Bert Gunter (Gunter, 1991). Unfortunately, as noted by Nelson (1992), much of the current interest in process capability indices is focused on determining competing estimators and their associated distributions, and little work has dealt with the more pressing problems associated with the practical shortcomings of the indices. Continuous monitoring of process capability represents a step toward more meaningful capability assessments. However, much work is needed in this area. In particular, as practitioners move to measures of process capability that assess clustering around the target, the effect of non-normality may be less problematic. Currently, however, meaningful process capability assessments in the presence of non-normal distributions remains a research problem.

REFERENCES AND SELECTED BIBLIOGRAPHY

Boyles RA. (1991). The Taguchi capability index and corrigenda. J Qual Technol 23:17–26.

Chan LK, Cheng SW, Spiring FA. (1988). A new measure of process capability: C_{pm}. J Qual Technol 20:162–175.

Chou YM, Owen DB, Borrego SA. (1990). Lower confidence limits on process capability indices. J Qual Technol 22:223–229.

Gunter BH. (1991). Statistics corner (A five-part series on process capability studies). Qual Prog.

Johnson T. (1992). The relationship of C_{pm} to squared error loss. J Qual Technol 24.211–215.

Juran JM. (1979). Quality Control Handbook. New York: McGraw-Hill.

Kane VE. (1986). Process capability indices and corrigenda. J Qual Technol 18:41–52, 265.

Kotz S, Johnson NL. (1993). Process Capability Indices. London: Chapman & Hall.

Nelson PR. (1992). Editorial. J Qual Technol 24:175.

Rodriguez RN. (1992). Recent developments in process capability analysis. J Qual Technol 24:176–187.

Spiring FA. (1995). Process capability: A total quality management tool. Total Qual Manage 6(1):21–33.
Spiring FA. (1996). A unifying approach to process capability indices. J Qual Technol 29:49–58.
Vännman K. (1995). A unified approach to capability indices. Stat Sin 5(2):805–820.

17

Experimental Strategies for Estimating Mean and Variance Function

G. Geoffrey Vining
Virginia Polytechnic Institute and State University, Blacksburg, Virginia

Diane A. Schaub
University of Florida, Gainesville, Florida

Carl Modigh
Arkwright Enterprises Ltd., Paris, France

1. INTRODUCTION

An important approach for optimizing an industrial process seeks to find operating conditions that achieve some target condition for the expected value for a quality characteristic (the response) and minimize the process variability. Vining and Myers (1990) suggest that the response and the process variance form a dual response system. They use the dual response methodology proposed by Myers and Carter (1973) to find appropriate operating conditions. This dual response approach allows the analyst to see where the process can achieve the target condition and where the process variability is acceptable. As a result, the engineer can make explicit compromises. Del Castillo and Montgomery (1993) extend this method by showing how to use the generalized reduced gradient, which is available in some spreadsheet programs such as Microsoft Excel, to find the appropriate operating conditions. Lin and Tu (1995) suggest a mean squared error approach within this context. Copeland and Nelson (1996) suggest a direct function minimiation of the mean squared error with a bound on how far the estimated response can deviate from the desired target value.

Vining and Myers (1990) advocate replicating a full second-order design. Such an approach is often prohibitively expensive in terms of the overall number of design runs. Vining and Schaub (1996) note that often the process variance follows a lower order model than the response. They suggest replicating only a first-order portion of standard response surface designs, which significantly reduces the overall design size. This chapter extends the work of Vining and Schaub by exploring alternative ways for choosing the portion of the design to replicate.

2. CRITERION FOR EVALUATING DESIGNS

Suppose we run an appropriate experiment with a total of n runs. Let n_v be the number of distinct settings that are replicated. Consider as our model for the response,

$$\boldsymbol{y} = \boldsymbol{X}\beta + \epsilon$$

where $\boldsymbol{y}$ is the $n \times 1$ vector of response, $\boldsymbol{X}$ is the $n \times p_r$ model matrix, β is the $p_r \times 1$ vector of unknown coefficients, and ϵ is the $n \times 1$ vector of normally distributed random errors. Similarly, consider as the model for the process variance,

$$\tau = \mathbf{Z}\gamma$$

where τ is the $n_v \times 1$ vector of linear predictors, $\boldsymbol{Z}$ is the $n_v \times p_v$ model matrix for the linear predictors, and γ is the $p_v \times 1$ vector of unknown coefficients. We relate the ith linear predictor, τ_i, to the ith process variance σ_i^2 by

$$\sigma_i^2 = h(\tau_i)$$

where h is a twice differentiable monotonic function. Define h_i' to be the first derivative of h with respect to the ith τ. Often, analysts use the **exp** function for h, which is similar to using a **log** transformation on the observed sample variances. Throughout this chapter, we follow this convention; thus,

$$\sigma_i^2 = \mathbf{exp}(\tau_i)$$

This approach guarantees that $\sigma_i^2 > 0$.

Consider the joint estimation of β and γ. The expected information matrix, $\boldsymbol{J}$, is

$$J = \begin{bmatrix} X'W_{11}X & 0 \\ 0 & Z'W_{22}Z \end{bmatrix}$$

where W_{11} and W_{22} are diagonal matrices with nonzero elements $1/\sigma_i^2$ and $(h_i'/\sigma_i^2)/2$, respectively. Vining and Schaub (1996) prefer to use M, the expected information matrix expressed on a per-unit basis, where

$$M = \frac{1}{n} J$$

In some sense, M represents a moment matrix. One problem with this approach, however, is that we use all n of the experimental runs to estimate the response, but we use only n_v distinct settings to estimate the process variance. In this chapter, we propose an alternative moment matrix, M^*, defined by

$$M^* = \begin{bmatrix} (1/n)X'W_{11}X & 0 \\ 0 & (1/n_v)Z'W_{22}Z \end{bmatrix}$$

which is a block diagonal matrix with separate moment matrices for each model on the diagonals.

One definition of an "optimal" design is that it is one that maximizes the information present in the data. Much of optimal design theory uses appropriate matrix norms to measure the size of the information matrix. The determinant is the most commonly used matrix norm in practice, which leads to D-optimal designs. In this particular case, we must note that M^* depends on the σ_i^2's, which in turn depend on γ through the τ_i's. However, we cannot know γ prior to the experiment; hence, we encounter a problem in determining the optimal design. One approach proposed by Vining and Schaub (1996) assumes that $\tau_i = \tau_0$ for $i = 1, 2, \ldots, n$. Essentially, this approach assumes that in the absence of any prior information about the process variance function, the function could assume any direction over the region of interest. By initially assuming that the process variance function is constant, the analyst does not bias the weights in any particular direction. With this assumption, we can establish that an appropriate D-optimality-based criterion for evaluating designs is

$$D = \left[\left(\frac{1}{n}\right)^{p_r} \left(\frac{1}{n_v}\right)^{p_v} |X'X| \cdot |Z'Z| \right]^{1/(p_r+p_v)}$$

The criterion provides an appropriate basis for comparing designs. By its definition, we are able to compare in a meaningful fashion designs that use different numbers of total runs and different replication schemes.

3. COMPUTER-GENERATED DESIGNS

We used this criterion within a modified DETMAX (Mitchell, 1974) algorithm to generate optimal finite-run designs. Figures 1–7 display the three-factor designs generated by this algorithm over a cuboidal region of interest for $n = 14$, $n = 15$, $n = 18$, $n = 22$, $n = 26$, $n = 32$, and $n = 59$, respectively. Taken together, these figures suggest how the optimal design evolves with additional design runs.

Figure 1 indicates that the computer starts with a Notz (1982) design with a resolution III fraction replicated. The Notz design is interesting because it uses seven out of the eight cube or factorial points. It adds three axial points in order to estimate the pure quadratic terms. Figure 2 shows what happens as we add the next point to the design. As one should expect, it brings in the other factorial point. Figure 3 shows the optimal design for $n = 18$ total runs. Interestingly, it starts adding the face centers of the cube defined by the factorial runs. The resulting design is a central composite design with a resolution III portion replicated, which Vining

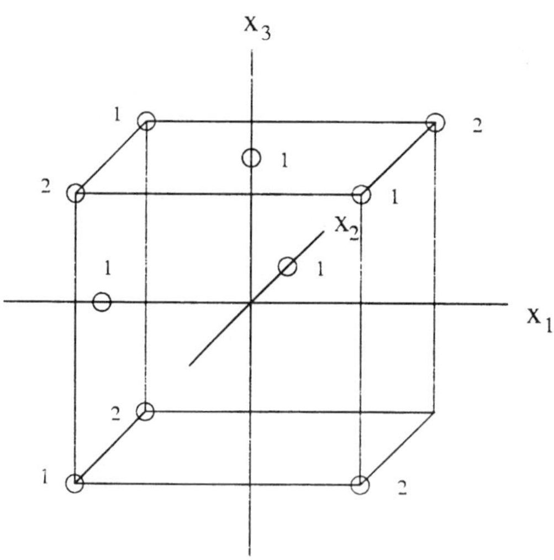

Figure 1 The three-factor *D*-optimal design for 14 runs over a cuboidal region.

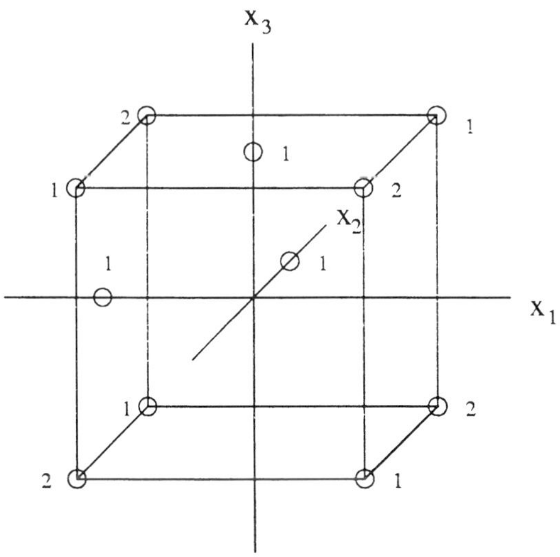

Figure 2 The three-factor *D*-optimal design for 15 runs over a cuboidal region.

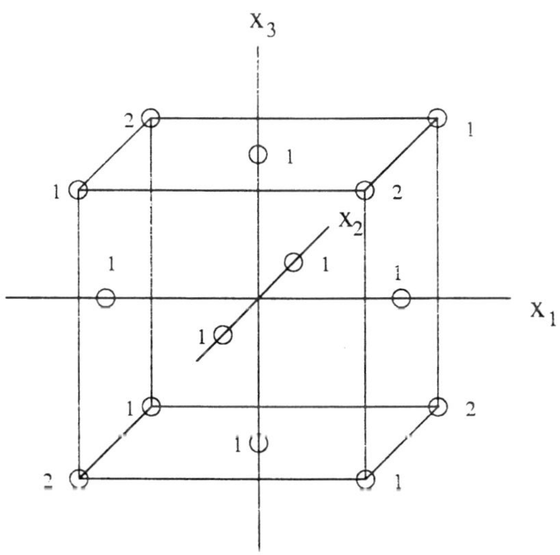

Figure 3 The three-factor *D*-optimal design for 18 runs over a cuboidal region.

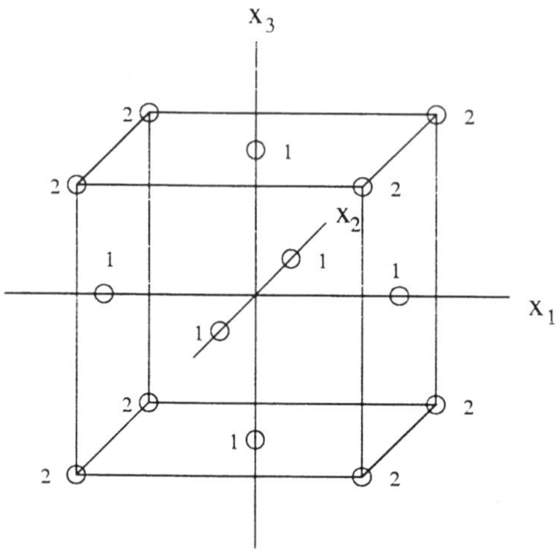

Figure 4 The three-factor D-optimal design for 22 runs over a cuboidal region.

and Schaub call a replicated factorial design. Figure 4 shows that at $n = 22$ the design replicates all of the cube points, as opposed to the replicated factorial, which would replicate only a resolution III fraction of the full factorial. Interestingly, Figure 5 shows that at $n = 26$, the computer adds midpoints of edges. Vining and Schaub recommend their replicated factorial design for this situation. The optimal design takes a slightly different strategy. Figures 6 and 7 show that as we continue to add runs, the computer moves to a 3^3 factorial with replicated cube points. It appears that the proposed criterion favors replicating the cube points and then augmenting with points from the full 3^3.

Figure 8 summarizes the D values for the three-factor computer-generated designs over a cuboidal region. Interestingly, the D value actually seems to peak around $n = 32$ total runs, with $D = 0.5851$. The initial increase in D with n makes a lot of sense because the extra runs provide necessary symmetries. As the cube points are replicated more and more, we presume that some imbalance in information results between the strict first-order terms and the strict second-order terms. This imbalance may explain why the D values drop slightly from $n = 32$ to the largest sample size studied, $n = 80$.

Figures 9 and 10 extend this study to the computer-generated designs for four and five factors, respectively. In each case, D increases as n increases. The largest values for D observed were 0.5806 for the four-factor case and 0.5816 for the five-factor case. These figures suggest either that D

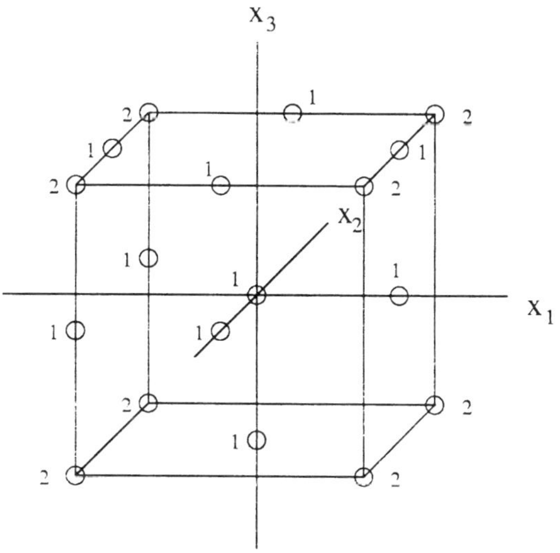

Figure 5 The three-factor *D*-optimal design for 26 runs over a cuboidal region.

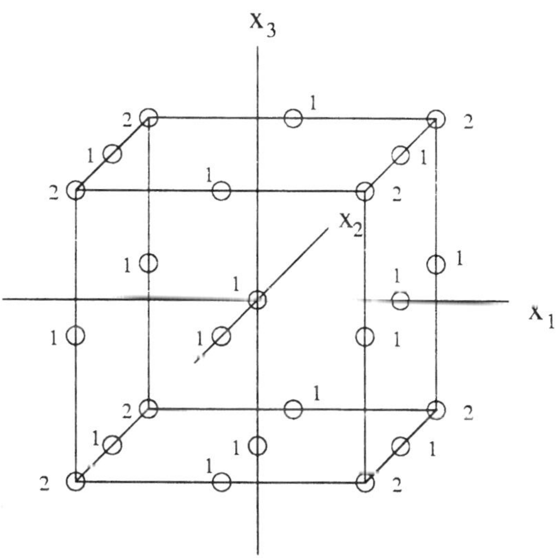

Figure 6 The three-factor *D*-optimal design for 32 runs over a cuboidal region.

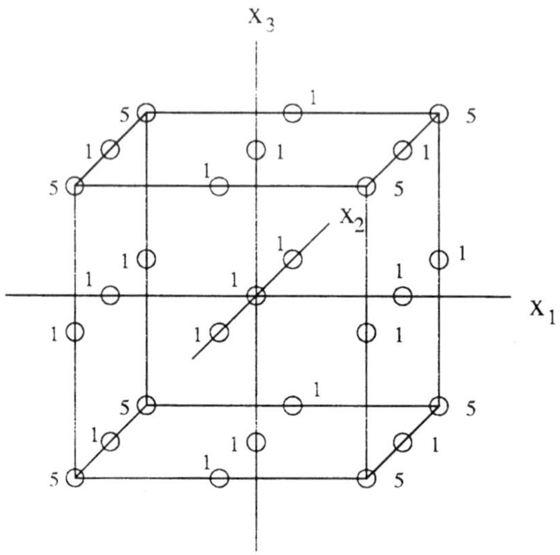

Figure 7 The three-factor *D*-optimal design for 59 runs over a cuboidal region.

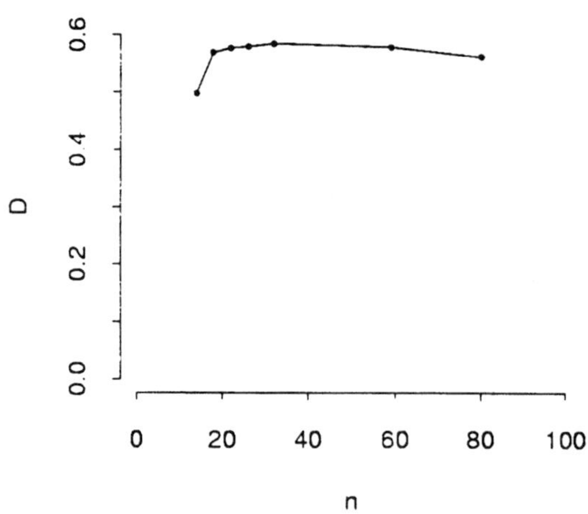

Figure 8 Plot of the value for *D* for the three-factor computer-generated design over a cuboidal region.

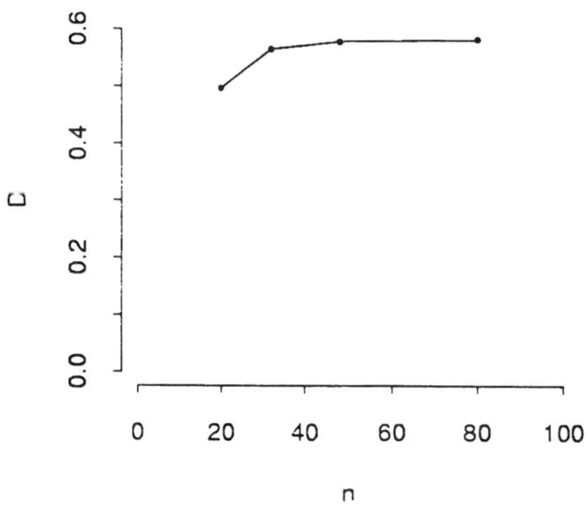

Figure 9 Plot of the value for D for the four-factor computer-generated design over a cuboidal region.

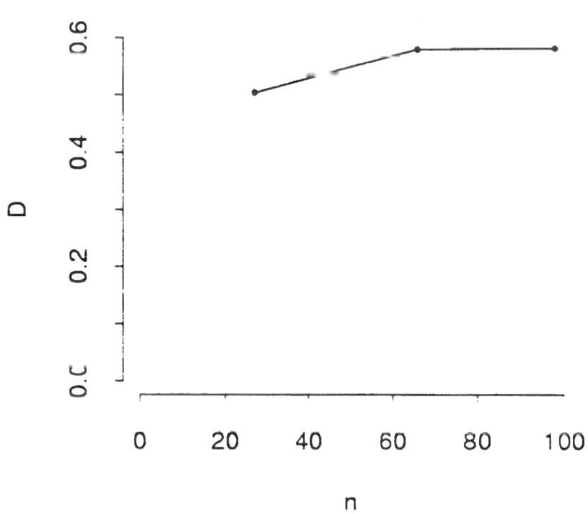

Figure 10 Plot of the value for D for the five-factor computer-generated design over a cuboidal region.

approaches some asymptote or that D may peak at some sample size larger than the ones studied.

4. COMPARISONS OF DIFFERENT REPLICATION STRATEGIES

Figures 11–13 use the D criterion to compare the following design strategies for three, four, and five factors over a cuboidal region:

A fully replicated central composite design (CCD)
A fully replicated Notz (1982) design
A replicated axial design (a CCD with only the axial points replicated)
A replicated factorial design (a CCD with only a resolution III fraction replicated)
A replicated 3/4 design (a CCD with only a 3/4 fraction replicated)
A replicated full factorial (a CCD with the entire factorial portion replicated)

The fully replicated CCD should always be a "near-optimal" design for each situation. In some sense, it provides a "gold standard" for comparisons. However, replicating a full CCD is rather expensive in terms of overall design size. The Notz design is a minimum run D optimal design for the second-order model over a cuboidal region. Replicating a minimal point design is one logical way to reduce the overall design size. Vining and Schaub (1996) note that the replicated Notz design performs surprisingly well in the joint estimation of the two models. Vining and Schaub proposed the replicated axial and the replicated factorial as alternative designs for reducing the total number of runs. The replicated 3/4 design is another possible alternative. The optimal designs generated in the previous section strongly suggest the replicated full factorial strategy.

Figure 11 summarizes the three-factor results. In this figure, m refers to the number of runs at each replicated setting. We evaluated each design using 4, 8, and 12 replicates. As expected, the replicated CCD appears to be the best overall design. Interestingly, the replicated full factorial actually was better for $m = 4$. The designs that replicated only a portion of their runs all became less efficient as the replication increased. We believe that this is due to an increase in the imbalance in these designs. The replicated full factorial performed slightly better than the other partially replicated designs. The replicated 3/4 and the replicated factorial performed very similarly. The replicated axial performed quite poorly. The replicated Notz performed almost as well as the replicated CCD.

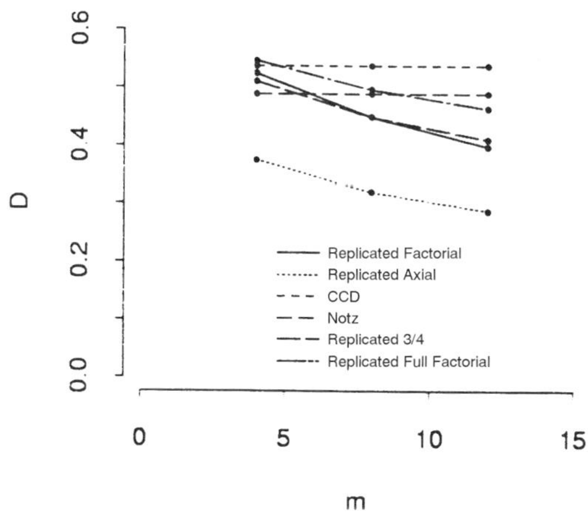

Figure 11 Comparisons of designs in terms of D for the three-factor cuboidal case.

Table 1 summarizes the number of runs required by each design. The replicated factorial requires the fewest, and the replicated CCD requires the most. Our D criterion takes the total sample size into account and thus provides a fair comparison for these designs. In many situations, the experimenter cannot afford large numbers of total runs due to either time or cost. The replicated factorial appears to be relatively competitive in terms of the D criterion while at the same time minimizing the total number of runs. In this light, the replicated factorial is often a very attractive design for this type of experimentation.

Figure 12 summarizes the four-factor results. Here, the replicated CCD performs uniformly best. Once again, the performance of all the

Table 1 Design Sizes for the Three-Factor Case

	m		
Design	4	8	12
Replicated factorial	26	42	58
Replicated axial	32	56	80
Replicated 3/4	32	56	80
Replicated full factorial	38	70	102
Notz	40	80	120
CCD	56	112	168

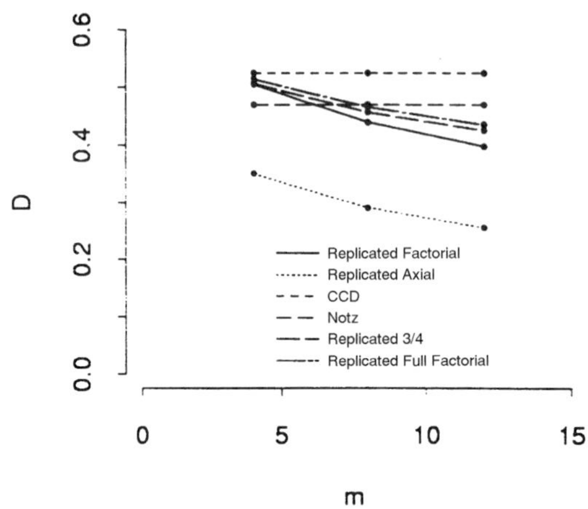

Figure 12 Comparisons of designs in terms of D for the four-factor cuboidal case.

designs that replicate only a portion of their runs decreases with greater replication. The replicated full factorial, replicated 3/4 factorial, and replicated factorial designs all perform similarly, with the replicated full factorial performing slightly better than the others and the replicated factorial performing slightly worse. The replicated axial performs very poorly. Once again, the replicated Notz performs similarly to the replicated CCD.

Table 2 summarizes the total number of runs for each design. In this case, the replicated factorial and the replicated axial require exactly the same number of runs. They in turn require fewer runs than any other design. Once again, taking into account Figure 12 and Table 2, the replicated factorial appears to be a reasonable design strategy in many situations.

Table 2 Design Sizes for the Four-Factor Case

	m		
Design	4	8	12
Replicated factorial	48	80	112
Replicated axial	48	80	112
Replicated 3/4	60	108	156
Replicated full factorial	72	136	200
Notz	60	120	180
CCD	96	192	288

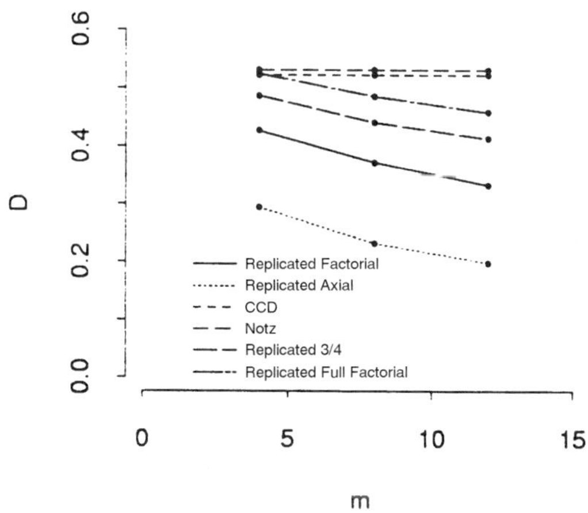

Figure 13 Comparisons of designs in terms of D for the five-factor cuboidal case.

Figure 13 summarizes the results for the five-factor case. Interestingly, the replicated Notz design performed best, edging out the replicated CCD. The replicated axial again performed worst. We see bigger differences in performance among the other three, with the replicated full factorial performing uniformly better than the replicated 3/4, which in turn uniformly outperformed the replicated factorial.

Table 3 summarizes the total number of runs required by each design. The replicated CCD here uses a resolution V fraction of the 2^5 factorial design. The replicated factorial, however, must use the full 2^5 factorial design in order to minimize the number of replicated points. Consequently, the replicated factorial is not always the smallest design.

Table 3 Design Sizes for the Five-Factor Case

	m		
Design	4	8	12
Replicated factorial	66	98	130
Replicated axial	56	96	136
Replicated 3/4	114	210	306
Replicated full factorial	74	138	202
Notz	84	168	252
CCD	104	208	312

The real message of Table 3 is that all of the design strategies require a large number of runs. In many situations, the total is prohibitive.

5. CONCLUSIONS

Our research suggests the following conclusions. First, the proposed D criterion suggests that if we fit a second-order model to the response and a first-order model to the process variance, then we need to replicate only a subset of the base second-order design. Second, this criterion appears to prefer replicating the full factorial as the sample size permits. Third, the replicated factorial and the replicated 3/4 factorial designs tend to perform well for small to moderate amounts of replication. Finally, for large amounts of replication, we may want to consider replicating at least a resolution V fraction (the replicated full factorial).

REFERENCES

Copeland KAF, Nelson PR. (1996). Dual response optimization via direct function minimization. J Qual Technol 28:331–336.

Del Castillo E, Montgomery DC. (1993). A nonlinear programming solution to the dual response problem. J Qual Technol 25:199–204.

Lin DKJ, Tu W. (1995). Dual response surface optimization. J Qual Technol 27:34–39.

Mitchell TJ. (1974). An algorithm for the construction of D-optimal experimental designs. Technometrics 16:211–220.

Myers RH, Carter WH Jr. (1973). Response surface techniques for dual response systems. Technometrics 15:301–317.

Notz W. (1982). Minimal point second order designs. J Stat Planning Inf 6:47–58.

Vining GG, Myers RH. (1990). Combining Taguchi and response surface philosophies: A dual response approach, J Qual Technol 22:38–45.

Vining GG, Schaub D. (1996). Experimental designs for estimating both mean and variance functions. J Qual Technol 28:135–147.

18
Recent Developments in Supersaturated Designs

Dennis K. J. Lin
The Pennsylvania State University, University Park, Pennsylvania

1. AGRICULTURAL AND INDUSTRIAL EXPERIMENTS

Industrial management is becoming increasingly aware of the benefits of running statistically designed experiments. Statistical experimental designs, developed by Sir R. A. Fisher in the 1920s, largely originated from agricultural problems. Designing experiments for industrial problems and designing experiments for agricultural problems are similar in their basic concerns. There are, however, many differences. The differences listed in Table 1 are based on the overall characteristics of all problems. Exceptions can be found in some particular cases, of course.

Industrial problems tend to contain a much larger number of factors under investigation and usually involve a much smaller total number of runs.

Industrial results are more reproducible; that is, industrial problems contain a much smaller replicated variation (pure error) than that of agricultural problems.

Industrial experimenters are obliged to run their experimental points in sequence and naturally plan their follow-up experiments guided by previous results; in contrast, agricultural problems harvest all results at one time. Doubts and complications can be resolved in industry by immediate follow-up experiments. Confirmatory experimentation is readily available for industrial problems and becomes a routine procedure to resolve assumptions.

Table 1 Differences Between Agricultural and Industrial Experiments

Subject	Agriculture	Industry
Number of factors	Small	Large
Number of runs	Large	Small
Reproducibility	Less likely	More likely
Time taken	Long	Short
Blocking	Nature	Not obvious
Missing values	Often	Seldom

The concept of blocking arose naturally in agriculture but often is not obvious for industrial problems. Usually, industrial practitioners need certain specialized training to recognize and handle blocking variables.

Missing values seem to occur more often in agriculture (mainly due to natural losses) than in industry. Usually, such problems can be avoided for industrial problems by carrying out well-designed experiments.

The supersaturated design method considered in this chapter suggests one kind of screening method for industrial problems involving a large number of potential relevant factors. It may not be an appropriate proposal for some agricultural problems.

2. INTRODUCTION

Consider the simple fact that where there is an effect, there is a cause. Quality engineers are constantly faced with distinguishing between the effects that are caused by particular factors and those that are due to random error. The "null" factors are then adjusted to lower the cost; the "non-null" (effective) factors are used to yield better quality. To distinguish between them, a large number of factors can often be listed as possible sources of effects. Preliminary investigations (e.g., using professional knowledge) may quickly remove some of these "candidate factors." It is not unusual, however, to find that more than 20 sources of effects exist and that among those factors only a small portion are actually active. This is sometimes called "effect sparsity." A problem frequently encountered in this area is that of how to reduce the total number of experiments. This is particularly important in situations where an individual run is expensive (e.g., with respect to money or time). With powerful statistical software readily available for data analysis, there is no doubt that data collection is the most important part of such problems.

To obtain an unbiased estimate of the main effect of each factor, the number of experiments must exceed (or at least be equal to) the number of factors plus one (for estimating the overall grand average). When the two numbers are equal, the design is called a saturated design; it is the minimum effort required to estimate all main effects. The standard advice given to users in such a screening process is to use the saturated design, which is "optimal" based on certain theoretical optimality criteria. However, the nonsignificant effects are not of interest. Estimating all main effects may be wasteful if the goal is simply to detect the few active factors. If the number of active factors is indeed small, then the use of a slightly biased estimate will still allow one to accomplish the identification of the active factors but significantly reduce the amount of experimental work. Developing such screening designs has long been a well-recognized problem, certainly since Satterthwaite (1959).

When all factors can be reasonably arranged into several groups, the so-called group screening designs can be used (see, e.g., Watson, 1961). Only those factors in groups that are found to have large effects are studied further here. The grouping scheme seems to be crucial but has seldom been discussed. The basic assumptions (such as assuming that the directions of possible effects are known), in fact, depend heavily on the grouping scheme. While such methods may be appropriate in certain situations (e.g., blood tests), we are interested in systematic supersaturated designs for two-level factorial designs that can examine k factors in $N < k + 1$ experiments in which no grouping scheme is needed. Recent work in this area includes, for example, that of Lin (1991, 1993a, 1993b, 1995, 1998), Tang and Wu (1997), Wu (1993), Deng and Lin (1994), Chen and Lin (1998), Cheng (1997), Deng et al. (1994, 1996a, 1996b), Yamada and Lin (1997) and Nguyen (1996).

3. SUPERSATURATED DESIGNS USING HADAMARD MATRICES

Lin (1993a) proposed a class of special supersaturated designs that can be easily constructed via half-fractions of the Hadamard matrices. These designs can examine $k = N - 2$ factors with $n = N/2$ runs, where N is the order of the Hadamard matrix used. The Plackett and Burman (1946) designs, which can be viewed as a special class of Hadamard matrices, are used to illustrate the basic construction method.

Table 2 shows the original 12 run Plackett and Burman design. If we take column 11 as the branching column, then the runs (rows) can be split into two groups: group I with the sign of +1 in column 11 (rows 2, 3, 5, 6, 7,

Table 2 A Supersaturated Design Derived from the Hadamard Matrix of Order 12

Run	Row	1	2	3	4	5	6	7	8	9	10	11
	1	+	+	−	+	+	+	−	−	−	+	−
1	2	+	−	+	+	+	−	−	−	+	−	+
2	3	−	+	+	+	−	−	−	+	−	+	+
	4	+	+	+	−	−	−	+	−	+	+	−
3	5	+	+	−	−	−	+	−	+	+	−	+
4	6	+	−	−	−	+	−	+	+	−	+	+
5	7	−	−	−	+	−	+	+	−	+	+	+
	8	−	−	+	−	+	+	−	+	+	+	−
	9	−	+	−	+	+	−	+	+	+	−	−
	10	+	−	+	+	−	+	+	+	−	−	−
6	11	−	+	+	−	+	+	+	−	−	−	+
	12	−	−	−	−	−	−	−	−	−	−	−

and 11) and group II with the sign of −1 in column 11 (rows 1, 4, 8, 9, 10, and 12). Deleting column 11 from group I causes columns 1–10 to form a supersaturated design to examine $N - 2 = 10$ factors in $N/2 = 6$ runs (runs 1–6, as indicated in Table 2). It can be shown that if group II is used, the resulting supersaturated design is an equivalent one. In general, a Plackett and Burman (1946) design matrix can be split into two half-fractions according to a specific branching column whose signs equal +1 or −1. Specifically, take only the rows that have +1 in the branching column. Then, the $N - 2$ columns other than the branching column will form a supersaturated design for $N - 2$ $N - 2$ factors in $N/2$ runs. Judged by a criterion proposed by Booth and Cox (1962), these designs have been shown to be superior to other existing supersaturated designs.

The construction methods here are simple. However, knowing in advance that Hadamard matrices entertain many "good" mathematical properties, the optimality properties of these supersaturated designs deserve further investigation. For example, the half-fraction Hadamard matrix of order $n = N/2 = 4t$ is closely related to a balanced incomplete block design with $(v, b, r, k) = (2t - 1, 4t - 2, 2t - 2, t - 1)$ and $\lambda = t - 1$. Consequently, the $E(s^2)$ value (see Section 4) for a supersaturated design from a half-fraction Hadamard matrix is $n^2(n - 3)/[(2n - 3)(n - 1)]$, which can be shown to be the minimum within the class of designs with the same size. Potentially promising theoretical results seem possible for the construction of a half-fraction Hadamard matrix. Theoretical implications deserve detailed scrutiny and are discussed below. For more details regarding this issue, please consult Cheng (1997) and Nguyen (1996).

Note that the interaction columns of Hadamard matrices are only partially confounded with other main-effect columns. Wu (1993) makes use of such a property and proposes a supersaturated design that consists of all main-effect and two-factor interaction columns from any given Hadamard matrix of order N. The resulting design has N runs and can accommodate up to $N(N-1)/2$ factors. When there are $k < N(N-1)/2$ factors to be studied, choosing columns becomes an important issue to be addressed.

4. CAPACITY CONSIDERATIONS

As mentioned, when a supersaturated design is used, the abandonment of perfect orthogonality is inevitable. The designs given in Lin (1993a) based on half-fractions of Hadamard matrices have a very nice mathematical structure but can be used only to examine $N-2$ factors in $N/2$ runs, where N is the order of the Hadamard matrix used. Moreover, these designs do not control the value of the maximal pairwise correlation r, and, in fact, large values of r occur in some cases.

Consider a two-level k-factor design in n observations with maximal pairwise correlation r. Given any two of the quantities (n, k, r), Lin (1995) presents the possible values that can be achieved for the third quantity. Moreover, designs given in Lin (1995) may be adequate to allow examination of many prespecified two-factor interactions. Some of the results are summarized in Table 3.

Table 3 shows the maximum number of factors, k_{max}, that can be accommodated when both n and r are specified for $3 \leq n \leq 25$ and $0 \leq r \leq 1/3$ (Table 3a for even n and Table 3b for odd n). We see that for $r \leq 1/3$, many factors can be accommodated. For fixed n, as the value of r increases, k_{max} also increases. That is, the larger the nonorthogonality, the more factors can be accommodated. In fact, k_{max} increases rapidly in this setting. Certainly the more factors accommodated, the more complicated are the biased estimation relationships that occur, leading to more difficulty in data analysis. On the other hand, for fixed r, the value of k_{max} increases rapidly as n increases. For $r < 1/3$, one can accommodate at most 111 factors in 18 runs or 66 factors in 12 runs; for $r \leq 1/4$, one can accommodate 42 factors in 16 runs; for $r \leq 1/5$, one can accommodate 34 factors in 20 runs. Provided that these maximal correlations are acceptable, this can be an efficient design strategy.

Table 3 Maximal Number of Factors Found, $k_{\max}$, as a Function of n and nr, for $3 \leq n \leq 25$ and $r \leq 1/3$

(a) Even n

Number of runs n	Maximum absolute cross product, $nr = \lvert c_i' c_j \rvert$				
	0	2	4	6	8
4	3				
6	—	10			
8	7	—			
10	—	12			
12	11	—	66		
14	—	13	—	113	
16	15	—	42	—	
18	—	17	—	111	
20	19	—	34		
22	—	20	—	92	—
24	23	—	33	—	276

(b) Odd n

Number of runs n	Maximum absolute cross product, $nr = \lvert c_i' c_j \rvert$			
	1	3	5	7
3	3			
5	4			
7	7	15		
9	7	12		
11	11	14		
13	12	14		
15	15	15	37	
17	15	17	50	
19	19	19	33	
21	19	19	34	92
23	23	23	33	94
25	23	23	32	76

5. OPTIMALITY CRITERIA

When a supersaturated design is employed, as previously mentioned, the abandonment of orthogonality is inevitable. It is well known that lack of orthogonality results in lower efficiency; therefore we seek a design that is as "nearly orthogonal" as possible. One way to measure the degree of nonorthogonality between two columns, c_i and c_j, is to consider their cross-product, $s_{ij} = c_i'c_j$; a larger $|s_{ij}|$ implies less orthogonality. Denote the largest $|s_{ij}|$ among all pairs of columns for a given design by s, and we desire a minimum value for s ($s = 0$ implies orthogonality). The quantity s can be viewed as the degree of orthogonality that the experimenter is willing to give up—the smaller, the better. This is by nature an important criterion. Given any two of the quantities (n, k, s), it is of interest to determine what value can be achieved for the third quantity. Some computational results were reported by Lin (1995). No theoretical results are currently available, however. It is believed that some results from coding theory can be very helpful in this direction. Further refinement is currently under investigation.

If two designs have the same value of s, we prefer the one in which the value of $|s_{ij}| = s$ is a minimum. This is intimately connected with the expectation of s^2, $E(s^2)$, first proposed by Booth and Cox (1962) and computed as $\sum s_i^2 f_i / \binom{k}{2}$, where f_i is the frequency of s_i among all $\binom{k}{2}$ pairs of columns.

Intuitively, $E(s^2)$ gives the increment in the variance of estimation arising from nonorthogonality. It is, however, a measurement for pairwise relationships only. More general criteria were obtained by Wu (1993) and Deng et al. (1994, 1996b). Deng and Lin (1994) outlined eight criteria useful for supersaturated designs: $s = \max |c_i'c_j|$; $E(s^2)$; ρ (Lin, 1995); D_f, A_f, E_f (Wu, 1993); B criterion (Deng et al., 1996a, 1996b); and r-rank (see Section 8). Further theoretical justification is currently under study. Optimal designs in light of these approaches deserve further investigations. In addition, the notion of multifactor (non)orthogonality is closely related to the multicollinearity in linear model theory.

6. DATA ANALYSIS METHODS

Several methods have been proposed to analyze the k effects, given only the $n(< k)$ observations from the random balance design contents (see, e.g., Satterthwaite, 1959). These methods can also be applied here. Quick methods such as these provide an appealing, straightforward comparison among factors, but it is questionable how much available information can be extracted using them; combining several of these methods provides a more satisfying result. In addition, three data analysis methods for data resulting from a supersaturated design are discussed in Lin (1995): (1) normal plotting, (2) stepwise selection, and (3) ridge regression.

To study so many columns in only a few runs, the probability of a false positive reading (type I error) is a major risk here. An alternative to the forward selection procedure to control these false positive rates is as follows. Let $\mathbf{N} = \{i_1, i_2, \ldots, i_p\}$ and $\mathbf{A} = \{i_p + 1, \ldots, i + k\}$ denote indexes of inert and active factors, respectively, so that $\mathbf{N} \cup \mathbf{A} = \{1, \ldots, k\} = \mathbf{S}$. If $\mathbf{X}$ denotes the $n \times p$ design matrix, our model is $\mathbf{Y} = \mu\mathbf{1} + \mathbf{X}\beta + \epsilon$, where $\mathbf{Y}$ is the $n \times 1$ observable data vector, μ is the intercept term, $\mathbf{1}$ is an n-vector of 1's, β is a $k \times 1$ fixed and unknown vector of factor effects, and ϵ is the noise vector. In the multiple hypothesis testing framework, we have null and alternative pairs $H_j : \beta_j = 0$ and $H_j^c : \beta_j \neq 0$ with H_j true for $j \in \mathbf{N}$ and H_j^c true for $j \in \mathbf{A}$.

Forward selection proceeds by identifying the maximum F statistics at successive stages. Let $F_j^{(s)}$ denote the F statistic for testing H_j at stage s. Consequently, define

$$j_t = \arg \max_{j \in \mathbf{S} - \{j_1, \ldots, j_{t-1}\}} F_j^{(t)}$$

where

$$F_j^{(t)} = \text{RSS}(j | j_1, \ldots, j_{s-1}) / \text{MSE}(j, j_1, \ldots, j_{s-1})$$

Letting $\max F_j^{(s)} = F^{(s)}$, the forward selection procedure is defined by selecting variables $j_1, \ldots, j_f$, where $F^{(f)} \leq \alpha$ and $F^{(f+1)} > \alpha$. If $F^{(1)} > \alpha$, then no variables are selected.

The type I (false positive) error rate may be controlled by using the adjusted p-value method (Westfall and Young, 1993). Algorithmically, at stage j, if $p^{(j)} > \alpha$, then stop; otherwise, enter X_j and continue. This procedure controls the type I error rate exactly at level α under the complete null hypothesis since

$$P(\text{Rejects at least one } H_i | \text{ all } H_i \text{ true}) = P(F^{(1)} \leq f_\alpha^{(1)}) = \alpha$$

In addition, if the first s variables are *forced* and the test is used to evaluate the significance of the next entering variable (of the remaining $k - s$ variables), the procedure is again exact under the complete null hypothesis of no effects among the $k - s$ remaining variables. The exactness disappears with simulated p values, but the errors can be made very small, particularly with control variates. The analysis of data from supersaturated designs along this direction can be found in Westfall et al. (1998).

7. EXAMPLES

Examples of supersaturated designs as real data sets can be found in Lin (1993, 1995). Here we apply the concept of supersaturated design to identify interaction effects from a main-effect orthogonal design. This example is adapted from Lin (1998). Consider the experiment in Hunter et al. (1982). A 12-run Plackett and Burman design was used to study the effects of seven factors (designated here as **A**, **B**, $\cdots$, **G**) on the fatigue life of weld-repaired castings. The design and responses are given in Table 4 (temporarily ignore columns 8–28). For the details of factors and level values, see Hunter et al. (1982).

Plackett and Burman designs are traditionally known as main-effect designs, because if all interactions can temporarily be ignored, they can be used to estimate all main effects. There are many ways to analyze such a main-effect design. One popular way is the normal plot [see Hamada and Wu (1992), Figure 1]. Using this method, it appears that factor **F** is the only significant main effect. Consequently a main-effect model is fitted as follows: $\hat{y} = 5.73 + 0.458\mathbf{F}$ with $R^2 = 44.5\%$.

Note that the low R^2 is not very impressive. Is it safe to ignore the interaction effects? Hunter et al. claim that the design did not generate enough information to identify specific conjectured interaction effects. If this is not the case here, is it possible to detect significant interaction effects? Hamada and Wu (1992) introduced the concept of effect heredity. After main effects were identified, they used forward selection regression to identify significant effects among a group consisting of (1) the effects already identified and (2) the two-factor interactions having at least one component factor appearing among the main effects of those already identified. In this particular example, a model for factor **F** and interaction **FG** was chosen:

$$\hat{y} = 5.7 + 0.458\mathbf{F} - 0.459\mathbf{FG}, \qquad R^2 = 89\% \tag{1}$$

Now, if we generate all interaction columns, **AB**, **AC**, ..., **FG**, together with all main effect columns, **A**, **B**, ..., **G**, we have $7 + 21 = 28$ columns. Treat all of those 28 columns in 12 runs as a supersaturated design (Lin, 1993) as shown in Table 4. The largest correlation between any pair of the design columns is $\pm 1/3$. The results from a regular stepwise regression analysis (with $\alpha = 5\%$ for entering variables) yields the model

Table 4 The Cast Fatigue Experiment Data with Interaction Columns

	1	2	3	4	5	6	7	8	9	10	11	12	13	14	15	16	17	18	19	20	21	22	23	24	25	26	27	28	
Run	**A**	**B**	**C**	**D**	**E**	**F**	**G**	**AB**	**AC**	**AD**	**AE**	**AF**	**AG**	**BC**	**BD**	**BE**	**BF**	**BG**	**CD**	**CE**	**CF**	**CG**	**DE**	**DF**	**DG**	**EF**	**EG**	**FG**	Responses
1	1	1	−1	1	1	1	−1	1	−1	1	1	1	−1	−1	1	1	1	−1	−1	−1	−1	1	1	1	−1	1	−1	−1	6.058
2	1	−1	1	1	1	−1	−1	−1	1	1	1	−1	−1	−1	−1	−1	1	1	1	1	−1	−1	1	−1	−1	−1	−1	1	4.733
3	−1	1	1	1	−1	−1	−1	−1	−1	−1	1	1	1	1	1	−1	−1	−1	1	−1	−1	−1	−1	−1	−1	1	1	1	4.625
4	1	1	1	−1	−1	−1	1	1	1	−1	−1	−1	1	1	−1	−1	−1	1	−1	−1	−1	1	1	1	−1	1	−1	−1	5.899
5	1	1	−1	−1	−1	1	−1	1	−1	−1	−1	1	−1	−1	−1	−1	1	−1	1	1	−1	1	1	−1	1	−1	1	−1	7.000
6	1	−1	−1	−1	1	−1	1	−1	−1	−1	1	−1	1	1	1	−1	1	−1	1	−1	1	−1	−1	1	−1	−1	1	−1	5.752
7	−1	−1	−1	1	−1	1	1	1	1	−1	1	−1	−1	1	−1	1	−1	−1	−1	1	−1	−1	−1	1	1	−1	−1	1	5.682
8	−1	−1	1	−1	1	1	−1	1	−1	1	−1	−1	1	−1	1	−1	−1	1	−1	1	1	−1	−1	−1	1	1	−1	−1	6.607
9	−1	1	−1	1	1	−1	1	−1	1	−1	−1	1	−1	−1	1	1	−1	1	−1	−1	1	−1	1	−1	1	−1	1	−1	5.818
10	1	−1	1	1	−1	1	1	−1	1	1	−1	1	1	−1	−1	1	−1	−1	1	−1	1	1	−1	1	1	−1	−1	1	5.917
11	−1	1	1	−1	1	1	1	−1	−1	1	−1	−1	−1	1	−1	1	1	1	−1	1	1	1	−1	−1	−1	1	1	1	5.863
12	−1	−1	−1	−1	−1	−1	−1	1	1	1	1	1	1	1	1	1	1	1	1	1	1	1	1	1	1	1	1	1	4.809

$$\hat{y} = 5.73 + 0.394\mathbf{F} - 0.395\mathbf{FG} - 0.191\mathbf{AE}, \qquad R^2 = 95\% \tag{2}$$

a significantly better fit to the data than Eq. (1). An application of the adjusted p-value method (Westfall, et al. 1998) reaches the same conclusion in this example.

Note that the **AE** interaction, in general, would never be chosen under the effect heredity assumption. Of course, most practitioners may consider adding main effects **A**, **E**, and **G** to the final model because of the significance of interactions **FG** and **AE**. The goal here is only to identify potential interaction effects. In general, for most main-effect designs, such as Plackett and Burman type designs (except for 2^{k-p} fractional factorials), one can apply the following procedure [see Lin (1998) for the limitations]:

Step 1. Generate all interaction columns and combine them with the main-effect columns. We now have $k(k+1)/2$ design columns.

Step 2. Analyze these $k(k+1)/2$ columns with n experimental runs as a supersaturated design. Data analysis methods for such a supersaturated design are available.

Note that if the interactions are indeed inert, the procedure will work well, and if the effect heredity assumption is indeed true, the procedure will end up with the same conclusion as that of Hamada and Wu (1992). The proposed procedure will always result in better (or equal) performance than that of Hamada and Wu's procedure.

8. THEORETICAL CONSTRUCTION METHODS

Deng et al. (1994) proposed a supersaturated design of the form $\mathbf{X}_c = [\mathbf{H}, \mathbf{RHC}]$, where $\mathbf{H}$ is a normalized Hadamard matrix, $\mathbf{R}$ is an orthogonal matrix, and $\mathbf{C}$ is an $n \times (n-c)$ matrix representing the operation of column selection. Besides the fact that some new designs with nice properties can be obtained this way, the $\mathbf{X}_c$ matrix covers many existing supersaturated designs. This includes the supersaturated designs proposed by Lin (1993a), Wu (1993), and Tang and Wu (1993). Some justifications of its optimal properties have been obtained as follows.

It can be shown that

$$\mathbf{X}_c'\mathbf{X}_c = \begin{pmatrix} n\mathbf{I}_n & \mathbf{H}'\mathbf{RHC} \\ \mathbf{C}'\mathbf{H}'\mathbf{R}'\mathbf{H} & n\mathbf{I}_{n-c} \end{pmatrix} = \begin{pmatrix} n\mathbf{I}_n & \mathbf{WC} \\ \mathbf{C}'\mathbf{W}' & n\mathbf{I}_{n-c} \end{pmatrix}$$

where $\mathbf{W} = \mathbf{H}'\mathbf{RH} = (w_{ij}) = (\mathbf{h}_i'\mathbf{Rh}_j)$ and $\mathbf{h}_j$ is the jth column of $\mathbf{H}$. Further, the following theorem can be demonstrated.

Theorem

Let **H** be a Hadamard matrix of order n and $\mathbf{B} = (b_1, \ldots, b_r)$ be an $n \times r$ matrix with all entries ± 1 and $\mathbf{V} = \mathbf{H}'\mathbf{B} = (v_{ij}) = \mathbf{h}_i' b_j$. Then

1. For any fixed $1 \le j \le r$, $n^2 = \sum_{i=1}^n v_{ij}^2$.
2. In particular, let $\mathbf{B} = \mathbf{RH}$ and $\mathbf{W} = \mathbf{H}'\mathbf{RH} = (w_{ij})$. We have
   a. $(1/n)\mathbf{W}$ is an $n \times n$ orthogonal matrix.
   b. $n^2 = \sum_{i=1}^n w_{ij}^2 = \sum_{j=1}^n w_{ij}^2$.
   c. w_{ij} is always a multiple of 4.
   d. If $\mathbf{H}'$ is column-balanced, then $\pm n = \sum_{i=1}^n w_{ij} = \sum_{j=1}^n w_{ij}$.

Corollary

For any **R** and **C** such that (1) $\mathbf{R}'\mathbf{R} = \mathbf{I}$ and (2) rank $(\mathbf{C}) = n - c$, all X_c matrices have an identical $E(s^2)$ value.

This implies that the popular $E(s^2)$ criterion used in supersaturated designs is invariant for any choice of **R** and **C**. Therefore, it is not effective for comparing supersaturated designs. In fact, following the argument in Tang and Wu (1993), the designs given here will always have the minimum $E(s^2)$ values within the class of designs of the same size. One important feature of the goodness of a supersaturated screening design is its projection property (see Lin 1993b). We thus consider the r-rank property as defined below.

Definition

Let **X** be a column-balanced design matrix. The resolution rank (or r rank, for short) of **X** is defined as $f = d - 1$, where d is the minimum number subset of columns that are linearly dependent.

The following results are provided by Deng et al. (1994).

1. If no column in any supersaturated design **X** is fully aliased, then the r rank of **X** is at least 3.
2. $n\mathbf{Rh}_j = \sum_{i=1}^n w_{ij}\mathbf{h}_i$.
3. Let $\mathbf{W} = \mathbf{H}'\mathbf{D}(\mathbf{h}_l)\mathbf{H}$, where $\mathbf{D}(\mathbf{h}_l)$ is the diagonal matrix associated with $\mathbf{h}_l$, namely, the lth column vector of **H**; and $n = 4t$. Then
   a. If t is odd, then there can be exactly three 0's in each row, or each column, of **W**. The rest of w_{ij} can only be of the form $\pm 8k + 4$, for some nonnegative integer k.
   b. If t is even, then every entry w_{ij} in **W** can be of the form $\pm 8k$, for some nonnegative integer k.

These results are only the first step. Extension of these results to a more general class of supersaturated designs in the form $\mathbf{S}_K = (\mathbf{R}_1\mathbf{HC}_1, \ldots, \mathbf{R}_K \mathbf{HC}_K)$ is promising.

9. COMPUTER ALGORITHMIC CONSTRUCTION METHODS

More and more researchers are benefiting from using computer power to construct designs for specific needs. Unlike some cases from the optimal design perspective (such as *D*-optimal design), computer construction of supersaturated designs is not well developed yet. Lin (1991) introduced the first computer algorithm to construct supersaturated designs. Denote the largest correlation in absolute value among all design columns by r, as a simple measure of the degree of nonorthogonality that can willingly be given up. Lin (1995) examines the maximal number of factors that can be accommodated in such a design when r and n are given.

Al Church at GenCorp Company used the projection properties in Lin and Draper (1992, 1993) to develop a software package named DOE0 to generate designs for mixed-level discrete variables. Such a program has been used at several sites in GenCorp. A program named DOESS is one of the results and is currently in a test stage. Dr. Nam-Ky Nguyen (CSIRO, Australia) also independently works on this subject. He uses an exchange procedure to construct supersaturated designs and near-orthogonal arrays. A commercial product called Gendex is available for sale to the public, as a result. Algorithmic approaches to constructing supersaturated designs seem to have been a hot topic in recent years. For example, Li and Wu (1997) developed a so-called columnwise–pairwise exchange algorithm. Such an algorithm seems to perform well for constructing supersaturated designs by various criteria.

10. CONCLUSION

1. Using supersaturated designs involves more risk than using designs with more runs. However, it is far superior to other experimental approaches such as subjectively selecting factors or changing factors one at a time. The latter can be shown to have unresolvable confounding patterns, though such confounding patterns are important for data analysis and follow-up experiments.
2. Supersaturated designs are very useful in early stages of experimental investigation of complicated systems and processes involving many factors. They are not used for a terminal experiment.

Knowledge of the confounding patterns makes possible the interpretation of the results and provides the understanding of how to plan the follow-up experiments.

3. The success of a supersaturated design depends heavily on the "effect sparsity" assumption. Consequently, the projection properties play an important role in designing a supersaturated experiment.
4. Combining several data analysis methods to analyze the data resulting from a supersaturated design is always recommended. Besides the stepwise selection procedure [and other methods mentioned in Lin (1993)], PLS (partial least squares), adjusted *p* value (see Westfall, et al. (1998)), and Bayesian approaches are promising procedures for use in identifying active factors.
5. Another particularly suitable use for these designs is in testing "robustness," where the objective is not to identify important factors but to vary all possible factors so that the response will remain within the specifications.

REFERENCES

Booth KHV, Cox DR. (1962). Some systematic supersaturated designs. Technometrics 4:489–495.

Chen JH, Lin DKJ. (1998). On identifiability of supersaturated designs. J Stat Planning Inference, 72, 99–107.

Cheng CS. (1997). $E(s^2)$-optimal supersaturated designs. Stat Sini, 7, 929–939.

Deng LY, Lin KJ. (1994). Criteria for supersaturated designs. Proceedings of the Section on Physical and Engineering Sciences, American Statistical Association, pp 124–128.

Deng LY, Lin DKJ, Wang JN. (1994). Supersaturated Design Using Hadamard Matrix. IBM Res Rep RC19470, IBM Watson Research Center.

Deng LY, Lin DKJ, Wang JN. (1996a). Marginally oversaturated designs. Commun Stat 25(11):2557–2573.

Deng LY, Lin DKJ, Wang JN. (1996b). A measurement of multifactor orthogonality. Stat Probab Lett 28:203–209.

Hamada M, Wu CFJ. (1992). Analysis of designed experiments with complex aliasing. J Qual Technol 24:130–137.

Hunter GB, Hodi FS, Eager TW. (1982). High-cycle fatigue of weld repaired cast Ti-6Al-4V. Metall Trans 13A:1589–1594.

Li WW, Wu CFJ. (1997). Columnwise-pairwise algorithms with applications to the construction of supersaturated designs. Technometrics 39:171–179.

Lin DKJ. (1991). Systematic supersaturated designs. Working Paper No. 264, College of Business Administration, University of Tennessee.

Lin DKJ. (1993a). A new class of supersaturated designs. Technometrics 35:28–31.

Lin DKJ. (1993b). Another look at first-order saturated designs: The *p*-efficient designs. Technometrics 35:284–292.

Lin DKJ. (1998). Spotlight interaction effects in main-effect designs. Quality Engineering 11(1), 133–139.

Lin DKJ. (1995). Generating systematic supersaturated designs. Technometrics 37:213–225.

Lin DKJ, Draper NR. (1992). Projection properties of Plackett and Burman designs. Technometrics 34:423–428.

Lin DKJ, Draper NR. (1993). Generating alias relationships for two-level Plackett and Burman designs. Comput Stat Data Anal 15:147–157.

Nguyen N-K. (1996). An algorithmic approach to constructing supersaturated designs. Technometrics 38:69–73.

Plackett RL, Burman JP. (1946). The design of optimum multifactorial experiments. Biometrika 33:303–325.

Satterthwaite F. (1959). Random balance experimentation. Technometrics 1:111–137 (with discussion).

Tang B, Wu CFJ. (1997). A method for construction of supersaturated designs and its E(s^2) optimality. Can J Stat 25:191–201.

Westfall PH, Young SS. (1993). Resampling-Based Multiple Testing. New York: Wiley.

Westfall, PH, Young SS, Lin DKJ. (1998). Forward selection error control in the analysis of supersaturated designs. Stat Sin, 8, 101–117.

Watson, GS. (1961). A study of the group screening methods. Technometrics 3:371–388.

Wu CFJ. (1993). Construction of supersaturated designs through partially aliased interactions. Biometrika 80:661–669.

Yamada S, Lin DKJ. (1997). Supersaturated designs including an orthogonal base. Can J Stat 25:203–213.

Youden WJ, Kempthorne O, Tukey JW, Box GEP, Hunter JS. (1959). Discussion on "Random balance experimentation" by Satterthwaite. Technometrics 1:157–184.

19
Statistical Methods for Product Development: Prototype Experiments

David M. Steinberg
Tel Aviv University, Tel Aviv, Israel

Søren Bisgaard
University of St. Gallen, St. Gallen, Switzerland

1. INTRODUCTION

Many authors have emphasized the importance of product development for long-term business survival [1–4]. The rapid pace of technological progress in today's economy makes it increasingly important to reduce development time and get new products to market quickly. Page [5] discovered that most of the development cycle was devoted to the physical development of the product. In our experience, much of that effort goes into experiments whose goals may include improving performance, comparing design alternatives, increasing reliability, or verifying that the product meets its stated goals and specifications. Thus efficient methods of experimentation can be of great value in ramping up the learning curve and accelerating the product development process [6, 7].

In this chapter we focus in particular on the use of factorial experiments for prototype testing, building on the ideas in Bisgaard and Steinberg [8]. Prototype tests provide design engineers with valuable information about the performance of products before they are sent further downstream for tooling and ramp-up for production. The knowledge acquired from these tests can be used to optimize and robustify products. Often a sequence of prototypes is built, beginning with computer-aided design (CAD) drawings and leading to the construction of a full-scale product. Since prototype tests can be run from early on in the development cycle, they can help eliminate potential quality problems without the large costs and delays that are usually incurred when problems are discovered in the later phases of the design-to-production cycle.

The common paradigm for prototype testing is to build and evaluate a single model at each stage. This approach is implicit in the excellent account by Wheelwright and Clark [3, Chapter 10] of the role of prototypes in product development.

It is our experience that great gains can be made by using factorial experiments to study and improve product design at the prototype stage. Several alternatives can be made, varying important design factors according to a factorial plan. The results of such experiments can substantially accelerate the path from concept development to finished product and can significantly lower the risk of discovering serious quality problems late in the development cycle.

A striking example of the importance of rapid feedback at early stages in the design process is presented by Clark and Fujimoto [9, Chapter 7] in their comprehensive study of auto manufacturers. They found that the lead time for developing a new car was about 25% less in Japan than in the United States. One major reason for this difference was that the Japanese companies were much more successful than their American counterparts at rapidly reducing the number of design problems early in the development process. Clark and Fujimoto credited this difference to the prototyping strategies that were prevalent in the two countries. The U.S. companies built few prototypes and treated them as master models; the Japanese companies built many prototypes and used them to provide information for finding and solving design problems. Our approach couples the power of statistical experiments with the Japanese strategy.

Prototype experiments have two interesting statistical features. First, it is typically much more expensive to build a prototype than to test it. Thus there is good reason, once a prototype is built, to test it extensively. The relevant test conditions, which can often be laid out in a factorial plan, will then be nested within the prototype configurations, in what is known as a split-plot structure. Second, interest often focuses on a performance curve rather than on a single number output. In motor testing, for example, the test might examine fuel consumption as a function of load or rpm, torque as a function of rpm, compression ratio as a function of a single 360° stroke, or the curve trace of the torque or power delivered through a gear shift cycle from forward through neutral to reverse and back again. Other examples include the hysteresis curve in the testing of transformers, the spectrum of the emitted light in the testing of light bulbs, the hardness as a function of depth in ion implantation of steel, the pressure versus time curve in a pyrotechnic chain, and the characteristic curve in the testing of transistors.

Experiments that include factors related to product design along with factors that reflect test settings have received some attention within the robust parameter design strategy of Taguchi [10]. The paradigm recommended by Taguchi is to use a factorial design to prepare product or system configurations and then to run each configuration at different settings (following a second factorial plan) of noise and signal factors. The noise factors might reflect possible variations in the production or use environment, and the signal factors represent adjustable inputs that the product user can control to produce a desired response (e.g., the force applied to a brake pedal). This paradigm is similar to the setting we have in mind, in particular what Taguchi has called "dynamic experiments," which study the performance curve of a product with respect to a signal factor. However, our method of analysis differs from that proposed by Taguchi. An approach similar to ours was proposed by Miller and Wu [11] for robust design experiments with signal factors.

In this chapter we describe the general statistical methodology proposed by Bisgaard and Steinberg [8] for prototype tests. We begin in the next section with a general discussion of the product design process and the role of prototype testing. Section 3 presents a number of examples of prototype experiments. Section 4 describes a simple two-stage analysis that is appropriate when the experiment focuses on a performance curve and the test conditions are nested within prototypes. Section 5 illustrates the analysis with an experiment to improve an engine starting system [12]. Some concluding remarks follow in Section 6.

2. THE PRODUCT DESIGN PROCESS

In Figure 1 we show a schematic representation of the product development process first introduced by Bisgaard [6]. The steps shown there are the same ones found in most traditional texts on product development, but we emphasize in Figure 1 that the development process is best viewed as one that is cyclical and ongoing, not a linear procession with a distinct beginning and end. Most products evolve from similar predecessors, go through a sequence of improvement cycles, and ultimately spawn new products. These cycles within the product development process have much in common with the Plan–Do–Check–Act cycle of Deming [13] and Shewhart [20].

One of the most important features of the development process shown in Figure 1 is the acquisition of new knowledge at each stage. Experiments often play a key role in unlocking these secrets of nature. Even when the source of insight is a theoretical breakthrough or comes from observational data, experiments will typically be needed to verify the theory. In our own

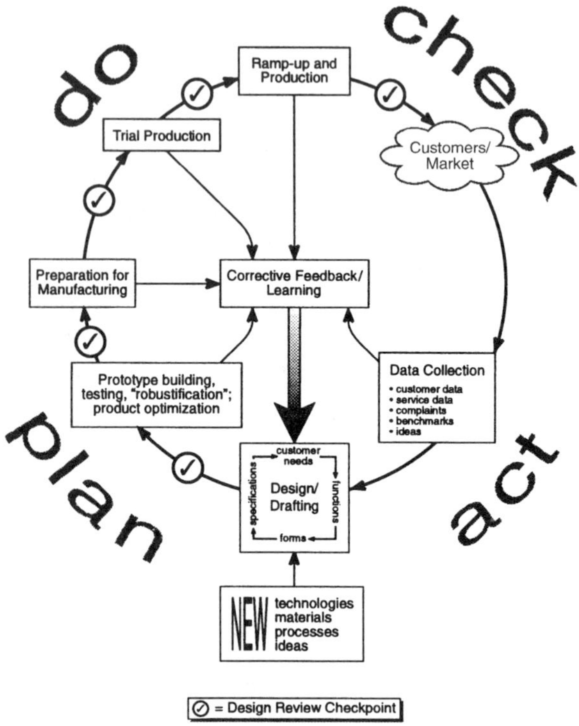

Figure 1 A conceptual view of the product development process as a cyclic learning cycle, analogous to Shewhart and Deming's Plan–Do–Check–Act cycle.

contacts with design engineers, we regularly see experiments used to test new concepts, compare designs, evaluate new materials, optimize performance, improve quality and reliability, and verify performance specifications. Efficient experimentation can be a crucial tool in the quest to bring high quality products to market ahead of the competition. Carefully planned factorial experiments can provide invaluable knowledge throughout the development cycle. See Bisgaard and Ellekjaer [7] for a broad conceptual account.

The prototype stage is especially well suited to experimental work. Typically prototypes are built fairly early in the development of a new product, when it is easiest to make design changes. Factorial experiments on prototypes can be an ideal method for comparing design alternatives and shaping the direction of future development. Once that direction is set and large amounts of time and money have been invested, it becomes increasingly difficult to make any fundamental changes to the product design. Thus the biggest payoff from additional knowledge, and hence from good experiments, is at the early stages in the development cycle when prototypes are being built and studied.

3. PROTOTYPE EXPERIMENTS: SOME EXAMPLES

3.1. Airplane Wing

Initial prototype development often takes the form of CAD drawings rather than actual physical mock-ups. Software that simulates the proposed operating environment can then be used to study the performance of the design on the computer. The experiment in question here was carried out by a team of engineers at the "concept design" stage. The two main goals were to improve the performance of the wing, as measured by thrust per unit weight, and to minimize the cost per unit performance. Five different aspects of the wing were studied: the sizes of three physical dimensions, the number of strength supports on the wing, and the type of material used in construction. Two possible values were considered for each of these factors, and eight prototypes were then defined, in accord with a standard 2^{5-2} fractional factorial experiment. Each prototype was carefully drawn by the design team using CAD software. The weight and cost of each prototype wing was then calculated and finite element analysis was used to compute the thrusts.

3.2. Engine Throttle Handle

Bisgaard [14] described an experiment to improve the performance of the throttle handle for an outboard motor. The goal of the experiment was to derive appropriate tolerances for seven physical dimensions by studying their effects on friction in the handle. The throttle handle is assembled from three parts: a knob, a handle, and a tube. Of the dimensions studied, three were related to the knob, three to the handle, and one to the tube. An interesting feature of this experiment is that separate experimental plans were set up for making prototypes of each of the three components (a 2^3 plan for the knobs, a 2^{3-1} plan for the handles, and a 2^1 plan for the tubes). All possible matchings of the prototype components were then assembled and tested for friction.

3.3. Engine Exhaust

Taguchi [10, p. 131] described an experiment to reduce the CO content of engine exhaust. Seven different characteristics of the engine design were studied using a saturated two-level design that specified eight prototype engines. Each engine was then run at three different driving modes, which constituted the test conditions for this study. Bisgaard and Steinberg [8] analyzed the results from this experiment and found that one of the factors had an interesting, and statistically significant, effect on the shape of the response curve, as shown in Figure 2. With this factor at its low level, the response curve was lower at the middle driving mode but higher at the high mode. The engineering significance of this effect depends on which driving modes will be encountered most often. The lower driving modes likely correspond to the sort of stop-and-start traffic common in large cities, and it might then be desirable to choose the factor at its low level to reduce the CO content at these modes.

3.4. Kitchen Mixer

Ott [15] described an experiment to improve a kitchen mixer. Each mixer was assembled from three components: a top unit, a bottom unit, and gears. An experiment was run to determine which of these three components was the cause of inefficient operation. Forty-eight mixers were used in the study, half of them efficient and half inefficient. Each mixer was disassembled, and

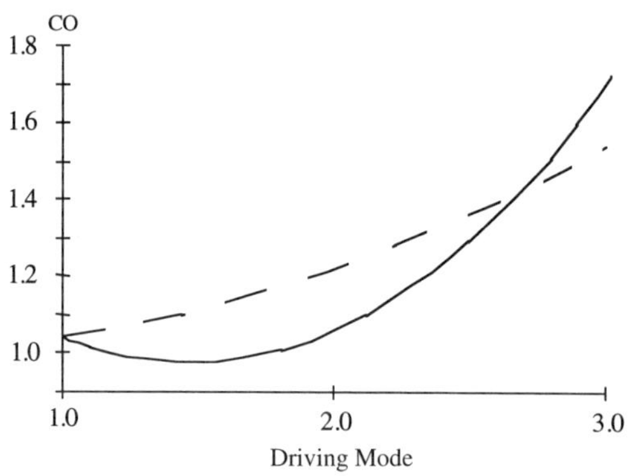

Figure 2 The estimated response curves for CO exhaust versus driving mode at the two levels of factor A for the engine exhaust experiment. The response curve with A at its high level (solid line) is lower than the curve with A at its low level (dashed line) across most of the driving modes but shows a sharp increase at high driving modes.

then 48 new mixers were assembled, swapping parts from the original mixers to form a 2^3 factorial design whose factors were the three components. The two levels for each factor were determined by the source of the component in an efficient (or inefficient) mixer. The experiment clearly pointed to the tops as the source of the problem.

3.5. Pyrotechnic Device

Milman et al. [16] reported on an experiment to improve the safety of a pyrotechnic device. It was known that the safety improvements could be achieved by using a new type of initiator, but there was concern that this change would adversely affect the performance of the device. An experiment was run to test 24 prototype devices, mating each of three safe initiators with four types of main charge and two types of secondary charge. The observed response for each prototype device was a trace of pressure against time.

3.6. Fluid Flow Controller

Bisgaard and Steinberg [8] described an experiment to study how prototype fluid flow control devices respond to changes in electrical input and flow rate. The controller was assembled from two components. Two experimental factors described dimensions of the first component, and a third factor described a dimension of the second component. As in the engine throttle experiment, the eight prototype controllers were formed by making four versions of the first component (following a 2^2 plan) and two versions of the second component and then mating all possible pairs of components. Each prototype was subjected to six test conditions formed by crossing three levels of the electrical input with two flow rates.

3.7. Hearing Aid

A remote control unit developed to permit easy control of a new, miniaturized hearing aid suffered from poor reception. A factorial experiment was carried out to test several conjectures as to the source of the problem, in particular that the difficult-to-control variation in the receptor coil was causing variations in the transmission frequency and that the type of cover used was affecting reception. The experiment showed that coil variation was the major problem and that it could be easily remedied by exploiting a large interaction between the coil and the transmission program (another factor in the experiment). The choice of cover was found to have no effect at all.

3.8. Bearing Manufacture

Although we have emphasized throughout this chapter the use of factorial experiments for prototype products, the same ideas can be applied to prototype process development. Hellstrand [17] described an experiment conducted at SKF, one of the world's largest manufacturers of ball bearings, to improve a production process. The goal of the experiment was to improve bearing life, and three factors were studied in a 2^3 plan: heat treatment, osculation, and cage design. The experiment uncovered a large interaction effect between heat treatment and osculation that led to a fivefold increase in bearing life.

4. ANALYSIS OF PROTOTYPE EXPERIMENTS

4.1. Standard Experimental Plans

Some prototype experiments are standard factorials or fractional factorials (e.g., the airplane wing and throttle handle experiments). No special methods are needed for the analysis of these experiments.

4.2. Two-Stage Analysis for Nested Test Conditions

Prototypes are typically much more expensive to make than to test, and it will then be advisable to apply a sequence of test conditions to each prototype. This scheme generates a split-plot structure in which the test conditions are nested within the prototype design. The analysis should correctly account for the nesting.

We suggest a simple, yet general, two-stage analysis method for experiments with nested test conditions:

1. Estimate the effects of the test factors for each prototype. We discuss below some useful ways to summarize these effects.
2. Use the effects found in stage 1 as "data" in a standard factorial analysis to study the effects of the design factors that guided the construction of the prototypes.

As an example, suppose there is a single test factor t and interest focuses on the performance curve that describes its relationship to an output y. For each prototype, fit a polynomial performance curve. The model equation for the ith prototype is

$$\beta_{i,0} + \sum_{l=1}^{m} \beta_{i,l} g_l(t) \qquad i = 1, \ldots, n \tag{1}$$

where $g_l(t)$ is a polynomial of degree l. We define the polynomials so that they are orthogonal with respect to the levels of the test factor. An advantage of this is that only the mean level effects involve interprototype ("whole plot") error. Any effects related to the slope or curvature or higher order properties of the performance curve will involve only intraprototype ("subplot") error. We also scale the orthogonal polynomials so that

$$\sum_{j=1}^{s} g_l^2(t_j) = 1, \qquad l = 1, \ldots, m$$

where the sum runs over all the test settings. The scaling guarantees that all the coefficients (except the constant) will have the same variance, a property that is important at the second stage of the analysis.

The use of orthogonal polynomials with our scaling convention leads to simple coefficient estimates. If we denote by $\mathbf{y}_i' = (y_{i1}, \ldots, y_{is})$ the observations on the ith prototype at each of the s test conditions, the least squares estimates of the coefficients are given by

$$\hat{\beta}_{i,0} = \bar{y}_i \tag{2a}$$

$$\hat{\beta}_{i,l} = \sum_{j=1}^{s} g_l(t_j) y_{i,j}, \qquad l = 1, \ldots, m \tag{2b}$$

The constant term is the average of the s observations, and the polynomial coefficients are simple linear contrasts.

At the second stage of the analysis, each of the polynomial coefficients found above is treated as a response variable and a separate analysis is carried out for the coefficients of each degree. The analysis of the constant terms reveals which factors affect the mean level of the performance curve, the analysis of the linear coefficients shows which factors affect slope, etc. Important effects that stand out from error can be identified with standard tools such as normal probability plots and analysis of variance (ANOVA). Note that the effects on the mean level include "whole plot" error, but effects on other aspects of the performance curve, including average coefficients, involve only "subplot" error. ANOVA can account for this situation by doing a split-plot analysis. For the graphical analysis, separate plots must be prepared for the two sets of effects. Our scaling convention from stage 1 implies that all the performance curve coefficients have the same variance.

We take similar care at the second stage to ensure that the effects have the same variance and can thus be combined on a single probability plot. We recommend computing the average value of each coefficient (for ease of interpretation) and then scaling all the design factor contrasts to have the same variance as the average. This property can be checked by setting up the regression matrix Z for the design factor effects with all elements in the first column equal to 1 and then verifying that $Z'Z = nI$, where I is the identity matrix. Each row of the matrix $(Z'Z)^{-1}Z' = (1/n)Z'$ then gives one of the factor effects.

Orthogonal polynomials are a convenient choice to describe a performance curve, but other sets of orthogonal functions could also be used. For some of the engine testing applications described in Section 1, we would naturally expect periodic behavior. In that case, trigonometric functions could be used to generate orthogonal contrasts in the test conditions.

Some experiments involve more than one test factor. Examples above are the fluid flow controller and the engine starting system studies. For these experiments, the natural approach is to estimate the effect of each test factor for each of the prototypes. Interactions among the test factors can also be included if the test array permits their estimation. The analysis will then reveal which product characteristics can be used to affect the dependence of the response on the various test factors. For example, in the fluid flow controller experiment, one important goal was to obtain accurate predictions of the relationship between the response and the test conditions so that controllers could be designed to meet any desired response pattern.

The two-stage analysis has an appealing simplicity. It can also be justified more formally using theory developed for growth curve models in our performance curve context. Bisgaard and Steinberg [8] showed that, for these models, the two-stage analysis actually computes generalized least squares estimates of the parameters (maximum likelihood estimates if the data are normally distributed). We refer interested readers to that article for details on the statistical model and its analysis.

Our analysis approach shares some common ground with that recommended by Taguchi [10] for robust design experiments, but there are some important differences that we would like to point out. The approach taken by Taguchi is to compute, for each prototype, a single summary measure across all the test conditions. This summary measure, which he calls a signal-to-noise ratio, is then taken as a response variable much as in our stage 2 analysis. The major difference between Taguchi's approach and ours is that we compute a complete, multicoefficient summary at our first stage, as opposed to Taguchi's use of a univariate summary. This difference may appear small but is in fact substantial. The single-number summary can throw away much valuable information that is captured by the complete summary. Steinberg and Bursztyn [18] and Bisgaard and Steinberg [8] showed that Taguchi's approach can miss important effects and identify spurious effects that are easily handled by the multicoefficient summary.

4.3. Analysis with Analog Traces

The observed response for each prototype may be a continuous analog trace against time, as in the pyrotechnic experiment. These curves can be analyzed by applying the methods of Section 4.2 to a digitized version of the response along a grid of time points.

An alternative strategy that is often useful is to take as response variables particular features of the observed performance curves that are of interest. In the experiment on the pyrotechnic device, an important feature was the delay time (i.e., the time from activation until the pressure first begins to increase). Feature analysis has the advantage of focusing attention on the most salient aspects of the performance curves. Most features will involve both whole plot and subplot error components and will have differing variances. So it will not in general be possible to combine estimated effects for different features (as we do above for the performance curve effects).

Feature analysis can also be applied when physical considerations suggest a nonlinear model that, modulo some unknown parameters, describes the response curve. The estimated parameters can then be taken as the first-stage summaries of the performance curves for the prototypes. Box and Hunter [19] applied for this approach for nonlinear models.

5. EXAMPLE: THE ENGINE STARTING SYSTEM EXPERIMENT

In this section we show how our two-stage analysis method can be applied to an experiment on engine starting systems that was described by Grove and Davis [12, p. 329]. For additional examples, we refer the interested reader to Bisgaard and Steinberg [8].

The goal of the engine starting system experiment was to reduce the sensitivity of the system to variations in ambient temperature. The performance of the system was evaluated via the relationship between the air-to-fuel (AF) ratio at the tip of the spark plug and the fuel mass pulse, which is controlled by the electronic engine management system. This measure was adopted because the automotive engineers knew that it was a key indicator of ignition success. The experiment studied seven components of the starting system: injector type, distance from injector tip to valve head, injection timing, valve timing, spark plug reach, spark timing, and fuel rail pressure. Six different injector types were used; three levels were used for each of the remaining factors. The L_{18} orthogonal array was used to define the experimental plan for the prototype starting systems. Each of the 18 systems was then tested at six conditions, formed by crossing three fuel mass pulses (30, 45, and 60 msec) with two temperatures (−15°C and +15°C). Two tests were run at each condition, so there are 12 results for each prototype.

The full data set, additional details on the experiment, and a number of alternative analyses can be found in Grove and Davis [12]. We proceed here only with our approach.

Increasing the fuel mass pulse (FMP) injects more fuel into the engine, and initial plots of the data for each prototype show, as expected, a negative correlation between the AF ratio and the FMP. They also show that the AF ratio is typically higher at −15°C than at +15°C. A number of possible models might be considered linking the AF ratio to the FMP, and there is not clear evidence in the experiment to prefer one model over another. For some prototypes, the AF ratio is almost a linear function of the FMP; for others the inverse of the AF ratio is nearly linear, and for others the log of the ratio is most nearly linear. We elected to work with the relationship between the logarithm of the AF ratio and the logarithm of the FMP, which seemed to be most appropriate for the full set of prototypes both for achieving linearity and for reducing the dependence of residual variation on the mean level of response. But we caution that other metrics could also be used and might lead to somewhat different conclusions.

The first stage of our analysis is to estimate for each prototype the effects of log FMP and temperature, including their interactions, on log AF ratio. The levels of FMP were equally spaced (30, 45, and 60 msec), and if we had kept FMP on its original scale we could have used standard polynomial contrasts to compute its linear and quadratic effects. For example, the linear effect would be proportional to the average of the results at 60 msec minus the average of the results at 30 msec. The logarithms of the FMP levels are 3.40, 3.81, and 4.09, and the resulting scaled contrasts are (−0.372, 0.040, 0.332) (linear) and (0.169, −0.406, 0.237) (quadratic). The main effect contrast for temperature is (−0.289, 0.289). The interaction contrasts are similar to the FMP contrasts, but multiplied by 1 or −1, according to whether the temperature is high or low, respectively. Each of the contrasts, when squared and summed over the 12 test points, gives a sum of 1, in accord with our scaling convention.

The second stage of our analysis estimates the effects of the design factors on each of the first-stage coefficients. Since there are 18 prototypes, the "average" contrast in the effects computation has each element equal to 1/18. All the remaining factor effect contrasts are scaled to have the same sum of squares. The linear contrast for each three level factor is (−0.068, 0, 0.068), and the quadratic contrast is (0.0393, −0.0786, 0.0393). Injector type, the 6 level factor, is represented by five orthogonal contrasts. These contrasts are formed by taking the main effects and interactions of the 2 and 3 level columns that were used at the design stage to assign the levels of this factor.

Figure 3 shows a normal probability plot of the effects on mean level (i.e., on the constant terms from the within-prototype regressions). None of the contrasts sharply deviates from a straight line through the origin. Only the two lowest values hint at statistical significance. The strongest contrast is one that corresponds to injector type and indicates that types 4, 5, and 6 have lower average AF ratios than do types 1, 2, and 3. The other large contrast is for the linear effect of fuel rail pressure and indicates lower average AF ratios with higher pressure.

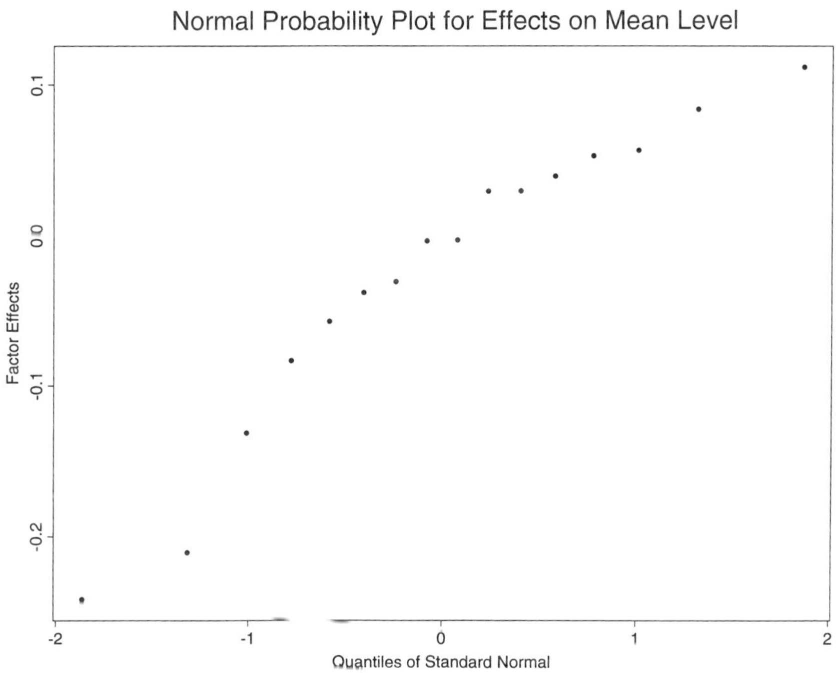

Figure 3 A normal probability plot of the factor effects on the mean level of response from our stage 2 analysis of the engine starting system experiment.

Figure 4 shows a normal probability plot for the effects related to the performance curve. The contrasts for the linear effect of log FMP and for the effect of temperature are clearly significant and dominate all the others. Figure 5 shows a normal probability plot without the two very large contrasts and helps to clarify which contrasts stand out from noise. The only contrasts that appear to be statistically significant are the three largest and the two smallest, all of which correspond to interaction effects with temperature. The factors that interact with temperature are the injector type (two significant contrasts), the distance from the injector tip to the valve head (both the linear and quadratic components), and the valve timing (the linear component). The next largest negative contrast is the interaction between temperature and the quadratic component of the valve timing, so it seems prudent to also take account of this effect in developing a model for the system.

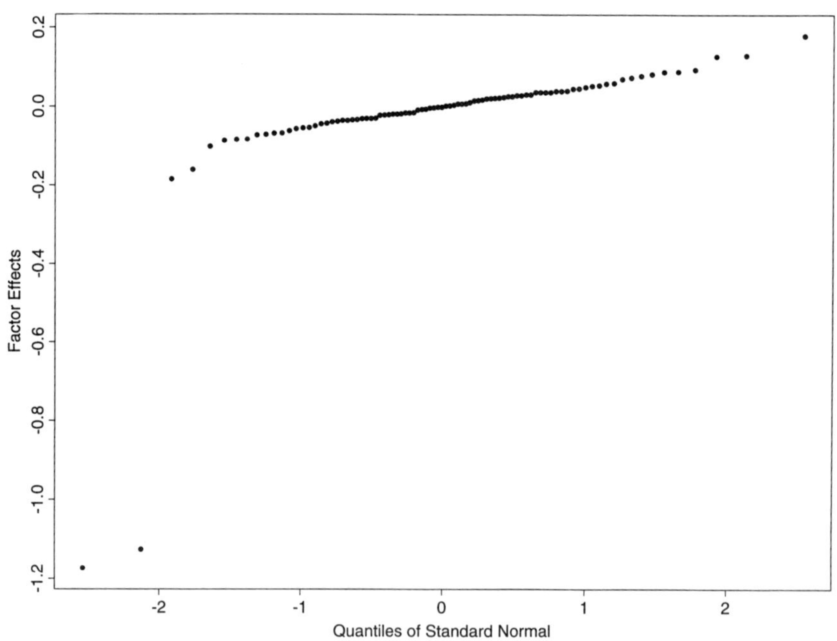

Figure 4 A normal probability plot of the factor effects on the performance curve from our stage 2 analysis of the engine starting system experiment.

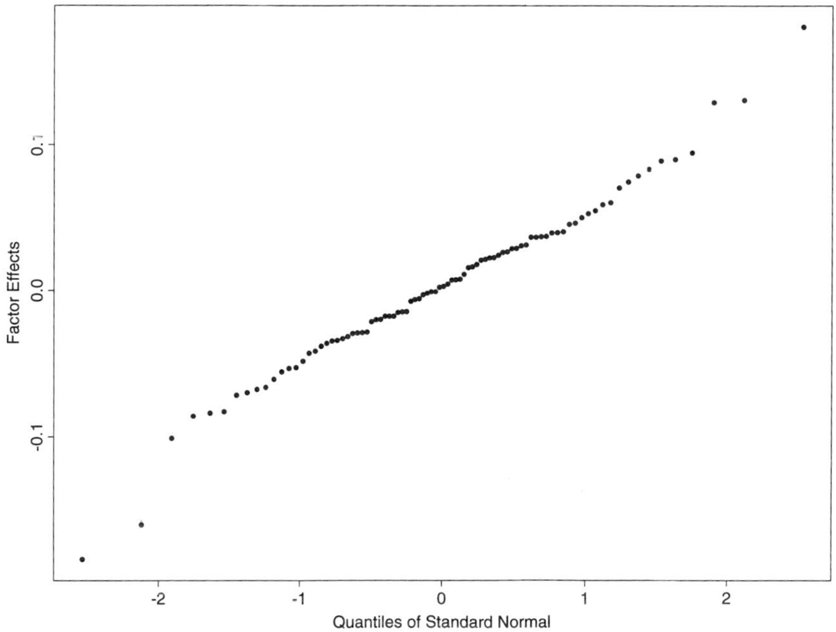

Figure 5 A normal probability plot of the factor effects on the performance curve from our stage 2 analysis of the engine starting system experiment, after deleting the two large effects due to the linear contrasts for fuel mass pulse and temperature.

We can now use the above information to compare different system configurations. First, we observe that the experiment has indeed borne out a clear linear relationship between log AFR and log FMP. The average relationship estimates log AFR by 3.83–1.13 (log FMP). For the three fuel mass pulses used in the experiment, the resulting estimates of log AFR are 4.25 (at 30 msec), 3.79 (at 45 msec), and 3.46 (at 60 msec). There is no detectable curvature in the log AFR–log FMP relationship, and the only possible dependence on the design factors is that the mean level of the line may decrease when injector 4, 5, or 6 is used and when fuel rail pressure is increased. The design factors have no effect on the slope of the line. Overall, we conclude that the relationship is quite consistent across the prototype conditions.

There is also a strong relationship between log AFR and temperature, but it is affected by interactions with three of the design factors. It is easiest to study and model those effects by computing the average stage 1 temperature effect at each level of the relevant factors, which are listed in Table 1. The average temperature effect was −1.173. Since the goal of the experiment was to reduce sensitivity to temperature variation, we seek levels of the three factors that make the temperature effect closer to 0. The best choice is to take an injector of type 6 and use the middle tip-to-head distance and the low level of valve timing (the middle level is almost equally good). If we assume that the design factors have additive effects on the temperature effect, the estimated increases in that effect from each of these choices are 0.296 (from injector type), 0.225 (from the tip-to-head distance), and 0.153 (from the valve timing). The estimated temperature effect is then −0.499, about 60% closer to 0 than its average value. Thus the experiment has identified factor settings that substantially reduce the sensitivity to temperature, resulting in less variation in product response and more uniform starting performance.

It is worth noting that if we place the mean level effects and the performance curve effects on the same probability plot (after appropriate scaling of the mean level effects), many of the mean level effects stand out from the line through the origin, contrary to our earlier conclusion that at most two contrasts are significant and then just barely. This finding suggests that the within-prototype error, on which we base the statistical significance of the performance curve contrasts, is too small for judging the mean level contrasts. That, in turn, implies that a substantial amount of the variability in the data may be at the interprototype level. This information could be valuable for future efforts to make the performance curves still more uniform.

Table 1 Average Estimated Temperature Effect from the Stage 1 Analysis at Each Level of the Three Factors That Had Significant Interactions with Temperature in the Stage 2 Analysis

	Level					
Factor	1	2	3	4	5	6
Injector type	−1.581	−1.058	−1.268	−1.202	−1.055	−0.877
Tip-to-head distance	−1.507	−0.948	−1.066			
Valve timing	−1.020	−1.030	−1.469			

6. CONCLUSIONS

Prototype testing is an important stage in the development of new products and production processes. Great gains are possible by exploiting factorial designs in prototype studies. Engineers can use these studies to compare design options, to increase the feedback from the prototypes, and to accelerate the design process.

Statistical methods for prototype experiments must take account of the fact that prototypes, being expensive to build but often cheap to test, may be run through a battery of test conditions, which themselves constitute a factorial design. Our two-stage analysis provides a simple scheme for modeling the ensuing performance curve and its dependence on the design factors. It correctly accounts for the split-plot error structure that arises when the test conditions are nested within the prototype design and permits quick identification of important effects from normal probability plots.

ACKNOWLEDGMENTS

The research of D. M. Steinberg was carried out in part while he was visiting the Center for Quality and Productivity Improvement, University of Wisconsin–Madison. He is grateful to the Center for providing excellent research facilities. The research of S. Bisgaard was carried out in part under grant number DMI 950014 from the U.S. National Science Foundation.

REFERENCES

1. GL Urban, JR Hauser. Design and Marketing of New Products. Englewood Cliffs, NJ: Prentice-Hall, 1993.
2. JW Wesner, JM Hiatt, DC Trimble. Winning with Quality. Applying Quality Principles in Product Development. Reading, MA: Addison-Wesley, 1994.
3. SC Wheelwright, KB Clark. Revolutionizing Product Development. New York: The Free Press, 1992.
4. WI Zangwill. Lightning Strategies for Innovation. New York: Lexington Books, 1993.
5. AL Page. Assessing new product development practices and performances: Establishing crucial norms. J Prod Innov Manag 10:273–290, 1993.
6. S Bisgaard. A conceptual framework for the use of quality concepts and statistical methods in product design. J Eng Design 3:31–47, 1992.

7. S Bisgaard, MR Ellekjaer. Designing quality into products during the design and development phase. Proc Eur Org Qual, Trondheim, Norway 2:285–296, 1997.
8. S Bisgaard. DM Steinberg. The design and analysis of $2^{k-p} \times s$ prototype experiments. Technometrics 39:52–62, 1997.
9. KB Clark, T Fujimoto. Product Development Performance: Strategy, Organization and Management in the World Auto Industry. Boston: Harvard Business School Press, 1991.
10. G Taguchi. Introduction to Quality Engineering. White Plains, NY: Kraus International Publications, 1986.
11. A Miller, CFJ Wu. Parameter design for signal-response systems: A different look at Taguchi's dynamic parameter design. Stat Sci. 1996. Vol. 11, 122–136.
12. DM Grove, TP Davis. Engineering Quality and Experimental Design. Burnt Mill, Harlow, UK: Longman Scientific and Technical, 1992.
13. WE Deming. Out of the Crisis. Cambridge, MA: Massachusetts Institute of Technology, Center for Advanced Engineering Study, 1986.
14. S Bigsaard. Designing experiments for tolerancing assembled products. Technometrics 38:142–152, 1997.
15. ER Ott. A production experiment with mechanical assemblies. Ind Qual Cont 9:124–130, 1953.
16. B Milman, I Sirota, DM Steinberg. Improving the safety of a pyrotechnic ignitor through a controlled experiment. Propel Explo Pyrotech 20:294–299, 1995.
17. C Hellstrand. The necessity of modern quality improvement and some experience with its implementation in the manufacturer of rolling bearings. Phil Trans Roy Soc (Lond) A 327:529–537, 1989.
18. DM Steinberg, D Bursztyn. Dispersion effects in robust design experiments with noise factors. J Qual Tech 26:12–20, 1994.
19. GEP Box, WG Hunter. A useful method for model-building. Technometrics 4: 301–318, 1962.
20. WA Shewhart. Statistical method from the viewpoint of quality control. Washington DC.: Graduate School, U.S. Department of Agriculture.

20

Optimal Approximate Designs for *B*-Spline Regression with Multiple Knots

Norbert Gaffke and Berthold Heiligers
Universität Magdeburg, Magdeburg, Germany

1. INTRODUCTION

Piecewise polynomial regression may serve as an alternative to nonlinear regression models in the case of a single real regressor variable, since polynomial splines possess excellent approximation properties. If the knots have been chosen appropriately, the spline model is linear in the parameters, and hence tools from linear model analysis and experimental design can be utilized. For an overview on the use of polynomial splines in regression modeling, the reader is referred to Ref. 1.

Let $[a, b]$ be a compact interval ($a, b \in \mathbb{R}$, $a < b$) with associated partition by given knots,

$$a = \kappa_0 < \kappa_1 < \cdots < \kappa_{\ell-1} < \kappa_\ell = b$$

where $\ell \geq 1$. A polynomial spline (with respect to the knots $\kappa_0, \kappa_1, \ldots, \kappa_\ell$) of degree at most $d \geq 1$ is a function on $[a, b]$ that coincides on each subinterval $[\kappa_i, \kappa_{i+1}]$ with some polynomial of degree at most d, $0 \leq i \leq \ell - 1$, and that satisfies some smoothness conditions at the interior knots $\kappa_1, \ldots, \kappa_{\ell-1}$, stated next. Let $s_1, \ldots, s_{\ell-1}$ be given integers with $0 \leq s_i \leq d - 1$ for all $i = 1, \ldots, \ell - 1$, where s_i denotes the desired degree of smoothness at knot κ_i of the spline functions considered. We abbreviate $\kappa = (\kappa_0, \ldots, \kappa_\ell)$ for the vector of knots and $s = (s_1, \ldots, s_{\ell-1})$ for the vector of smoothness degrees. Let $\mathcal{S}_d(\kappa, s)$ be the set of all polynomial splines of degree at most d with respect to the knots κ being s_i times continuously differentiable at κ_i for all $i = 1, \ldots, \ell - 1$. Note that $s_i = 0$ means simply continuity at κ_i, and $\ell = 1$ describes ordinary dth degree polynomial regression. Obviously, $\mathcal{S}_d(\kappa, s)$ is a linear space, and its dimension is known to be (cf. Ref. 2, Theorem 5)

$$k = \ell d + 1 - \sum_{i=1}^{\ell-1} s_i \tag{1}$$

To define the particular B-spline basis $B_1, \ldots, B_k$ of $\mathcal{S}_d(\kappa, s)$ to be employed, we assign multiplicity $d - s_i$ to each interior knot $\kappa_i, i = 1, \ldots, \ell - 1$, and multiplicity $d + 1$ to both boundary knots. Consider the extended knot vector t having the knots $\kappa_0, \ldots, \kappa_\ell$ as components where each knot is repeated according to its multiplicity, i.e.,

$$t = (t_1, \ldots, t_{k+d+1}) = (\underbrace{a, \ldots, a}_{d+1}, \underbrace{\kappa_1, \ldots, \kappa_1}_{d-s_1}, \ldots, \underbrace{\kappa_{\ell-1}, \ldots, \kappa_{\ell-1}}_{d-s_{l-1}}, \underbrace{b, \ldots, b}_{d+1}) \tag{2}$$

Now a family $B_{i,q}, i = 1, \ldots, k + d - q; q = 0, 1, \ldots, d$, of functions on $[a, b]$ is recursively defined as follows.

$$B_{i,0}(x) = \begin{cases} 1 & \text{if } t_i \le x < t_{i+1} \\ 0 & \text{otherwise} \end{cases}$$

and for $q \ge 1$,

$$B_{i,q}(x) = w_{i,q}(x) B_{i,q-1}(x) + [1 - w_{i+1,q}(x)] B_{i+1,q-1}(x) \tag{3}$$

where

$$w_{j,q}(x) = \begin{cases} (x - t_j)/(t_{q+j} - t_j) & \text{if } t_j < t_{q+j} \\ 0 & \text{otherwise} \end{cases}; \; 1 \le j \le k + d - q + 1; \; 1 \le q \le d$$

Then the B-spline basis $B_1, \ldots, B_k$ of $\mathcal{S}_d(\kappa, s)$ is given by

$$B_i = B_{i,d}, \qquad i = 1, \ldots, k \tag{4}$$

(cf. Ref. 2, Theorems 10 and 11).

It is not difficult to see that the basis enjoys the properties

$$0 \le B_i(x) \le 1 \qquad \text{for all } i = 1, \ldots, k \text{ and all } x \in [a, b] \tag{5a}$$

$$B_1(x) = 1 \qquad \text{if and only if } x = a \tag{5b}$$

$$B_k(x) = 1 \qquad \text{if and only if } x = b \tag{5c}$$

$$\sum_{i=1}^{k} B_i(x) = 1 \qquad \text{for all } x \in [a, b] \tag{5d}$$

$$\{x \in [a, b] : B_i(x) > 0\} = \begin{cases} [a, t_{d+2}) & \text{if } i = 1 \\ (t_i, t_{i+d+1}) & \text{if } i = 2, \ldots, k-1 \\ (t_k, b] & \text{if } i = k \end{cases} \tag{5e}$$

We note that the small support property (5e) is a particular feature of the basic splines B_i. Figure 1 shows the *B*-splines for a special case.

A further favorable property of the *B*-spline basis, Eq. (4), is its *equivariance* under affine-linear transformation of the knot vector κ. That is, if the interval $[a, b]$ (and its knots κ_i, $i = 0, \ldots, \ell$) are transformed to another

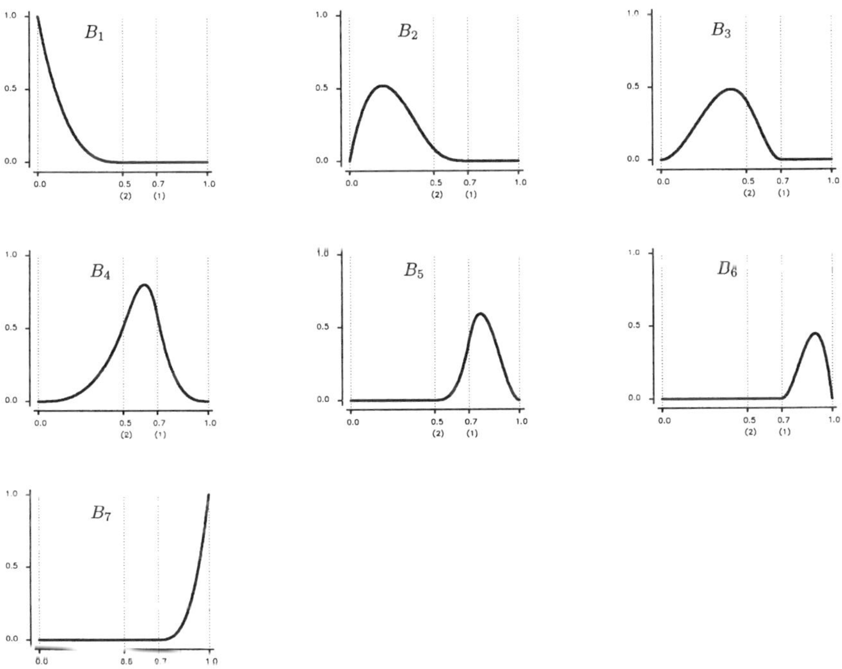

Figure 1 *B*-Splines for $d = 3$, $l = 3$, $\kappa = (0, 0.5, 0.7, 1)$, and $s = (2, 1)$.

interval $[\tilde{a}, \tilde{b}]$ with knots $\tilde{\kappa}_i = L(\kappa_i)$, $i = 0, \ldots, \ell$, by the affine-linear transformation L, then the B-spline basis $\tilde{B}_1, \ldots, \tilde{B}_k$ of $\mathcal{S}_d(\tilde{\kappa}, s)$ defined correspondingly by Eqs. (3) and (4) is

$$\tilde{B}_i(\tilde{x}) = B_i\big(L^{-1}(\tilde{x})\big) \qquad \text{for all } i = 1, \ldots, k \text{ and all } \tilde{x} \in [\tilde{a}, \tilde{b}] \tag{6}$$

Hence Eq. (6) allows us to standardize the interval $[a, b]$, e.g., to $[0, 1]$.

The spline regression model states that a regression function y is a member of the space $\mathcal{S}_d(\kappa, s)$, i.e.,

$$y(x) = \sum_{i=1}^{k} \theta_i B_i(x) \qquad \text{all } x \in [a, b]$$

for some coefficient vector $\theta = (\theta_1, \ldots, \theta_k)'$, which has to be estimated from the data (the prime denotes transposition). Under the standard statistical assumptions that the observations of the regression function at any x values are uncorrelated and have equal (but possibly unknown) variance σ^2, the ordinary least squares estimator of θ will be used. So for designing the experiment, i.e., for choosing the x values at which the observations of $y(x)$ are to be taken, the concepts of optimal linear regression design apply. For mathematical and computational tractability we restrict ourselves to the approximate theory. An approximate design ξ consists of a finite set of distinct support points $x_1, \ldots, x_r \in [a, b]$ (where the support size $r \geq 1$ may depend on ξ) and corresponding weights $\xi(x_1), \ldots, \xi(x_r) > 0$ with $\sum_{i=1}^{r} \xi(x_i) = 1$. The design ξ calls for $\xi(x_i) \times 100\%$ of all observations of the regression function at x_i for all $i = 1, \ldots, r$. The moment matrix (or information matrix) of ξ is given by

$$M(\xi) = \sum_{i=1}^{r} \xi(x_i) B(x_i) B(x_i)' \tag{7}$$

where $B(x) = (B_1(x), \ldots, B_k(x))'$. Note that, by Eq. (5e), for all $x \in [a, b]$ the matrix $B(x)B(x)'$ has a principal block of size $(d+1) \times (d+1)$ outside which all the entries of $B(x)B(x)'$ vanish. Hence the moment matrix $M(\xi)$ of the design ξ is a band matrix with d diagonals above and below the main diagonal, i.e., the (i, j)th entries of $M(\xi)$ are zero whenever $|i - j| > d$.

Under a design ξ, all coefficients θ_i, $i = 1, \ldots, k$, are estimable if and only if the moment matrix of ξ is nonsingular, or equivalently if and only if it is positive definite. Among those designs ξ [with $M(\xi)$ being positive definite], an optimal design is one that minimizes $\Phi(M(\xi))$, where Φ is a given (real-valued) optimality criterion defined on the set $\mathrm{PD}(k)$ of all positive definite $k \times k$ matrices. We are concerned here with Kiefer's Φ_p criteria ($-\infty \leq p \leq 1$) including the most popular D, A, and E criteria through $p = 0, -1, -\infty$, respectively. These are defined by

$$\Phi_p(M) = \left(\frac{1}{k} \sum_{i=1}^{k} \lambda_i(M)^p \right)^{-1/p} \qquad \text{if } p \notin \{0, -\infty\}$$

$$\Phi_0(M) = [\det(M)]^{-1/k}, \qquad \Phi_{-\infty}(M) = 1/\lambda_1(M)$$

where $\lambda_1(M) \leq \lambda_2(M) \leq \cdots \leq \lambda_k(M)$ denote the eigenvalues of $M \in \mathrm{PD}(k)$, arranged in ascending order. We note that $\Phi_p(M)$ is continuous as a function of p. In particular, $\Phi_{-\infty}(M) = \lim_{p \to -\infty} \Phi_p(M)$ for all $M \in \mathrm{PD}(k)$, and hence the non-smooth E criterion can be approximated by a smooth Φ_p (with, e.g., $p = -50$).

In Section 2 we describe the algorithm and discuss the numerical results. Some results on the support of optimal designs for special cases are proved in Section 3, providing thus a first step toward a theoretical explanation of the numerical results.

2. COMPUTING NUMERICALLY OPTIMAL DESIGNS

The basic algorithm we used is that of Gaffke and Heiligers [3], with necessary adaptations to the present situation of polynomial spline regression as are described in detail in Ref. 4. So we only briefly outline the method.

A sequence of moment matrices M_n, $n = 1, 2, \ldots$, is computed, corresponding to some approximate designs ξ_n, $n = 1, 2, \ldots$. The current design ξ_n, however, is *not* computed (except for the final iteration when the algorithm terminates). Thus an increasing set of support points calling for some clustering or elimination rules is avoided. For twice continuously differentiable optimality criteria Φ having compact level sets (as, e.g., the Φ_p criteria with $-\infty < p < 1$), the generated sequence of moment matrices M_n have been shown to converge to an optimal solution to

$$\text{Minimize} \quad \Phi(M) \tag{8a}$$

$$\text{Subject to } M \in \mathrm{Conv}\{B(x)B(x)' : x \in [a, b]\} \cap \mathrm{PD}(k) \tag{8b}$$

where Φ is the optimality criterion under consideration and Conv S denotes the convex hull of a set S of matrices (cf. Ref. 3, Theorem 2.2). Note that, by Eq. (7), restriction (8b) just expresses that a feasible matrix M is nonsingular and is the moment matrix of some approximate design.

So the algorithm solves problem (8) numerically. Additionally, for the final iterate M^*, say, a decomposition is computed (see below),

$$M^* = \sum_{i=1}^{r} w_i^* B(x_i^*) B(x_i^*)' \tag{9}$$

with $r \in \mathbb{N}$, $a \le x_i^* < \cdots < x_r^* \le b$, and $w_i^*, \ldots, w_r^* > 0$ such that $\sum_{i=1}^{r} w_i^* = 1$. A numerically optimal design is then given by ξ^* having support points x_i^* and weights $\xi^*(x_i^*) = w_i^*$, $i = 1, \ldots, r$.

Any starting point M_1 is chosen from the feasible set (8b), e.g., $M_1 = M(\xi_1)$ with an initial design ξ_1 whose support contains k distinct points $x_1 < \cdots < x_k$ such that $B_i(x_i) > 0$ for all $i = 1, \ldots, k$ (see Lemma 1 in Section 3). Given $n \in \mathbb{N}$ and the current (feasible) iterate M_n, a feasible search direction $\overline{M}_n$ is computed as the optimal solution of a quadratic convex problem

$$\text{Minimize} \quad g_n'(m - m_n) + \tfrac{1}{2}(m - m_n)' H_n (m - m_n) \tag{10a}$$

$$\text{Subject to } m \in \operatorname{Conv}\{m(x_1), \ldots, m(x_r), m_n\} \tag{10b}$$

Here we have denoted by lowercase letters m_n, m, and $m(x_i)$ the moment *vectors* obtained from M_n, M, and $M(x_i) = B(x_i) = B(x_i)B(x_i)'$, respectively, by a usual vector operation turning matrices to column vectors. Owing to the symmetry and the band structure of the moment matrices it suffices to apply the vector operation to the main diagonal and the d diagonals above the main diagonal. So the vector operator considered here selects that part of a symmetric matrix A and arranges the entries in some fixed order, resulting in a vector $\operatorname{vec}(A) \in \mathbb{R}^K$, where $K = (d+1)(k - d/2)$. In Eqs. (10) we have

$$m_n = \operatorname{vec}(M_n), \qquad m(x_i) = \operatorname{vec}(B(x_i)B(x_i)'), \qquad g_n = V \operatorname{vec}(G_n)$$

where $x_1, \ldots, x_r$ are certain points from $[a, b]$ to be described next (note that these points including their total number r depend on n, but this dependence is dropped here to simplify the notation), and G_n denotes the gradient of Φ at M_n in the space of symmetrical $k \times k$ matrices endowed with the scalar product $\langle A, B\rangle = \operatorname{tr}(AB)$. The matrix V occurring when vectorizing the gradient is a fixed $K \times K$ diagonal matrix with diagonal entries equal to 1 or 2, such that those components of $\operatorname{vec}(G_n)$ coming from the diagonal of G_n receive weight 1 while the off-diagonal elements are weighted by 2. This is to ensure that g_n is the gradient at m_n of the function

$$\phi(m) = \Phi(\operatorname{vec}^{-1}(m)) \tag{11}$$

where vec^{-1} is the inverse operation of vec, converting an K-dimensional vector into a band matrix [m being restricted in Eq. (11) to the set of all vectors obtained by vectorizing positive definite band matrices]. Note that although M_n is a band matrix, this is not true in general for the gradient G_n, e.g., for the Φ_p criteria with $-\infty < p < 1$ we have

$$G_n = -\frac{1}{k}[\Phi_p(M_n)]^{p+1} M_n^{p-1}$$

The points $x_i, i = 1, \ldots, r$, in (10b) are most crucial for obtaining a good search direction by solving the quadratic problem. Their choice is guided by the equivalence theorem, i.e., the first-order optimality conditions for problem (8). A feasible moment matrix M^* is an optimal solution if and only if

$$B(x)'(-G^*)B(x) \leq \mathrm{tr}(-G^* M^*) \qquad \text{for all } x \in [a, b] \tag{12}$$

where G^* is the gradient of Φ at M^*. Moreover, if M^* *is* an optimal solution, then for any representation of M^* as

$$M^* = \sum_{i=1}^{r} w_i^* B(x_i^*) B(x_i^*)'$$

with some

$$r \in \mathbb{N}, \qquad x_1^*, \ldots, x_r^* \in [a, b], \qquad w_1^*, \ldots, w_r^* > 0, \quad \sum_{i=1}^{r} w_i^* = 1$$

one has

$$B(x_i^*)'(-G^*)B(x_i^*) = \mathrm{tr}(-G^* M^*) \qquad \text{for all } i = 1, \ldots, r$$

From this it appears reasonable to choose in (10b) the *local* maximum points $x_1, \ldots, x_r$ of the function

$$B(x)'(-G_n)B(x), \qquad x \in [a, b] \tag{13}$$

(including, of course, its *global* maximum points). In fact, computing these is not too difficult, since $B(x)'(-G_n)B(x)$ is a polynomial spline of degree at most $2d$, i.e., a polynomial of degree at most $2d$ on each subinterval $[\kappa_i, \kappa_{i+1}]$, $i = 0, \ldots, \ell - 1$. Thus, standard routines for computing all zeros of polynomials can be used. Figure 2 shows an example of the function (13) for an early iterate and for the final one.

The matrix H_n in (10a) is a positive definite $K \times K$ matrix, which should be an approximation of the Hessian matrix of ϕ from (11) at m_n. A good job is done by the Broyden–Fletcher–Goldfarb–Shanno (BFGS) update

$$H_n = H_{n-1} + (\gamma_n'\delta_n)^{-1}\gamma_n\gamma_n' - (\delta_n'H_{n-1}\delta_n)^{-1}(H_{n-1}\delta_n)(H_{n-1}\delta_n)'$$

where

$$\delta_n = m_n - m_{n-1}, \qquad \gamma_n = g_n - g_{n-1}, \qquad n \geq 2$$

with any positive definite initial choice of H_1.

The quadratic minimization problem (10) can be solved by the Higgins–Polak method as described in Ref. 3. Let $\overline{m}_n$ be the solution

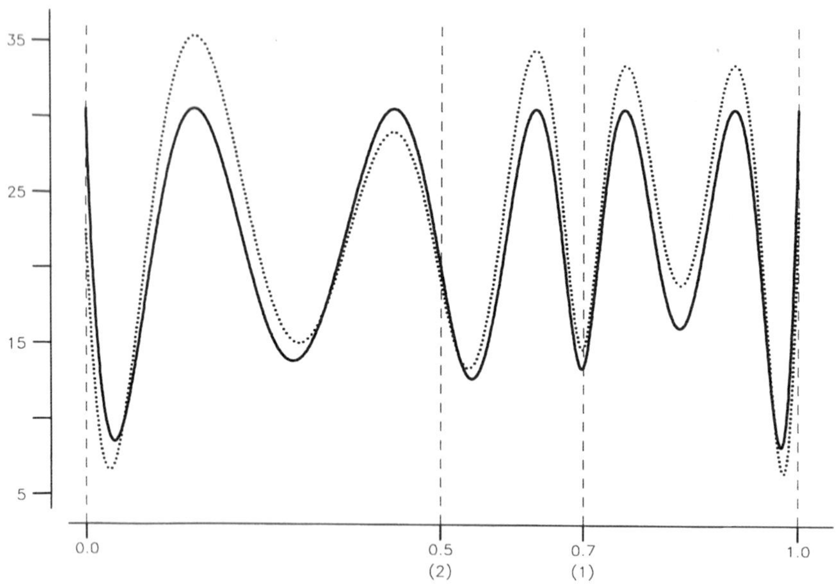

Figure 2 The function (13) for iterate $n = 10$ (dotted line) and for the final iterate $n = 43$ (solid line). Under consideration is the cubic spline model as in Figure 1, and the optimality criterion is the A criterion ($p = -1$).

obtained. We note that the Higgins–Polak method also provides weights $w_0, w_1, \ldots, w_r \geq 0$ summing up to 1 and such that

$$\overline{m}_n = w_0 m_n + \sum_{i=1}^{r} w_i m(x_i) \tag{14}$$

but this is used only in the final step (see below). Let $\overline{M}_n = \text{vec}^{-1}(\overline{m}_n)$. Now, a search along the line segment

$$(1 - \alpha)M_n + \alpha \overline{M}_n, \qquad 0 \leq \alpha \leq \overline{\alpha}$$

(with some fixed $\overline{\alpha} < 1$, usually close to 1) is performed to obtain the next iterate M_{n+1}.

To summarize, the method for solving (8) is a modified quasi-Newton method. The search direction is based on a local second-order approximation of the objective function Φ. The constraint set in (10b) over which the quadratic approximation is minimized may be viewed as a polyhedral neighborhood of the current vector iterate m_n. It may appear more natural to minimize that quadratic approximation over the set of *all* moment vectors

$$m = \text{vec}(M), \qquad M \in \text{Conv}\{B(x)B(x) : x \in [a, b]\}$$

This, however, is practically impossible.

After termination of the algorithm with a final iterate M_n (for stopping criteria see Ref. 3), a corresponding numerically optimal design ξ^* is computed by applying the Higgins–Polak method to the problem of minimizing the final quadratic approximation (10a) over the slightly smaller set

$$\text{Conv}\{m(x_1), \ldots, m(x_r)\}$$

that is, the final vector iterate m_n is removed from the generator set in (10b). This has proved to be favorable, since otherwise a positive w_0 may occur in (14) that could prevent the identification of a corresponding design. We thus obtain an optimal solution m^*, say, to that quadratic problem, a non-empty subset I of indices from $\{1, \ldots, r\}$, and positive weights $w_i^*, i \in I$, summing up to 1 and such that

$$m^* = \sum_{i \in I} w_i^* m(x_i) \tag{15}$$

In all our numerical experiments we observed that m^* is very close to the final vector iterate m_n and shares numerically the same value of ϕ. Hence, a numerically optimal design is given by ξ^* supported by x_i, $i \in I$, and weights $\xi^*(x_i) = w_i^*$.

The algorithm shows good convergence behavior, in particular a good *local* convergence rate as it is usually observed by a quasi-Newton method. For instance, the D-optimal designs for spline degree $d = 2$ and one single interior knot (i.e., $\ell = 2$, $s_1 = 1$) derived theoretically in Ref. 5, page 43, and in Ref. 6, Theorem 2, are found very accurately by the algorithm. For degrees $d = 3, 4, 5$ and one single interior knot, D-optimal designs within the class of designs with minimum support size k were found numerically by Lim [6]. The present algorithm computed precisely these designs as the numerically D-optimal ones in the class of *all* designs (up to two printing errors in the tables on page 176 of Ref. 6).

Table 1 Numerically Optimal Designs in the Spline Model (Fig. 1)

D		*A*		*E*	
Support	Weight	Support	Weight	Support	Weight
0.00000	0.14286	0.00000	0.08848	0.00000	0.07361
0.00000	*0.14286*	*0.00000*	*0.09128*	*0.00000*	*0.07361*
0.16329	0.14286	0.15315	0.16875	0.14473	0.17424
0.14473	*0.14286*	*0.14473*	*0.16962*	*0.14473*	*0.17424*
0.43037	0.14286	0.43415	0.18454	0.43418	0.20559
0.43418	*0.14286*	*0.43418*	*0.18134*	*0.43418*	*0.20559*
0.62989	0.14286	0.63363	0.14444	0.63316	0.17364
0.63316	*0.14286*	*0.63316*	*0.14328*	*0.63316*	*0.17364*
0.75929	0.14286	0.75807	0.15269	0.75720	0.14992
0.75720	*0.14286*	*0.75720*	*0.14820*	*0.75720*	*0.14992*
0.90894	0.14286	0.91179	0.17049	0.91907	0.15293
0.91907	*0.14286*	*0.91907*	*0.17121*	*0.91907*	*0.15293*
1.00000	0.14286	1.00000	0.09062	1.00000	0.07008
1.00000	*0.14286*	*1.00000*	*0.09506*	*1.00000*	*0.07008*

Note: Under consideration are the D, A, and E criteria. The numbers in italics give the optimal designs supported by the Chebyshev points.

Table 1 shows a few numerical results for the D and A criteria and the approximate E criterion Φ_{-50} in the cubic spline model as in Figures 1 and 2. The designs addressed in Table 1 by italics are the $D-$, $A-$, and E-optimal designs within the subclass of those designs concentrated on the Chebyshev points, i.e., supported by the k extremal points of the equioscillating spline in $\mathcal{S}_d(\kappa, s)$ (cf. Ref. 7, Section 2). For the D and A criteria these are computed by a simplified variant of the above algorithm, fixing $x_1, \ldots, x_r (r = k)$ to those Chebyshev points, while the E-optimal design is from Ref. 7, Theorem 4. By that theorem the E-optimal design (among *all* designs) is supported by the Chebyshev points. We see from Table 1 that the Φ_{-50}-optimal design numerically coincides with the E-optimal design. For the D and A criteria the Chebyshev restricted designs do not differ much from the numerically optimal designs. The D efficiency of the former with respect to the latter is 0.99335, and the A efficiency is 0.99476. Similar results hold true for other spline setups.

In all the cases we considered, the numerically optimal design has minimum support size and the boundary points a and b are support points. For D optimality, the minimum support size property has been conjectured in Ref. 5, page 45, Conjecture 1. In our final section we present some first results toward a theoretical foundation of the observed phenomena.

3. SOME RESULTS ON OPTIMAL *B*-SPLINE REGRESSION DESIGNS

The B-spline basis $B_1, \ldots, B_k$ of $\mathcal{S}(\kappa, s)$ defined by (4) enjoys the fundamental property of *total positivity*; i.e., for any points $x_1, \ldots, x_k$ such that $a \leq x_1 < \cdots < x_k \leq b$ the collocation matrix

$$(B_i(x_j))_{i,j=1,\ldots k} \tag{16}$$

is totally positive. Recall that a $k \times k$ matrix $A = (a_{i,j})_{i,j=1,\ldots,k}$ is said to be totally positive if and only if all its minors are nonnegative, i.e., if and only if for any $p \in \{1, \ldots, k\}$ and all p row and column indices $1 \leq i_1 < \cdots < i_p \leq k$ and $1 \leq j_1 < \cdots < j_p \leq k$ one has

$$\det(a_{i_\mu, j_\nu})_{\mu,\nu=1,\ldots,p} \geq 0$$

Moreover, by Ref. 2, Theorem 12, the collocation matrix (16) is nonsingular if and only if its diagonal elements are positive. From this we obtain

Lemma 1

For any design ξ, the moment matrix of ξ from Eq. (7) is nonsingular (and hence positive definite) if and only if there are support points $z_1 < \cdots < z_k$ of ξ such that $B_i(z_i) > 0$ for all $i = 1, \ldots, k$.

Proof. Arrange the support points of ξ in increasing order, $a \leq x_1 < \cdots < x_r \leq b$, say. We may write

$$M(\xi) = N(\xi)W(\xi)N(\xi)' \tag{17}$$

where

$$N(\xi) = (B_i(x_j))_{\substack{i=1,\ldots,k \\ j=1,\ldots,r}} \qquad \text{and} \qquad W(\xi) = \mathrm{diag}(\xi(x_1), \ldots \xi(x_r))$$

Obviously, $M(\xi)$ is nonsingular if and only if the rows of $N(\xi)$ are linearly independent, or equivalently, if and only if there exist k column incides $1 \leq j_1 < \cdots < j_k \leq r$ such that the submatrix

$$(B_i(z_\nu))_{i,\nu=1,\ldots,k} \qquad \text{where } z_\nu = x_{j_\nu};\ \nu = 1, \ldots, k$$

is nonsingular. As noted above, this is equivalent to $B_i(z_i) > 0$ for all $i = 1, \ldots, k$, and the lemma is proved. □

A design ξ is said to be *admissible for* $S_d(\kappa, s)$, if and only if there is no design $\tilde{\xi}$ such that $M(\xi) \leq M(\tilde{\xi})$ and $M(\xi) \neq M(\tilde{\xi})$. That is, the admissible designs are precisely those whose moment matrices are maximal with respect to the Loewner partial ordering in the set of all moment matrices of designs. The Loewner partial ordering in the set of all symmetrical $k \times k$ matrices is defined by

$$A \leq B \quad \text{if and only if } B - A \text{ is positive semidefinite}$$

Note that admissiblity of a design does not depend on the particular choice of the basis of the spline space $S_d(\kappa, s)$. For, if we choose another basis $f = (f_1, \ldots, f_k)'$ (e.g., the truncated power basis as in Ref. 8), then this is related to our B-spline basis $B = (B_1, \ldots, B_k)'$ by a linear transformation, i.e., $f = TB$, for some nonsingular $k \times k$ matrix T. Hence the resulting moment matrices of designs under basis f,

$$M_f(\xi) = \sum_{i=1}^{r} \xi(x_i) f(x_i) f(x_i)'$$

($x_1, \ldots, x_r$ being the support points of ξ) are related to the moment matrices $M(\xi)$ under the B-spline basis by

$$M_f(\xi) = TM(\xi)T' \qquad \text{for all designs } \xi \tag{18}$$

Obviously, for the Loewner partial ordering we have

$$A \leq B \Longleftrightarrow TAT' \leq TBT'$$

for any symmetrical $k \times k$ matrices A and B.

Any reasonable optimality criterion Φ is decreasing with respect to the Loewner partial ordering, i.e.,

$$\text{If } A, B \text{ are positive definite and } A \leq B, \text{ then } \Phi(A) \geq \Phi(B). \tag{19}$$

Many optimality criteria Φ are *strictly* decreasing; i.e., if additionally $A \neq B$ in (19), then $\Phi(A) > \Phi(B)$. Examples are the Φ_p criteria for finite p we used in Section 2. If Φ is strictly decreasing, then obviously any Φ-optimal design is admissible.

The result of Ref. 8, Theorem 1.1, states that a design ξ is admissible for $S_d(\kappa, s)$ if and only if

$$\#(\text{supp}(\xi) \cap (\kappa_i, \kappa_j)) \leq d - 1 + \sum_{t:i<t<j} \lfloor d - \frac{1}{2} s_t \rfloor \text{ for all } 0 \leq i < j \leq \ell, \tag{20}$$

where $\text{supp}(\xi)$ denotes the support of ξ and $\lfloor x \rfloor$ is the largest integer $\leq x$. For the case that $s_i \in \{0, 1\}$ for all $i = 1, \ldots, \ell - 1$, the observed minimum support size property of Φ_p-optimal designs (where $p < \infty$) is explained by the following result (cf. also Ref. 8, pp. 1558–1559).

Lemma 2

Let $s_i \in \{0, 1\}$ for all $i = 1, \ldots, \ell - 1$. If ξ is admissible for $\mathcal{S}_d(\kappa, s)$ and the moment matrix $M(\xi)$ is nonsingular, then the support size of ξ is equal to k [the dimension of $\mathcal{S}_d(\kappa, s)$], and the boundary knots κ_0, κ_ℓ and all the interior knots κ_i with smoothness $s_i = 0$ are in the support of ξ.

Proof. By (1), $k = \ell d + 1 - \alpha$, where α denotes the number of interior knots κ_i with $s_i = 1$. Consider the $\beta = \ell + 1 - \alpha$ knots

$$\kappa_{i_1} < \cdots < \kappa_{i_\beta}$$

which are the end knots of the interval and the interior knots with smoothness zero. By Eq. (20), for all $\nu = 1, \ldots, \beta - 1$,

$$\#(\text{supp}(\xi) \cap (\kappa_{i_\nu}, \kappa_{i_{\nu+1}})) \leq d - 1 + \alpha_\nu(d-1)$$

where α_ν denotes the number of knots of smoothness 1 in the interval $(\kappa_{i_\nu}, \kappa_{i_{\nu+1}})$. Hence,

$$\begin{aligned}\#(\text{supp}(\xi) \backslash \{\kappa_{i_1}, \ldots, \kappa_{i_\beta}\}) &\leq (\beta-1)(d-1) + (d-1)\sum_{\nu=1}^{\beta-1} \alpha_\nu \\ &= (\beta - 1 + \alpha)(d-1) = \ell(d-1) = k - \beta\end{aligned}$$

Since $M(\xi)$ is nonsingular, we have $\#\text{supp}(\xi) \geq k$, and thus $\kappa_{i_1}, \ldots, \kappa_{i_\beta} \in \text{supp}(\xi)$ and $\#\text{supp}(\xi) = k$. □

For polynomial spline regression with higher smoothness, a theoretical explanation of the minimum support size property of Φ_p-optimal designs is still outstanding. It has not even been proved that the support of a Φ_p-optimal design necessarily includes the boundary points of the interval $[a, b]$. However, for D optimality ($p = 0$) the latter can be proved (see Lemma 3 below; see also Ref. 6, Lemma 1).

For the rest of the chapter we will be concerned with D-optimal designs for polynomial spline regression. As is well known [and is obvious from Eq. (18)], D optimality of a design (within any class of designs) does not depend on the particular choice of the basis of the space $S_d(\kappa, s)$; thus, we will use the notion of a D-optimal design for $\mathcal{S}_d(\kappa, s)$. The following result has been stated by Kim [6, Lemma 1]. However, the proof given in that paper is not convincing, in our view, and we give different proof here.

Lemma 3

The D-optimal design for $\mathcal{S}_d(\kappa, s)$ (with arbitrary degree, knots, and associated multiplicities) has both boundary points $\kappa_0 = a$ and $\kappa_\ell = b$ among its support points.

Proof. Let ξ be any design with nonsingular moment matrix $M(\xi)$, and let $x_1 < \cdots < x_r$ be the support points of ξ. Consider the representation (17) of $M(\xi)$. In the following we denote by $N_\xi\binom{i_1, \ldots, i_p}{j_1, \ldots, j_p}$ the submatrix of $N(\xi)$ with respective row and column indices $1 \leq i_1 < \cdots < i_p \leq k$ and $1 \leq j_1 < \cdots < j_p \leq r$ (where $1 \leq p \leq k$), i.e.,

$$N_\xi\begin{pmatrix} i_1, \ldots, i_p \\ j_1, \ldots, j_p \end{pmatrix} = \left(B_{i_\mu}(x_{j_\nu})\right)_{\mu,\nu=1,\ldots,p}$$

By (17) and the Cauchy–Binet formula, we have

$$\det M(\xi) = \sum_{1 \le j_1 < \cdots < j_k \le r} \left(\prod_{\nu=1}^{k} \xi(x_{j\nu})\right) \det^2 N_\xi\begin{pmatrix} 1, \ldots, k \\ j_1, \ldots, j_k \end{pmatrix} \tag{21}$$

Suppose that $x_1 > a$. We will prove that ξ cannot be D optimal.

Let $\tilde{\xi}$ be the design obtained from ξ by replacing the support point x_1 by a and $\tilde{\xi}(a) = \xi(x_1)$, $\tilde{\xi}(x_i) = \xi(x_i)$ for all $i = 2, \ldots, r$. By (21) and its version for $\tilde{\xi}$, we obtain

$$\det M(\xi) - \det M(\tilde{\xi})$$
$$= \sum_{2 \le j_2 < \ldots < j_k \le r} \xi(x_1)\left(\prod_{\nu=2}^{k} \xi(x_{j_\nu})\right)\left[\det^2 N_\xi\begin{pmatrix} 1, 2, \ldots, k \\ 1, j_2, \ldots, j_k \end{pmatrix} - \det^2 N_{\tilde{\xi}}\begin{pmatrix} 1, 2, \ldots, k \\ 1, j_2, \ldots, j_k \end{pmatrix}\right] \tag{22}$$

From (5a)–(5e) we see that the first column of $N_{\tilde{\xi}}\begin{pmatrix} 1,2,\ldots,k \\ 1,j_2,\ldots,j_k \end{pmatrix}$ is the first unit vector in $\mathbb{R}^k$; thus

$$\det N_{\tilde{\xi}}\begin{pmatrix} 1, 2, \ldots, k \\ 1, j_2, \ldots, j_k \end{pmatrix} = \det N_\xi\begin{pmatrix} 2, \ldots, k \\ j_2, \ldots, j_k \end{pmatrix} \tag{23}$$

Since the collocation matrix $N_\xi\begin{pmatrix} 1,2,\ldots,k \\ 1,j_2,\ldots,j_k \end{pmatrix}$ is totally positive, we have by the Hadamard-type inequality (cf. Ref. 9, p. 191) and by (5a),

$$0 \le \det N_\xi\begin{pmatrix} 1, 2, \ldots, k \\ 1, j_2, \ldots, j_k \end{pmatrix} \le B_1(x_1) N_\xi\begin{pmatrix} 2, \ldots, k \\ j_2, \ldots, j_k \end{pmatrix} \le \det N_\xi\begin{pmatrix} 2, \ldots, k \\ j_2, \ldots, j_k \end{pmatrix} \tag{24}$$

Moreover, by (5b), the last inequality in (24) is strict whenever the matrix $N_\xi\begin{pmatrix} 2,\ldots,k \\ j_2,\ldots,j_k \end{pmatrix}$ is nonsingular. In fact, such indices $2 \le j_2 < \cdots < j_k \le r$ exist. For, by (21), since $M(\xi)$ is nonsingular, there exist indices $1 \le j_1 < j_2 < \cdots < j_k \le r$ such that $N_\xi\begin{pmatrix} 1,2,\ldots,k \\ j_1,j_2,\ldots,j_k \end{pmatrix}$ is nonsingular, and again by applying the Hadamard type inequality to the latter totally positive matrix we obtain

$$0 < \det N_\xi \begin{pmatrix} 1, 2, \ldots, k \\ j_1, j_2, \ldots, j_k \end{pmatrix} \leq B_1(x_{j_1}) \det N_\xi \begin{pmatrix} 2, \ldots, k \\ j_2, \ldots, j_k \end{pmatrix}$$

Together with Eqs. (22)–(24), it follows that

$$\det M(\xi) < \det M(\tilde{\xi})$$

and thus ξ canot be D optimal.

The case for $x_r < b$ is treated analogously.

Some results on D-optimal designs for $S_d(\kappa, s)$ *within the class of minimum support designs* were derived in Ref. 10. Actually, in that paper different polynomial degrees d_i on each subinterval $[\kappa_i, \kappa_{i+1}]$, $i = 0, \ldots, \ell - 1$, were admitted, but we will not follow this extension here. For short, a design with support size $k = \dim \mathcal{S}_d(\kappa, s)$ that is D optimal within the subclass of all designs with support size k will be called a *D-optimal minimum support design for $\mathcal{S}_d(\kappa, s)$*. As is well known, a D-optimal minimum support design assigns equal weights $1/k$ to each of its support points. As the proof of Lemma 3 shows, the result of that lemma pertains also to a D-optimal minimum support design. Hence, by (17), a D-optimal minimum support design for $S_d(\kappa, s)$ is determined by its support points

$$x_1^* = a < x_2^* < \cdots < x_{k-1}^* < x_k^* = b$$

where $\mathbf{x}^* = (x_1^*, \ldots, x_k^*)$ is an optimal solution to the problem

$$\text{Maximize} \quad \det N(\boldsymbol{x}) \tag{25a}$$

$$\text{Subject to } x_1 = a < x_2 < \cdots < x_{k-1} < x_k = b \tag{25b}$$

where $\mathbf{x} = (x_1, \ldots, x_k)$ and $N(\mathbf{x})$ denotes the collocation matrix to $x_1, \ldots, x_k$, i.e.,

$$N(\boldsymbol{x}) = \big(B_i(x_j)\big)_{i,j=1,\ldots,k}$$

For the case of merely continuous polynomial spline regression (that is, $s_i = 0$ for all $i = 1, \ldots, \ell - 1$) it was claimed by Park [10, p. 152] by somewhat heuristic arguments, that the D-optimal minimum support design in $\mathcal{S}_d(\kappa, 0)$ is obtained by putting together the support points of the D-optimal designs for ordinary dth degree polynomial regression on the subintervals $[\kappa_i, \kappa_{i+1}]$, $i = 0, \ldots, \ell - 1$. We give a proof thereof in Corollary 5. We start with a more general result.

Lemma 4

Let $i_0 \in \{1, \ldots, \ell - 1\}$ be such that the interior knot κ_{i_0} has smoothness $s_{i_0} = 0$. Then the support of the D-optimal minimum support design for $\mathcal{S}_d$ (κ, s) is the union of the supports of the D-optimal minimum support designs for $\mathcal{S}_d(\kappa^{(1)}, s^{(1)})$ and for $\mathcal{S}_d(\kappa^{(2)}, s^{(2)})$, respectively, where

$$\begin{aligned} \kappa^{(1)} &= (\kappa_0, \kappa_1, \ldots, \kappa_{i_0}), & s^{(1)} &= (s_1, \ldots, s_{i_0-1}) \\ \kappa^{(2)} &= (\kappa_{i_0}, \kappa_{i_0+1}, \ldots, \kappa_\ell) & s^{(2)} &= (s_{i_0+1}, \ldots, s_{\ell+1}) \end{aligned}$$

(If $i_0 = 1$ or $i_0 = \ell - 1$, the sets $\mathcal{S}_d(\kappa^{(1)}, s^{(1)})$ or $\mathcal{S}_d(\kappa^{(2)}, s^{(2)})$ are to be understood as the space of all dth-degree polynomials over $[\kappa_0, \kappa_1]$ or $[\kappa_{\ell-1}, \kappa_\ell]$, respectively.)

Proof. Consider the vector $t = (t_1, \ldots, t_{k+d+1})$ of multiple knots from Eq. (2). Let k_0 be the index for which

$$t_{k_0+1} = \cdots = t_{k_0+d} = \kappa_{i_0}$$

i.e.,

$$k_0 = d + 1 + \sum_{i=1}^{i_0-1} (d - s_i) = i_0 d + 1 - \sum_{i=1}^{i_0-1} s_i \tag{26}$$

From (5e) we see that

$$B_1, \ldots, B_{k_0-1} \text{ vanish on } [\kappa_{i_0}, b] \tag{27a}$$

and

$$B_{k_0+1}, \ldots, B_k \text{ vanish on } [a, \kappa_{i_0}] \tag{27b}$$

Also, observing (5a) and (5d), we have

$$B_{k_0}(\kappa_{i_0}) = 1 \tag{27c}$$

Let $\boldsymbol{x} = (x_1, \ldots, x_k)$ satisfy (25) and such that the collocation matrix $N(\boldsymbol{x})$ is nonsingular, i.e, $B_i(x_i) > 0$ for all $i = 1, \ldots, k$. By (27a)–(27c), $x_{k_0-1} < \kappa_{i_0}$ and $x_{k_0+1} > \kappa_{i_0}$. Hence the Hadamard type inequality for totally positive matrices entails, using notations for submatrices as in the proof of Lemma 3,

$$
\begin{aligned}
0 &< \det N(\boldsymbol{x}) \\
&\le \det N_{\boldsymbol{x}}\begin{pmatrix}1,\dots,k_0\\1,\dots,k_0\end{pmatrix}\det N_{\boldsymbol{x}}\begin{pmatrix}k_0+1,\dots,k\\k_0+1,\dots,k\end{pmatrix} && \text{(28a)}\\
&\le \det N_{\boldsymbol{x}}\begin{pmatrix}1,\dots,k_0-1\\1,\dots,k_0-1\end{pmatrix}B_{k_0}(x_{k_0})\det N_{\boldsymbol{x}}\begin{pmatrix}k_0+1,\dots,k\\k_0+1,\dots,k\end{pmatrix} && \text{(28b)}\\
&\le \det N_{\boldsymbol{x}}\begin{pmatrix}1,\dots,k_0-1\\1,\dots,k_0-1\end{pmatrix}\det N_{\boldsymbol{x}}\begin{pmatrix}k_0+1,\dots,k\\k_0+1,\dots,k\end{pmatrix} && \text{(28c)}
\end{aligned}
$$

If $x_{k_0} = k_{i_0}$, then the column vector $(B_1(\kappa_{i_0}, \dots, B_k(\kappa_{i_0}))'$ is the k_0th unit vector in $\mathbb{R}^k$, as follows from by (27c), (5a), and (5d). Hence, if $x_{k_0} = \kappa_{i_0}$, then there is equality in (28a)–(28c), but otherwise there is strict inequality throughout. Consequently, an optimal solution $\boldsymbol{x}^*$ to (25a) and (25b) must satisfy $x^*_{k_0} = \kappa_{i_0}$. Now, for any $\boldsymbol{x}$ satisfying (25b) and $x_{k_0} = \kappa_{i_0}$, we may write

$$
\det N_{\boldsymbol{x}}\begin{pmatrix}1,\dots,k_0-1\\1,\dots,k_0-1\end{pmatrix} = \det N_1(\boldsymbol{x}^{(1)})
$$

$$
\det N_{\boldsymbol{x}}\begin{pmatrix}k_0+1,\dots,k\\k_0+1,\dots,k\end{pmatrix} = \det N_2(\boldsymbol{x}^{(2)})
$$

Hence

$$
\det N(\boldsymbol{x}) = \det N_1(\boldsymbol{x}^{(1)}) \det N_2(\boldsymbol{x}^{(2)}) \tag{29}
$$

where we have denoted

$$
\begin{aligned}
\boldsymbol{x}^{(1)} &= (x_1, \dots, x_{k_0}), \qquad & N_1(\boldsymbol{x}^{(1)}) &= \big(B_i(x_j)\big)_{i,j=1,\dots,k_0}\\
\boldsymbol{x}^{(2)} &= (x_{k_0}, \dots, x_k), \qquad & N_2(\boldsymbol{x}^{(2)}) &= \big(B_i(x_j)\big)_{i,j=k_0,\dots,k}
\end{aligned}
$$

Equation (29) ensures that an optimal solution $\boldsymbol{x}^*$ of (25) must be such that $\boldsymbol{x}^{*(1)}$ is an optimal solution to the problem

Maximize $\quad \det N_1(\boldsymbol{x}^{(1)})$

Subject to $a = x_1 < x_2 < \cdots < x_{k_0-1} < x_{k_0} = \kappa_{i_0}$

and $\boldsymbol{x}^{*(2)}$ is an optimal solution to the problem

Maximize $\quad \det N_2(\boldsymbol{x}^{(2)})$

Subject to $\kappa_{i_0} = x_{k_0} < x_{k_0+1} < \cdots < x_{k-1} < x_k = b$

Now the assertion follows by observing that the matrices $N_1(\boldsymbol{x}^{(1)})$ and $N_2(\boldsymbol{x}^{(2)})$ are the collocation matrices to $\boldsymbol{x}^{(1)}$ and $\boldsymbol{x}^{(2)}$ under special bases of the spline spaces $\mathcal{S}_d(\kappa^{(1)}, s^{(1)})$ and $\mathcal{S}_d(\kappa^{(2)}, s^{(2)})$, respectively. For, note that by (26) and (1), k_0 is the dimension of the space $\mathcal{S}_d(\kappa^{(1)}, s^{(1)})$. The B-splines $B_1, \ldots, B_{k_0}$ restricted to the interval $[a, \kappa_{i_0}]$ are clearly members of $\mathcal{S}_d(\kappa^{(1)}, s^{(1)})$; they are linearly independent [since by (29) there is a nonsingular collocation matrix in these splines], and hence they are a basis of the space $\mathcal{S}_d(\kappa^{(1)}, s^{(1)})$. Similarly, it can be seen that the dimension of $\mathcal{S}_d(\kappa^{(2)}, s^{(2)})$ is equal to $k - k_0 + 1$, and the B-splines $B_{k_0}, \ldots, B_k$ restricted to the interval $[\kappa_{i_0}, b]$ form a basis of the space $\mathcal{S}_d(\kappa^{(2)}, s^{(2)})$. □

Repeated application of Lemma 4 for merely continuous polynomial spline regression yields

Corollary 5

The support of the D-optimal minimum support design for $\mathcal{S}_d(\kappa, 0)$ is the union of the supports of the D-optimal designs for ordinary dth-degree polynomial regressions over the subintervals $[\kappa_i, \kappa_{i+1}]$, $i = 0, \ldots, \ell - 1$.

REFERENCES

1. RL Eubank. (1984). Approximate regression models and splines. Commun Stat Theor Methods 13:433–484.
2. K Morken. (1996). Total positivity and splines. In: M Gasca, CA Micchelli, eds. Total Positivity and Its Application. Dordrecht: Kluwer, pp 47–84.
3. N Gaffke, B Heiligers. (1996). Second order methods for solving extremum problems from optimal linear regression design. Optimization 36:41–57.
4. N Gaffke. (1998). Numerical computation of optimal approximate designs in polynomial spline regression. Journal of Combinatorics, Information and System Sciences, Special Volume "Design of Experiments and Related Combinatorics." 23: 85–94.
5. VK Kaishev. (1989). Optimal experimental designs for the *B*-spline regression. Comput Stat Data Anal 8:39–47.
6. YB Lim. (1991). *D*-Optimal design in polynomial spline regression. Korean J Appl Stat 4:171–178.
7. B Heiligers. (1998). *E*-Optimal designs for polynomial spline regression. Journal of Statistical Planning and Inference, 75: 159–172.
8. WJ Studden, DJ VanArman (1969). Admissible designs for polynomial spline regression. Ann Math Stat 40:1557–1569.
9. T Ando. (1987). Totally positive matrices. Linear Algebra Appl 90:165–219.
10. SH Park. (1978). Experimental designs for fitting segmented polynomial regression models. Technometrics 20:151–154.

21

On Dispersion Effects and Their Identification

Bo Bergman
Linköping University, Linköping, and Chalmers University of Technology, Gothenburg, Sweden

Anders Hynén
ABB Corporate Research, Västerås, Sweden

1. INTRODUCTION

Understanding variation is fundamental to quality improvement and customer satisfaction. That was realized early by Shewhart (1931) and later emphasized by, for example, Deming (1986, 1993). While Shewhart and Deming mainly concentrated on the reduction of variation by removing so-called assignable or special causes of variation, Taguchi (1986) suggested a systematic way to make products and processes insensitive to sources of variations (see also Taguchi and Wu (1980)). This strategy is usually called robust design methodology or robust design engineering; see, for instance, Kackar (1985), Phadke (1989), and Nair (1992). An important step is to identify factors, controllable by the designer or process developer, that affect the dispersion of a response variable y of interest.

Let $\mathbf{x}$ denote a vector of control factors, and let $\mathbf{z}$ be a vector of environmental variables that vary in a way usually not controllable by the designer, although some of its components might be controllable during the course of an experiment. A quite general way to describe the outcome y is

$$y = f(\mathbf{x}) + g(\mathbf{z}) + h(\mathbf{x}, \mathbf{z}) + \epsilon(\mathbf{x}, \mathbf{z}) \tag{1}$$

where $f(\mathbf{x})$ is the expectation of y, $f(\mathbf{x}) + g(\mathbf{z}) + h(\mathbf{x}, \mathbf{z})$ is the conditional expectation of y given $\mathbf{z}$; here $h(\mathbf{x}, \mathbf{z})$ corresponds to the interaction between

$\mathbf{x}$ and $\mathbf{z}$. In robust design methodology we want to determine levels of the factors, i.e., components of $\mathbf{x}$, such that the effect on y of the variation of ϵ and $\mathbf{z}$ is made as small as possible while $f(\mathbf{x})$ is kept on target. Assume that it is possible to vary all components of $\mathbf{z}$ in an experiment. Then the interaction between $\mathbf{x}$ and $\mathbf{z}$ is important in order to identify a robust design; see, for example, Box et al. (1988) and Bergman and Holmquist (1988). Very often, however, we cannot vary all components of $\mathbf{z}$; we have to find factors (components of $\mathbf{x}$) that affect the dispersion of y, i.e., variables having a dispersion effect. To clarify this approach we expand the variance of y by conditioning on the environmental factors $\mathbf{z}$:

$$\mathrm{Var}[y] = E[\mathrm{Var}[y|\mathbf{z}]] + \mathrm{Var}[E[y|\mathbf{z}]] \tag{2}$$

The two terms in the variance of y can be interpreted as follows. The first term on the right, $E[\mathrm{Var}[y|\mathbf{z}]]$, portrays how the variance of y, given $\mathbf{z}$, is affected by dispersion effects, i.e., factors affecting the spread of the data. The second term on the right, $\mathrm{Var}[E[y|\mathbf{z}]]$, portrays how the variance of y is affected by parameters in the location model—including fixed effects of $\mathbf{z}$ such as design by environmental interaction effects. The approach is motivated by the incorporation of dispersion effects, since direct location modeling of both design factors and environmental factors is allowed; thus this standpoint reduces the risk of confounding location effects and dispersion effects. Theoretical justification for the approach is also provided by Shoemaker et al. (1991), Box and Jones (1992), and Myers et al. (1992).

In this chapter we discuss how to identify control factors, i.e., product or process parameters, having dispersion effects; in particular, we discuss how dispersion effects can be identified using unreplicated experimental designs in the 2^{k-p} series of fractional factorial designs (see Bergman and Hynén, 1997). For some extensions to more general designs, see Blomkvist et al. (1997) and Hynén and Sandvik Wiklund (1996).

2. IMPROVING ROBUSTNESS THROUGH DISCOVERY OF DISPERSION EFFECTS

When it is possible to vary environmental (noise) factors in an experiment, robustness improvement is possible through location effect modeling if interaction effects are found. However, in this section improving robustness through minimization of the first variance term in (2) is considered. It was not until fairly recently that dispersion effect modeling became a central issue in parameter design, originally not emphasized even by Taguchi. Historically, there are many anecdotes associated with dispersion effect modeling, but many of these are merely anecdotes or aimed at making the estimation of location effects as efficient as possible. During the past decade this problem area experienced a rapid growth of interest, as shown by the number of applications and published papers. In general, there are two approaches for dispersion effects modeling: Either the experiment is replicated, or it is not. Major emphasis in this chapter is placed on the latter case; however, for the sake of completeness both approaches are considered.

In a replicated experiment, identification of dispersion effects is fairly straightforward. Depending on the error structure of the experiment, e.g., on whether or not the replicates are carried out fully randomized, the identified dispersion effects are effects measuring variability either between or within trials. Some may use the terms *replicates* and *duplicates*, or *genuine* and *false* replicates, respectively. If we compute sample variances, under each treatment combination, on which new effects can be computed, the analysis is rather uncomplicated. Taking the logarithm prior to computing the effects improves estimation (see Bartlett and Kendall, 1946). The new effects, which can be seen as dispersion effect estimates, can be plotted on normal probability paper to discriminate between large and small effects or analyzed with other techniques such as analysis of variance. For more background on this topic, see Nair and Pregibon (1988) and Bisgaard and Fuller (1995).

If the problem of dispersion effect modeling is a fresh arrival, identification of dispersion effects from unreplicated experiments is of even more recent date. Rather pioneering, Box and Meyer (1986b) published a paper addressing dispersion effect identification from unreplicated two-level fractional factorial experiments in the 2^{k-p} series. Their contribution was not entirely unique; Daniel (1976), Glejser (1969), and many others had touched upon the subject earlier, but Box and Meyer were the first to propose dispersion effect identification from unreplicated experiments as an important aspect of parameter design. In a paper by Bergman and Hynén (1997), in which the problem area is surveyed and a new method is introduced; dispersion effects from unreplicated designs in the 2^{k-p} series can now be identified with well-known statistical significance testing techniques (see also Section 3). It is still too early to judge the significance of the new method, but compared to existing methods the new proposal does not rely on distributional approximations or model discrimination procedures that are entirely ad hoc. There is, however, an assumption of normality that is rather critical (see Hynén, 1996). Moreover, the method proposed in Bergman and Hynén (1997) is generalized to experimental designs other than the two-level designs from the 2^{k-p} series by Blomkvist et al. (1997) and to the inner and outer array setup by Hynén and Sandvik Wiklund (1996). The use of normal probability plotting and transformations in combination with the method of Bergman and Hynén (1997) is considered by Blomkvist et al. (1997).

A different approach to the same problem was taken by Nelder and Lee (1991) and by Engel and Huele (1996). In both papers a generalized linear model approach is taken; see also McCullagh and Nelder (1989).

Overall, even if many problems remain to be solved, the contribution provided by the papers cited above constitutes a technique that can be useful for many purposes. It can be used to identify general heteroscedasticity, to relate heteroscedasticity to certain factors studied in the experiment, or simply to provide an additional component to the design engineer's toolbox useful for identifying the most robust design solution. Also note that the techniques used for unreplicated designs may be used in conjunction with duplicated designs to identify different components. The method suggested by Bergman and Hynén (1997) is discussed in the following section.

3. DISPERSION EFFECTS IN TWO-LEVEL FRACTIONAL FACTORIAL DESIGNS

Let i denote one of the factors in an unreplicated two-level fractional factorial design. Define σ_{i+}^2 as the average variance of the observations when factor i is at its high level, and let σ_{i-}^2 be defined correspondingly. Factor i is said to have a dispersion effect if $\sigma_{i+}^2 \neq \sigma_{i-}^2$. Natural but naive indicators for σ_{i+}^2 and σ_{i-}^2 are the sample variances based on all observations when factor i is at its high and low level, respectively, i.e., $s^2(i+)$ and $s^2(i-)$. Box and Meyer (1986b) suggested the use of ratios $F_i = s^2(i+)/s^2(i-)$ to identify dispersion effects. However, despite the notation, they noted that the F ratios did not belong to an F distribution owing to the presence of dispersion and location effect aliasing. The location effects had to be eliminated before estimating dispersion effects. Therefore, estimates were calculated from residuals obtained after eliminating suspected location effects. Later, some alternatives to Box and Meyer's approach were given. Nair and Pregibon (1988) extended the method to the case with replications. Furthermore, essential contributions are given by Wang (1989) and Wiklander (1994), who propose alternatives to the unreplicated case.

3.1. Location Effects

Let $\mathbf{y}$ be the $(n \times 1)$ response vector from a complete or fractional factorial experiment with an $(n \times n)$ design matrix $\mathbf{X}$ with column vectors $\mathbf{x}_0, \ldots, \mathbf{x}_{n-1}$. Column $\mathbf{x}_0$ is a column of 1's, and the remaining columns represent contrasts for estimating the main and interaction effects. We assume that the observations $y_1, \ldots, y_n$ are independent with variances $V[y_u] = \sigma_u^2$, $u = 1, \ldots, n$. Possibly, σ_u depends on the factors varied in the experiment. Note that, for example,

$$\sigma_{i+}^2 = \frac{2}{n} \sum_{x_{ui}=+1} \sigma_u^2$$

Let $\mathbf{z} = (1/n)\mathbf{X}'\mathbf{y}$ be the vector of estimated mean response, main, and interaction effects. As usual, we denote $E[\mathbf{z}] = \boldsymbol{\beta}$, whereupon $z_0, \ldots, z_{n-1}$ are independent with equal variances

$$\text{Var}[z_k] = \frac{1}{n^2} \sum_{u=1}^{n} \sigma_u^2, \qquad k = 0, \ldots, n-1$$

As noted by Box and Meyer (1993), the "vital few and trivial many" principle suggested by Juran (the Pareto principle) ensures that in most cases only a few β's are nonnegligible. Therefore, we can use the normal plotting technique suggested by Daniel (1976) to find these β's (see also Daniel, 1959). Of course, there may be problems due to confoundings when highly fractionated designs are used, but this issue is not discussed further here. See, for example, Box and Meyer (1986a, 1993), who give an interesting approach to these problems using Bayesian techniques.

Under the Pareto principle, only a few degrees of freedom are used to estimate nonnegligible β values. Therefore, the remainder of the contrasts can be used to estimate the variance σ_0^2, i.e.,

$$\sigma_0^2 = \frac{1}{n} \sum_{u=1}^{n} \sigma_u^2$$

In order to identify dispersion effects, we shall use additional contrasts that are based on linear combinations of those column vectors in $\mathbf{X}$ associated with negligible location effects.

3.2. Dispersion Effects

Box and Meyer (1986b) created new column vectors based on columns from the original design matrix $\mathbf{X}$:

$$\mathbf{x}_{j|i+} = \frac{1}{2}(\mathbf{x}_j + \mathbf{x}_{i \cdot j}) \qquad \text{and} \qquad \mathbf{x}_{j|i-} = \frac{1}{2}(\mathbf{x}_j - \mathbf{x}_{i \cdot j}) \tag{1}$$

where $\mathbf{x}_{i \cdot j}$ is the column vector corresponding to the row-wise (Hadamard) product of $\mathbf{x}_i$ and $\mathbf{x}_j$; i.e., if $\mathbf{x}_i$ and $\mathbf{x}_j$ correspond to main location effects, then $\mathbf{x}_{i \cdot j}$ corresponds to the $i \times j$ interaction effect. Note, for example, that the uth element of $\mathbf{x}_{j|i+}$ is equal to $\mathbf{x}_{uj}$ if $\mathbf{x}_{ui} = +1$ and zero otherwise.

Let us now introduce the set Γ_i of nonordered *pairs* of column vectors $\{\mathbf{x}_j, \mathbf{x}_{i \cdot j}\}$ from $\mathbf{X}$, such that the pair $\{\mathbf{x}_0, \mathbf{x}_{i \cdot 0}\}$ is excluded and neither of the corresponding contrasts $z_j = \mathbf{x}_j'\mathbf{y}$ and $z_{i \cdot j} = \mathbf{x}_{i \cdot j}'\mathbf{y}$ has been judged to estimate nonnegligible location effects, i.e., their expected values are judged to be zero:

$$E[\mathbf{x}_j'\mathbf{y}] = E[\mathbf{x}_{i \cdot j}'\mathbf{y}] = 0 \tag{2}$$

Note that there are $(n-1)/2$ members of Γ_i if all location effects are judged to be negligible, i.e., if we have $E[\mathbf{x}_j'\mathbf{y}] = 0$ for all j. It is straightforward to show that the contrasts corresponding to (1), $z_{j|i+} = \mathbf{x}_{j|i+}'\mathbf{y}$ and $z_{j|i-} = \mathbf{x}_{j|i-}'\mathbf{y}$, have variances

$$Var[z_{j|i+}] = \frac{n}{2}\sigma_{i+}^2 \quad \text{and} \quad Var[z_{j|i-}] = \frac{n}{2}\sigma_{i-}^2, \quad \text{respectively} \tag{3}$$

Now, let $\mathbf{x}_i$ be associated with a studied factor, i.e., let $\mathbf{x}_i'\mathbf{y}$ estimate one of the main effects. If σ_{i+}^2 and σ_{i-}^2 are different, this factor has a dispersion effect. Therefore, the difference between $z_{j|i+}^2$ and $z_{j|i-}^2$ gives information about the magnitude of this dispersion effect. If we can find many indices j such that $\{\mathbf{x}_j, \mathbf{x}_{i \cdot j}\}$ belongs to Γ_i, then all the corresponding $z_{j|i+}^2$ and $z_{j|i-}^2$ can be used to estimate the difference between σ_{i+}^2 and σ_{i-}^2. Moreover, since the column vectors $\mathbf{x}_{j|i+}$ are orthogonal, the contrasts $z_{j|i+}^2$ are independent. Therefore, we can use an F test for testing $H_{i0} : \sigma_{i+}^2 = \sigma_{i-}^2$ against $H_{i1} : \sigma_{i+}^2 \neq \sigma_{i-}^2$ with the test statistic

$$F_i = \frac{\sum_{\{j|(x_j, x_{i \cdot j}) \in \Gamma_i\}} z_{j|i+}^2}{\sum_{\{j|(x_j, x_{i \cdot j}) \in \Gamma_i\}} z_{j|i-}^2} \tag{4}$$

Under H_{i0}, the distribution of F is an F distribution with (m, m) degrees of freedom, where m is the number of elements $\{\mathbf{x}_j, \mathbf{x}_{i \cdot j}\}$ in Γ_i. Note that the test is double-sided; i.e., both large and small values of F_i supply evidence against the null hypothesis.

3.3. Alternative Expressions

The intuitive understanding of the above expressions might be somewhat vague. However, more intuitive expressions exist. Compute new "residuals," $\tilde{r}_u$, $u = 1, \ldots, n$, based on a location model including the active location effects expanded with the effects associated with column i and all interaction terms between i and the active location effects. Then the statistic D_i^{BH} may be computed as

$$D_i^{BH} = \frac{\sum_{x_{ui}=+1} \tilde{r}_u^2}{\sum_{x_{ui}=-1} \tilde{r}_u^2}$$

A third interpretation is the following. Given the identified location model, fit separate regressions to the two sets of data associated with the high and low levels of column i, respectively; i.e., use column i as a branching column. Compute the corresponding residual vectors, $\breve{\mathbf{r}}_{i+}$ and $\breve{\mathbf{r}}_{i-}$, and calculate D_i^{BH} as

$$D_i^{BH} = \frac{\sum_{u=1}^{n/2} \breve{r}_{ui+}^2}{\sum_{u=1}^{n/2} \breve{r}_{ui-}^2}$$

This alternative is, in fact, a generalization of the parametric test suggested by Goldfeld and Quandt (1965). They proposed a similar approach for identifying heteroscedasticity in a more general regression model.

Regarding the three alternatives, we see that the second one intuitively explains the differences between D_i^{BH} and the methods based directly on residuals. That is, it is necessary to adjust the original residuals to obtain independence between the two sets of residuals. This independence is, of course, conditional on the judgments made in (2) but provides the sufficient requirements for D_i^{BH} being F-distributed. The third alternative is the most natural way forward to generalize the proposed method to designs other than the 2^{k-p} series, e.g., to nongeometric Plackett and Burman designs and to factorial designs with more than two levels (see Blomkvist et al., 1997).

4. AN ILLUSTRATION FROM DAVIES (1956)

In Davies (1956), data from an improvement study concerning the quality of dyestuff was presented. The outcome was also given an interesting reanalysis by Wiklander (1994). We use the same data set to illustrate our method. The improvement study was carried out as a 2^{5-1} fractional factorial experiment without replicates involving five factors, labeled *A*–*E*. The defining relation was chosen as $I = -ABCDE$. The quality of the dyestuff was measured by a photoelectric spectrometer, which gave a quality characteristic of "the smaller the better" type; i.e., the lower the value recorded, the better the quality. Responses and all 15 orthogonal columns concerning location main and interaction effects are given in Table 1.

Since no independent error estimate is available, the normal probability plot of contrasts suggested by Daniel (1976) is a convenient tool for analysis (see Fig. 1). From this plot it appears that factor *D* is the only location effect present in the data; hence columns other than *D* can be used for estimating dispersion effects. In Davies (1956) only location effects were considered, but Wiklander (1994) detected and showed evidence of a dispersion effect from factor *E*. Further investigations will be conducted using our method.

An estimate of the dispersion effect from factor *E* becomes available on combining certain columns according to Eq. (1). The pairs of columns included must be judged to belong to the set Γ_E, i.e., judged not to correspond to active location effects. These new contrasts and their calculated values appear in Table 2.

For illustration, the contrast $z_{A|E+}$ is derived by combining columns *A* and *AE*, i.e.,

$$z_{A|E+} = (1/2)(\mathbf{x}_A + \mathbf{x}_{A \cdot E})'\mathbf{y}$$

Furthermore, testing $H_{i0} : \sigma_{i+}^2 = \sigma_{i-}^2$ against $H_{i1} : \sigma_{i+}^2 \neq \sigma_{i-}^2$ for other factors will require calculations analogous to those in Table 2 but based on other contrasts. The results from such a procedure are presented in Table 3. Note however, that the five *F* tests are not independent.

From Table 3, we see that factor *E* has a dispersion effect that is difficult to disregard. Wiklander (1994) detected this dispersion effect and found it significant. However, she used only (3, 3) degrees of freedom in a similar test. Furthermore, even factor *D* might have a dispersion effect that was not detected by Wiklander (1994). However, a complete analysis of data should always involve residual analysis, which here reveals a possible abnormality in observation 11. Treating y_{11} as a missing observation and recalculating it by setting some negligible contrast to zero (see Draper and Stoneman, 1964) shows that the dispersion effect from *D* becomes insignificant. Furthermore, the dispersion effect from *E* is fairly insensitive to changes in y_{11}, and it is therefore reasonable to consider *E* as the only active dispersion effect on the dyestuff data. Of course, there is also always the risk of overestimating the significance due to the multiple test effect.

Table 1 Design Matrix, Responses, and Confounding Structure up to Two-Factor Interactions for the Dyestuff Data

u	A	B	C	D	AB	AC	AD	BC	BD	CD	−DE	−CE	−BE	−AE	−E	y_u
1	−	−	−	−	+	+	+	+	+	+	−	−	−	−	+	201.5
2	−	−	−	−	−	−	−	+	+	+	+	+	+	−	−	178.0
3	−	+	−	−	−	+	+	−	−	+	+	+	−	+	−	183.5
4	−	+	−	−	+	−	−	−	−	+	−	−	+	+	+	176.0
5	−	−	+	−	+	−	+	−	+	−	+	−	+	+	−	188.5
6	+	−	+	−	−	+	−	−	+	−	−	+	−	+	+	178.5
7	−	+	+	−	−	−	+	+	−	−	−	+	+	−	+	174.5
8	+	+	+	−	+	+	−	+	−	−	+	−	−	−	−	196.5
9	−	−	−	+	+	+	−	+	−	−	−	+	+	+	−	255.5
10	+	−	−	+	−	−	+	+	−	−	+	−	−	+	+	240.5
11	−	+	−	+	−	+	−	−	+	−	+	−	+	−	+	208.5
12	+	+	−	+	+	−	+	−	+	−	−	+	−	−	−	244.0
13	−	−	+	+	+	−	−	−	−	+	+	+	−	−	+	274.0
14	+	−	+	+	−	+	+	−	−	+	−	−	+	−	−	257.5
15	−	+	+	+	−	−	−	+	+	+	−	−	−	+	−	256.0
16	+	+	+	+	+	+	+	+	+	+	+	+	+	+	+	274.5

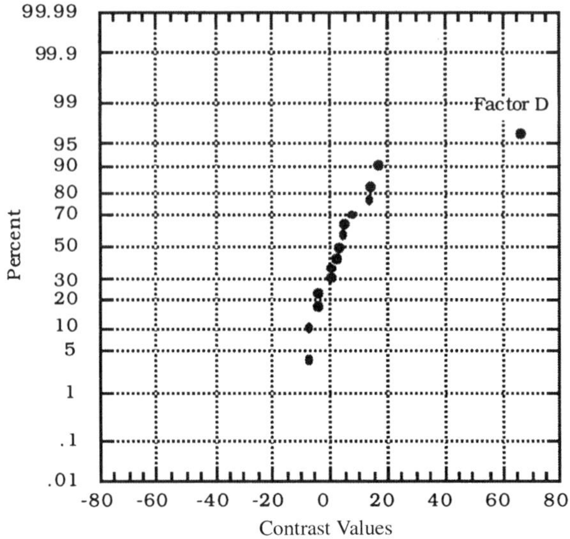

Figure 1 Normal probability plot of contrasts for the dyestuff data.

Table 2 Contrasts of Use for Estimating the Dispersion Effect from Factor E

Contrast	E = "+"	E = "−"
$z_{A\|E\pm}$	−7.5	11.0
$z_{B\|E\pm}$	0.5	−61.0
$z_{C\|E\pm}$	37.5	75.0
$z_{AB\|E\pm}$	9.5	124.0
$z_{AC\|E\pm}$	26.5	−2.0
$z_{AD\|E\pm}$	35.0	6.5

Table 3 F Ratios for the Five Factors from the Dyestuff Data

H_{i0}	F ratio	d.f.	P value
$\sigma^2_{A+} = \sigma^2_{A-}$	0.36	(6,6)	0.239
$\sigma^2_{B+} = \sigma^2_{B-}$	2.83	(6,6)	0.231
$\sigma^2_{C+} = \sigma^2_{C-}$	0.37	(6,6)	0.252
$\sigma^2_{D+} = \sigma^2_{D-}$	4.47	(7,7)	0.0665
$\sigma^2_{E+} = \sigma^2_{E-}$	0.14	(6,6)	0.0305

5. GENUINE REPLICATES AND SPLIT-PLOT DESIGNS

Genuine replicates require full randomization both between runs and within replications, which entails a large amount of experimental work (see, e.g., Box et al., 1978, p. 319). When experiments are expensive, as is often the case in industry, the randomization procedure within replicates is sometimes neglected and the experiment is given a split-plot structure. As seen in one of the examples provided by Bergman and Hynén (1997), this does not have to be a disadvantageous property but can instead be used to estimate two different variance components. Earlier analytic techniques did not support this special property, for which reason split-plot designs have received some criticism. However, some constructive remarks were made by Box and Jones (1992), Lucas and Ju (1992), and Anbari and Lucas (1994).

The method presented in this chapter is applicable to experiments with both genuine and split-plot replicates. Genuine replicates simply increase the degrees of freedom associated with the test statistic, Eq. (4), while split-plot replicates enable estimation of one additional variance component. Therefore, the latter of these two techniques ought to give the greatest increase in knowledge of how the system really works.

6. ON THE PLANNING OF ROBUST DESIGN EXPERIMENTS

The area of robust design methodology is constantly developing; thus a routine for planning experiments is very difficult to establish. In particular, developments enabling new methods for dispersion effect estimation will require changes in existing robust design techniques. We do not claim that the method presented in this chapter is the final step within this area. On the contrary, further research is necessary to fully understand the impact of dispersion estimation on experimental work. In this chapter, we have focused mainly on identification, although the success of an experiment is dependent on thorough planning. Therefore, some effects on the planning phase are worth mentioning.

Finding new techniques for testing and estimating dispersion effects from unreplicated experiments is a large step toward improving design economy. For instance, at the screening stage of sequential experimentation, replicates for identifying dispersion effects will not be necessary. Furthermore, it becomes possible to estimate additional variance components, which gives new perspectives on the use of some special designs such as split-plot designs as well as Taguchi's cross-product designs. Finally, and probably the most important issue to keep in mind, no technique is so perfect that sequential experimentation becomes unimportant. Problem solving is an iterative learning process, where "all-encompassing" solutions seldom come instantaneously. The Plan–Do–Study–Act cycle, or the Deming cycle (see Deming, 1993), is a model for every learning process, even the experimental one.

ACKNOWLEDGMENTS

This study has ben financed by the Swedish Research Council for Engineering Sciences. We also wish to thank the participants in a project on design of experiments and robust design methodology supported by a number of Swedish industrial firms. We are also grateful to our colleagues at the Division of Quality Technology for their valuable support.

REFERENCES

Anbari FT, Lucas JM. (1994). Super-efficient designs: How to run your experiment and higher efficiency and lower cost. In 1994 ASQC 48th Annual Quality Congress Proceedings. May 24–16, 1994, Las Vegas, Nevada.

Bartlett MS, Kendall DG. (1946). The statistical analysis of variance—Heterogeneity and the logarithmic transformation. J Roy Stat Soc Ser B 8:128–138.

Berman B, Hynén A. (1997). Dispersion effects from unreplicated designs in the 2^{k-p} series. Technometrics 39(2).

Bergman B, Holmqvist L. (1988). A Swedish programme on robust design and Taguchi methods. In: Bendell T, (ed.) Taguchi Methods. Proceedings of the 1988 European Conference. London, 13–14 July 1988. Amsterdam: Elsevier Applied Science.

Bisgaard S, Fuller HT. (1995). Quality quandaries—Reducing variation with two level factorial experiments. Qual Eng 8(2):373–377.

Blomkvist O, Hynén A, Bergman B. (1997). A method to identify dispersion effects from unreplicated multilevel experiments. Qual Reliab Eng Int 13(2).

Box GEP, Jones S. (1992). Split-plot designs for robust product experimentation. J Appl Stat 19(1):3–26.

Box GEP, Meyer RD. (1986a). An analysis for unreplicated fractional factorials. Technometrics 28(1):11–18.

Box GEP, Meyer RD. (1986b). Dispersion effects from fractional designs. Technometrics 28(1):19–27.

Box GEP, Meyer RD. (1993). Finding the active factors in fractionated screening experiments. J Qual Technol 25(2):94–105.

Box GEP, Hunter WG, Hunter JS. (1978). Statistics for experimenters—An Introduction to Design, Data Analysis, and Model Building. New York: Wiley.

Box GEP, Bisgaard S, Fung C. (1988). An explanation and critique of Taguchi's contribution to quality engineering. Qual Reliab Eng Int 4(2):123–131.

Cook RD, Weisberg S. (1983). Diagnostics for heteroscedasticity in regression. Biometrika 70(1):1–10.

Daniel C. (1959). Use of half-normal plots in interpreting fctorial two-level experiments. Technometrics 1(4):311–341.

Daniel C. (1976). *Application of Statistics to Industrial Experimentation*. New York: Wiley.

Davies OL, (ed.) (1956). *Design and Analysis of Industrial Experiments*. London: Oliver and Boyd.

Deming WE. (1986). Out of the Crisis. Cambridge, MA: Cambridge University Press.

Deming WE. (1993). The New Economics, for Industry, Government, Education. Cambridge, MA: Massachusetts Institute of Technology.

Draper, NR, Stoneman DM. (1964). Estimating missing values in unreplicated two-level factorial and fractional factorial designs. Biometrics 20(3):443–458.

Engel J, Huele AF. (1996). A generalized linear modeling approach to robust design. Technometrics 38(4):365–373.

Glejser H. (1969). A new test for heteroscedasticity. No 6, RQT&MR Rep Ser Division of Quality Technology and Management, Linköping University, Sweden.

Goldfeld SM, Quandt RE. (1965). Some tests for homoscedasticity. J Am Stat Assoc 60(310):539–547.

Hynén A. (1996). A note on non-normality and dispersion effect identification in unreplicated factorial experiments. No 6, RQT&M Res Rep Ser, Division of Quality Technology and Management, Linköping University, Sweden.

Hynén A, Sandvik Wiklund P. (1996). On dispersion effects from inner and outer array experiments. No 9, RQT&M Res Rep Ser, Division of Quality Technology and Management, Linköping University, Sweden.

Kackar RN. (1985). Off-line quality control, parameter design, and the Taguchi method (with discussion). J Qual Technol 17(4):176–209.

Lucas JM, Ju HL (1992). Split plotting and randomization in industrial experiments. 1992 ASQC 46th Annual Quality Congress Transactions. May 18–20, 1992, Nashville, TN. Milwaukee, WI: ASQC.

McCullagh P, Nelder JA. (1989). Generalized Linear Models. 2nd ed. London: Chapman and Hall.

Myers R, Khuri AI, Vining G. (1992). Response surface alternatives to the Taguchi robust parameter design approach. Am Stat 46(2):131–139.

Nair VN. (1992). Taguchi's parameter design: A panel discussion. Technometrics 34(2):127–161.

Nair VN, Pregibon D. (1988). Analyzing dispersion effects from replicated factorial experiments. Technometrics 30(3):247–257.

Nelder JA, Lee Y. (1991). Generalized linear models for the analysis of Taguchi type experiments. Appl Stochastic Models Data Anal 7:107–120.

Phadke MS. (1989). Quality Engineering Using Robust Design. Englewood Cliffs, NJ: Prentice-Hall.

Shewhart WA. (1931). *Economic Control of Quality of Manufactured Product*. New York: Van Nostrand. (A 1981 reprint is available from the American Society for Quality Control.)

Shoemaker AC, Tsui K-L, Wu CFJ. (1991). Economical experimentation methods for robust design. Technometrics 33(4):415–427.

Taguchi G. (1981). On-Line Quality Control During Production. Tokyo: Japanese Standards Association.

Taguchi G. (1986). Introduction to Quality Engineering. Tokyo: Asian Productivity Organisation.

Taguchi G, Wu Y. (1980). Introduction to Off-Line Quality Control. Nagoya, Japan: Central Japan Quality Control Association.

Wang PC. (1989). Tests for dispersion effects from orthogonal arrays. Comput Stat Data Anal 8:109–117.

Wiklander K. (1994). Models for dispersion effects in unreplicated two-level factorial experiments. Thesis No. 1994: 1/ISSN 1100-2255, The University of Gothenburg, Sweden.

22

A Graphical Method for Model Fitting in Parameter Design with Dynamic Characteristics

Sung H. Park
Seoul National University, Seoul, Korea

Je H. Choi
Samsung Display Devices Co., Ltd., Suwon, Korea

ABSTRACT

Detecting the relationship between the mean and variance of the response and finding the control factors with dispersion effects in parameter design and analysis for dynamic characteristics are important. In this paper, a graphical method, called multiple mean-variance plot, is proposed to detect the relationship between the mean and variance of the response. Also to find the control factors with dispersion effects, the analysis of covariance method is proposed, and its properties are studied compared with the dynamic signal-to-noise ratio. A case study is presented to illustrate the proposed methods.

I. INTRODUCTION

Achieving high product quality at low cost is a very important goal in modern industry. One of the most popular statistical methods using an experimental design approach to reach this goal is parameter design, which is often called robust parameter design. Parameter design was proposed by Taguchi (1986, 1987) and explained further by Box (1988), Leon et al. (1987), Nair (1992), Phadke (1989), and Park (1996), among many others. The main idea of parameter design is to determine the setting of control factors (or design parameters) of a product or process in which the response characteristic is robust to the uncontrollable variations caused by the noise factors and hence has a small variability.

The parameter design uses the S/N (signal to noise) ratio 10 log (μ^2/σ^2), for the static characteristics (hereafter it will be called the static S/N ratio), where μ and σ^2 are the mean and variance of the response, respectively. S/N is the ratio of the power of the signal to the power of the noise. The S/N ratio for the dynamic characteristics, which will be hereafter called the dynamic S/N ratio, is defined as 10 log (β^2/σ^2) under the model $y = \alpha + \beta M + \varepsilon$ or $y = \beta M + \varepsilon$, where y is the response, M is the signal factor, and σ^2 is the variance of the error ε. Here β^2 implies the power of the signal, and σ^2 implies the power of the noise.

The usefulness of the dynamic S/N ratio has been proved, since many engineering systems can be adequately described as dynamic characteristic problems. See, for instance, many case studies presented in the American Supplier Institute (1991) symposium on Taguchi methods.

2. DESCRIPTION OF THE DYNAMIC CHARACTERISTICS SYSTEM

In the parameter design, the experimental factors are classified according to their roles into the following three classes.

1. Signal factor (M). This factor influences the average value but not the variability of the response. It is also called the target-control factor.
2. Noise factor (N). This factor has an influence over the response variability but cannot be controlled in actual applications.
3. Control factor ($\mathbf{x}$). This factor can be controlled and manipulated by the engineer, and its level is selected to make the product's response robust to noise factors. It is the goal of the experiment to determine the best levels of the control factors that are robust to noise factors under the existence of a signal factor.

Parameter design systems are classified into two categories according to the nature of the target value of the response. One is the static system, which has a fixed target value, and the other is the dynamic system, which has varied target values according to the levels of the signal factor. The dynamic system is shown in Figure 1. In this section the dynamic characteristic problem which has a continuous signal input and a continuous output with some control factors and noise factors is considered.

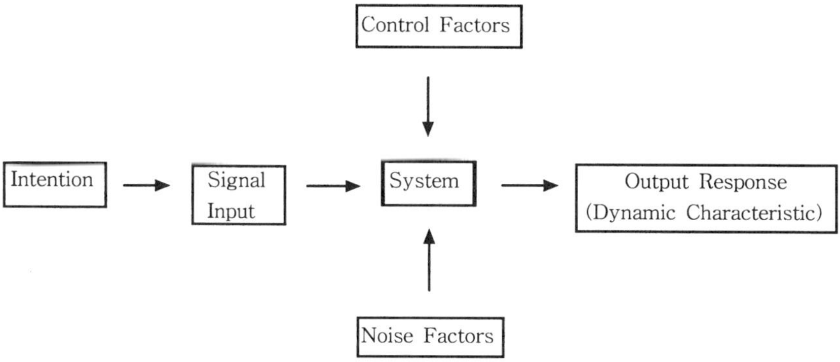

Figure 1 Dynamic system of parameter design.

3. UNKNOWN VARIANCE FUNCTION AND DETECTION

Let y_{ijk} denote the response corresponding to the ith setting of the control factors, jth level of the signal factor, and kth noise factor or repetition, for $i = 1, \ldots, l$; $j = 1, \ldots, m$; and $k = 1, \ldots, n$. Then the data structure of the response in the dynamic system is assumed to be expressed as

$$y_{ijk} = f_i(M_j) + \varepsilon_{ijk} \tag{1}$$

The data structure in this section has the following assumptions:

1. The error ε_{ijk} has zero expectation and variance σ_{ij}^2.
2. The expectation of y_{ijk} is $f_i(M_j)$ for all k and can be expressed as a polynomial, especially the first-order polynomial $\alpha_i + \beta_i M_j$ or $\beta_i M_j$.
3. The effect of the noise factors is included in the error variance σ_{ij}^2, so the subscript of the variance term σ_{ij}^2 does not contain k.
4. The variance of the response depends on its expected value and can be expressed as $\sigma_i^2 \times V[E(y_{ijk})]$, where the variance function $V(\cdot)$ represents the relationship of the variance of the response to its mean, and the term σ_i^2 represents the remaining part, which depends on the ith control factor setting.

The experimenter is interested in finding the control factor setting that makes σ_i^2 small and minimizes $\sigma_i^2 \times V[E(y_{ijk})]$. In general the relationship between the mean and variance is unknown, and the detection and modeling of the variance function $V(\cdot)$ is important.

For the detectiong of $V(\cdot)$ and model fitting, the following three-step optimization procedure is proposed.

Step 1. Detect the relationship between mean and variance by constructing a multiple mean–variance plot.

Step 2. Find the control factors with dispersion effects by the analysis of covariance (ANCOVA) method.

Step 3. Fit the response as a function $f_i(M_j)$ of the signal factor M to adjust the sensitivity of the response to the signal factor M.

3.1 Detecting the Relationship Between Mean and Variance by Using a Multiple Mean–Variance Plot

To detect the relationship $V(\cdot)$ between the variance and the mean of the response, a multiple mean–variance plot (MMVP) is suggested. Nair and Pregibon (1986) proposed the mean–variance plot, and Lunani et al. (1995) proposed the sensitivity–standard deviation (SS) plot for the dynamic characteristic problems. Lunani et al. considered the model where the variance structure satisfies the relationship

$$\mathrm{Var}\ (y_{ijk}) = \beta_i^{\theta}\sigma_i^2 \qquad \text{for some} \quad \theta$$

Under this model there is a logarithmic relationship between the sensitivity measure $(\hat{\beta}_i)$ and the standard deviation (s_i),

$$\log(s_i) = \log(\sigma_i) + \frac{\theta}{2}\log(\hat{\beta}_i) \tag{2}$$

where $\hat{\beta}_i$ and s_i are obtained from the regression fitting for each control factor setting i. Lunani et al. plotted $[\log(\hat{\beta}_i), \log(s_i)]$ for each control factor and visually examined the plots to check the nature of the relationship. They noticed that when some control factors have dispersion effects, the intercepts $\log(\sigma_i)$ can vary from one control factor setting to another, making it possible to have several parallel lines with a common slope $\theta/2$ in the SS plot under model (2).

The MMVP is proposed for model (1). It is the combination of the mean–variance plot and the multiple SS plot. Under model (1), there is a logarithmic relationship between $\bar{y}_{ij}$ and s_{ij}^2,

$$\log(s_{ij}^2) = \log(\sigma_i^2) + \log[V(\bar{y}_{ij})] \tag{3}$$

where $s_{ij}^2 = \Sigma_k (y_{ijk} - \bar{y}_{ij})^2/(n-1)$. Note that the expected values of $\bar{y}_{ij}$ and s_{ij}^2 are $E(\bar{y}_{ij}) = fi(M_j) = \mu_{ij}$ and $E(s_{ij}^2) = V[f_i(M_j)]\sigma_i^2 = V(\mu_{ij})\sigma_i^2$, where $\mu_{ij} = E(y_{ijk})$. The procedure used for the MMVP is as follows.

1. Get $l \times m$ data of pairs $(\bar{y}_{ij}, s_{ij}^2)$ for each control factor setting i and each signal factor level j.
2. Plot these paired data [log $(\bar{y}_{ij})$, log (s_{ij}^2)] on the scatter plot for each control factor.
3. Identify the points of the frame of each control factor according to its levels.
4. Detect the variance relationship $V(\cdot)$.

For example, if the orthogonal array L_{18} as the inner array and a three-level signal factor are used for experiments, a total of 54 (= 18 × 3) paired data [log $(\bar{y}_{ij})$, log (s_{ij}^2)] are obtained. By plotting [log $(\bar{y}_{ij})$, log (s_{ij}^2)], detection of the form of $V(\cdot)$ is possible. If the points are scattered like an exponential function, the exponential function taken on log $(\bar{y}_{ij})$, $\bar{y}_{ij}$ would make the points linear. If that is the case, then $V(\mu) = \exp(\theta\mu)$ is selected as the proper variance function. For an example see Figure 2 in Section 4.

Like the SS plot of Lunani et al., if the variance function is properly selected and the assumption of model (1) holds, then the points on the frame of the control factor with dispersion effects are identified on separate lines. Then the control factors with dispersion effects can be easily found.

Note that when the objective of the analysis is focused on the variance of the response, the term log (s^2) is usually used rather than s^2 for certain statistical reasons. One reason is that the effect on dispersion may be reasonably considered as a multiplicative effect rather than an additive effect. Moreover, a linear model on log (s^2) can be easily used without constraint. In addition, the performance of log (s^2) is stable when the hteroscedasticity problem occurs. Logothetis (1989) showed that the mean of log (s^2) depends on log (σ^2) and n, and the variance of log (s^2) is stable depending on only n, and furthermore log (s^2) converges to approximate normality as n increases.

3.2. Finding the Control Factors with Dispersion Effects by the ANCOVA Method

In the second step, finding the control factors with dispersion effects, the implementation of the ANCOVA method is proposed where the covariate is determined from the selected variance function at step 1. This method is an extension of the models of Logothetis (1989) and Engel (1992).

Logothetis (1989) thought that the relationship could be detected by using the regression model on log (s_i^2) with an independent variable log $(\bar{y}_i)$:

$$\log(s_i^2) = \log(\alpha) + \theta \log(\bar{y}_i) + \varepsilon_i \tag{4}$$

Engel (1992) noticed that the parameter log (α) is a nonconstant term in the Logothetis model and replaced it by the term log (α_i), which is a linear function of the control factors.

$$\log(s_i^2) = \log(\alpha_i) + \theta\ \log(\bar{y}_i) + \varepsilon_i \tag{5}$$

When the logarithm is taken on the variance term in model (1), the following equation is obtained.

$$\log[\mathrm{Var}(y_{ijk})] = \mathbf{x}_i\gamma + \log[V(E(y_{ijk}))] \tag{6}$$

Here $\mathbf{x}_i$ is the row vector of the control factors, and γ is a parameter vector. When this model is applied to practical applications, the fitting model (7) is used as the form of ANCOVA. Here the sample variance s_{ij}^2 on log is the dependent variable, the control factors are the factor given in the vector $\mathbf{x_i}$, and the sample mean $\bar{y}_{ij}$ or its function $h(\bar{y}_{ij})$ is the covariate, where $h(\cdot)^{\theta_i} = V(\cdot)$:

$$\log(s_{ij}^2) = \mathbf{x}_i\gamma + \theta_i\ \log[h(\bar{y}_{ij})] + \text{error} \tag{7}$$

The control factors of significance are selected to have dispersion effects from the analysis of the model.

Note that this model has two main differences from Engel's model:

1. The variance function $V(\mu)$ is a general function instead of μ^6.
2. The coefficient θ_i is considered a nonconstant parameter.

The variance function $V(\mu)$ cannot be easily detected in the static system, so the power of the mean model (5) is mainly used in Engel's paper. But in the dynamic system $V(\mu)$ can be detected, and it can have a general form. When several $V(\mu)$'s are candidates, for example, $V(\mu) = \mu^{\theta_i}$ or $V(\mu) = \exp(\theta_i\mu)$, the variance function $V(\mu)$ can also be detected at step 1, and the selection of $V(\mu)$ can be done by some variable selection technique of the regression with log (s_{ij}^2) as the dependent variable. The variance function $V(\mu)$ is preferred that separates the plotted points into parallel lines with a common slope and different intercepts, because the control factors with dispersion effects need to be detected and well selected there. Taking $V(\mu)$ as μ^θ is the direct generalization of the model of Engel.

A nonconstant θ_i has the practical meaning that as the mean increases the variance can increase at a different rate at each level of some control factors. If the coefficients of the lines look identical from the search at step 1, then fitting model (7) is the ANCOVA method without interaction between the covariate and the factors. At the beginning of the analysis, model (7) with constant term θ is used. If the points are separated into lines with different slopes in the frame of some control factors, then changing the variance function $V(\mu)$ or extending the term θ into θ_t may be considered.

When $V(\mu) = \mu^{\theta}$ and the parameter θ is taken as equal to 2 beforehand, the ANCOVA method is equivalent to the procedure for finding the control factors to maximize the static S/N ratio. We can observe that the following model is derived from model (7):

$$\log(s_{ij}^2) = 2\ \log(\bar{y}_{ij}) + \mathbf{x}_i\gamma + \text{error} \tag{8a}$$

$$\log(s_{ij}^2) - 2\log(\bar{y}_{ij}) = -\log\left(\frac{\bar{y}_{ij}^2}{s_{ij}^2}\right) = -\frac{1}{10}(S/N)_i = \mathbf{x}_i\gamma + \text{error} \tag{8b}$$

The dynamic characteristic approach has some merits compared to the static characteristic approach for detecting the variance function. One is that it has a large numer of degrees of freedom when the dispersion effects of control factors are checked. In the example of Engel (1992), the inner array is saturated, so there is no degree of freedom allowed for covariate log $(\bar{y}_i)$. However, the ANCOVA method has a large number of degrees of freedom when the level-of-signal factor is large. Another merit is the distribution of the mean response. As the mean value is more widely spread, the precision of estimation of the variance function increases (Davidian and Carroll, 1987). In the static system the response is usually distributed around the fixed target value and less spread. But in the dynamic system the target value varies according to the signal input value, and the response is widely spread according to the signal factor level. Taguchi's optimization procedure with the dynamic S/N ratio does not enjoy these merits. The sample variances of each signal factor level are combined into one quantity, the dynamic S/N ratio. Here the ANCOVA method is proposed to utilize these merits by taking the sample mean and the sample variance at each signal factor level.

3.3. Fitting the Response as a Function of the Signal Factor

When the variance of the response is a function of the levels of the signal factor, the use of weighted least squares (WLS) is recommended to estimate $f_i(M)$ for each i. After the control factors with dispersion effects are chosen at step 2, the variance of response y_{ijk} is estimated as $\widehat{\sigma}_i^2\ \widehat{V(\bar{y}_{ij})}$. Then the weights are the inverse of the estimated variance of each response, and the

WLS method is applied for each control factor setting i to adjust the sensitivity of the response to the signal factor M.

4. AN EXAMPLE: CHEMICAL CLEANING EXPERIMENT

In this section, the data set from the chemical cleaning process for Kovar metal components (American Supplier Institute, 1991) is reanalyzed to show how to use the ANCOVA method and multiple mean–variance plot proposed in Section 3 to find the control factors with dispersion effects and the functional relationship between the mean and the variance.

The response y is the amount of the material removed as a result of the chemical cleaning process. The inner array is L_{18} including a two-level factor A and three-level factors B, C, D, E, F, and G. The outer array consists of a three-level signal factor M crossed with L_4 for a compound array of three two-level noise factors X, Y, Z. The signal factor M is the acid exposure time, which is known to have a linear impact on the expected value of the response. By imposing the linearity of the signal factor, the process becomes predictable and more controllable from the engineering knowledge. The information about the experimental factors and the raw data are given in Tables 1 and 2.

Table 1 Experimental Factors and Levels for Chemical Cleaning Experiment

	Factor level		
Factor label and description	0	1	2
Control factor			
A Part status at Brite-dip	Dry	Wet	
B Descale acid exposure time	B_0	B_1	B_2
C Descale acid strength	C_0	C_1	C_2
D Descale acid temperature	Low	Med	High
E Nitric/acetic (ratio)	E_0	E_1	E_2
F Percent in Brite-dip acid	Low	Med	High
G Brite-dip acid temperature	standard	Remachine	
Noise factor			
X Descale acid age	New	Used	
Y Brite-dip acid age	New	Used	
Z Part type	Stamped	Machined	
Signal factor			
M Exposure time in Brite-dip	M_1	M_2	M_3

Table 2 Experimental Layout and Raw Data for Chemical Cleaning Experiment

Signal factor								M_1				M_2				M_3				Noise factor
								0	0	1	1	0	0	1	1	0	0	1	1	X
Control factor								0	1	0	1	0	1	0	1	0	1	0	1	Y
	A	B	C	D	E	F	G	0	1	1	0	0	1	1	0	0	1	1	0	Z
Col.	1	2	3	4	5	6	7	1	2	3	4	1	2	3	4	1	2	3	4	
1	0	0	0	0	0	0	0	9	11	15	11	14	17	22	14	19	29	26	18	
2	0	0	1	1	1	1	1	27	31	30	25	43	55	63	43	67	63	88	43	
3	0	0	2	2	2	2	2	26	36	38	26	40	57	71	44	59	82	92	59	
4	0	1	0	0	1	1	2	27	39	50	24	51	86	92	48	78	113	123	68	
5	0	1	1	1	2	2	0	11	3	27	14	24	33	37	22	32	34	51	31	
6	0	1	2	2	0	0	1	14	20	23	16	27	30	44	20	38	50	59	28	
7	0	2	0	1	2	2	2	12	18	23	13	26	31	32	23	29	42	46	29	
8	0	2	1	2	0	0	0	25	33	45	36	34	45	95	55	42	66	127	80	
9	0	2	2	0	1	1	1	19	24	27	17	33	46	46	27	40	61	70	33	
10	1	0	0	2	2	1	0	25	43	40	32	38	53	62	41	56	73	95	51	
11	1	0	1	0	0	2	1	25	42	36	21	42	49	63	36	47	58	81	56	
12	1	0	2	1	1	0	2	16	20	17	8	22	32	36	20	34	43	53	33	
13	1	1	0	1	2	2	1	27	39	28	19	46	62	55	36	58	84	78	48	
14	1	1	1	2	0	0	2	26	48	77	37	42	84	104	60	52	111	109	78	
15	1	1	2	0	1	1	0	31	56	59	32	58	88	98	50	77	115	128	71	
16	1	2	0	2	2	0	1	17	17	34	18	23	31	56	30	32	42	67	40	
17	1	2	1	0	0	1	2	21	25	61	20	30	43	95	36	40	60	124	60	
18	1	2	2	1	1	2	0	13	26	22	10	23	27	27	20	26	52	41	28	

The response data y_{ijk} ($i = 1, \ldots, 18; j = 1, 2, 3; k = 1, 2, 3, 4$) are summarized into 18 × 3 paired data [log ($\bar{y}_{ij}$), log (s_{ij}^2)] for each control factor setting i and signal factor level j. These paired data [$\bar{y}_{ij}$), log (s_{ij}^2)] and [log ($\bar{y}_{ij}$), log (s_{ij}^2)] are plotted in Figures 2a and 2b. These figures definitely show that a function relationship exists between the variance and mean of the response and can be explained by a linear function on the log–log scale. We can assume that $h(\mu) = \mu$ rather than $h(\mu) = \exp(\mu)$ for model (7).

In Figure 3, the multiple mean–variance plots of [log ($\bar{y}_{ij}$), log (s_{ij}^2)] show which control factors have dispersion effects. In the frame of factor A, the points for level 1 (symbol +) are shown along with the points for level 0 (∘). Two separate fitting lines can be drawn, with a common slope and different intercepts. When the level of factor A is 0, the response has a smaller variance. By similar work, level 0 of factor B can be selected. For factor C the difference between levels 0 (∘) and 2 (◇) does not look large, and

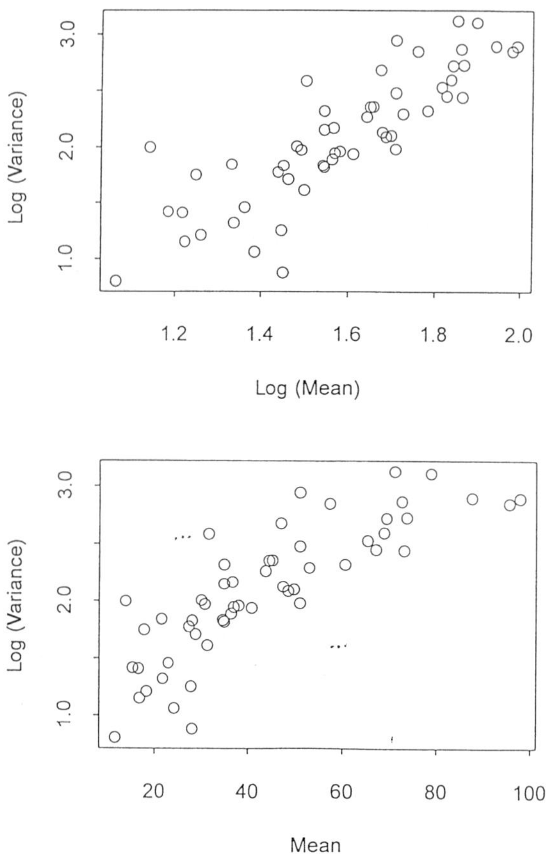

Figure 2 Plots of [log ($\bar{y}_{ij}$), log (s_{ij}^2)] and [$\bar{y}_{ij}$, log (s_{ij}^2)] for chemical cleaning experiment.

either of those may be selected. In the other frames of Figure 3, the points are not divided into separate lines according to the levels of factors *D*, *E*, *F*, and *G*.

The results from the analysis of the dynamic S/N ratio are presented in Table 4. These results show that *A*, *B*, *C*, and *D* are the important factors with respect to the S/N ratio. The best level selected is A_0, B_0, C_0 and D_1, which is similar to the selected level in the results from the ANCOVA method except for factor *D*. But in the ANCOVA method, factors *C* and *D* are not very significant (their *p* values are 0.053 and 0.057, respectively), and other levels of these factors may be selected.

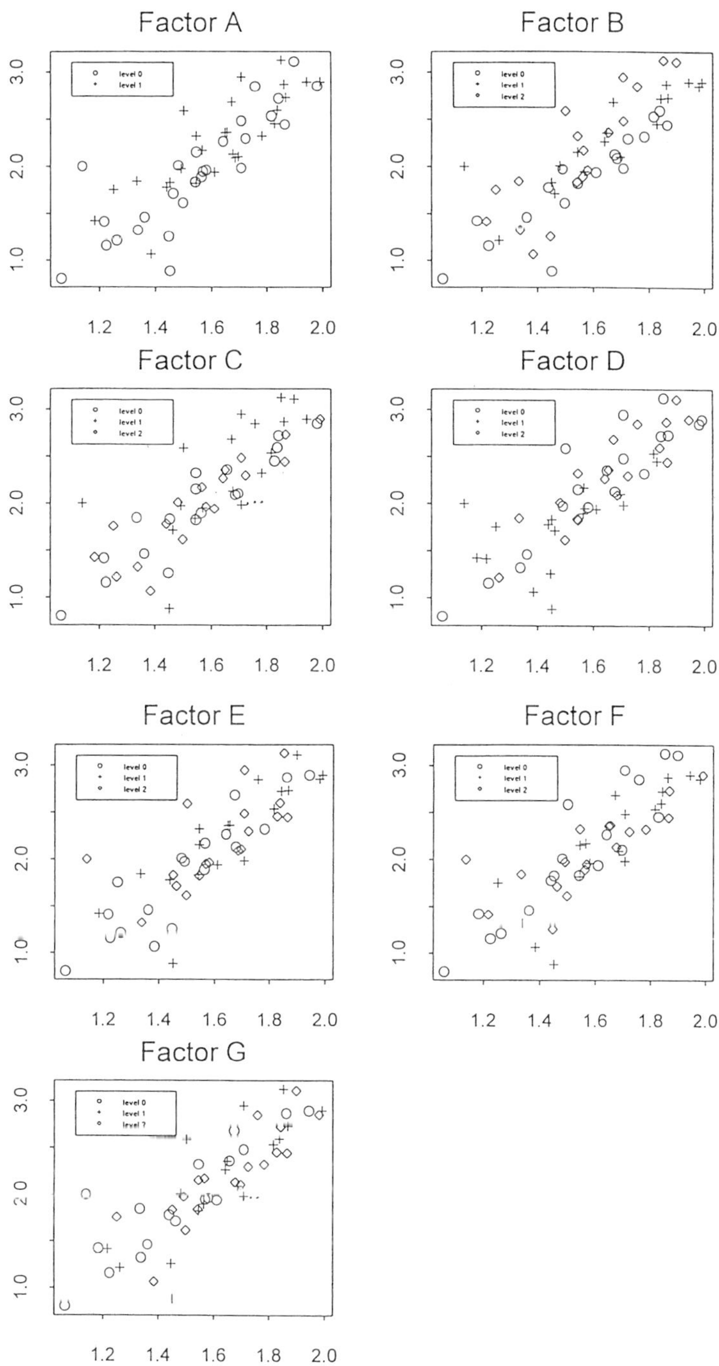

Figure 3 Multiple plots of [log ($\bar{y}_{ij}$), log (s_{ij}^2)] for chemical cleaning experiment.

Table 3 ANOVA Table for the ANCOVA Method with Covariate log ($\bar{y}_{ij}$)

Source	DF	Adjusted SS	F	p Value
Covariate	1	4.81013	100.12	0.000
A	1	0.32313	5.66	0.022
B	2	0.91742	7.01	0.002
C	2	0.35064	3.15	0.053
D	2	0.30866	3.06	0.057
E	2	0.09540		
F	2	0.14475		
G	2	0.05400		
$A \times B$	2	0.18098		
(e)	37	2.59806		
Pooled error	45	3.09112		
T	53	17.92377		

Table 4 ANOVA Table for the Dynamic Signal-to-Noise Ratio

Source	DF	Adjusted SS	F	$\rho(\%)$
A	1	14.010	14.84	12.13
B	2	30.166	15.98	26.27
C	2	32.417	17.17	28.34
D	2	23.101	12.24	19.84
E	2	0.002		
F	2	2.620		
G	2	1.667		
$A \times B$	2	3.542	2.42	1.93
(e)	2	0.803		
Pooled error	8	5.907		11.49
T	17	109.132		100.00

REFERENCES

American Supplier Institute. (1991). Taguchi Symposium: Case Studies and Tutorials, Dearborn, MI: ASI.

Box GEP. (1988). Signal-to-noise ratios, performance criteria, and transformations (with discussion). Technometrics 30:1–40.

Davidian M, Carrol RJ. (1987). Variance function estimation. Am Stat Assoc 82:1079–1091.

Engel J. (1992) Modelling variation in industrial experiments. Appl Stat 41(3): 579–593.

Leon R, Shoemaker AC, Kackar RN. (1987). Performance measures independent of adjustment: An explanation and extension of Taguchi's signal-to-noise ratios (with discussion). Technometrics 9:253–285.

Logothetis N. (1989). Establishing a noise performance measure. Statistician 38: 155–174.

Lunani M, Nair V, Waserman GS. (1995). Robust Design with Dynamic Characteristics: A Graphical Approach to Identifying Suitable Measures of Dispersion. Tech Rep 253, University of Michigan.

Nair VN, ed. (1992). Taguchi's parameter design: a panel discussion. Technometrics 34: 127–160.

Nair VN, Pregibon D. (1986). A data analysis strategy for quality engineering experiments. AT&T Tech J 65:73–84.

Park SH. (1996). Robust Design and Analysis for Quality Engineering. London: Chapman & Hall.

Phadke MS. (1989). Quality Engineering Using Robust Design. Englewood Cliffs, NJ: Prentice-Hall.

Taguchi G. (1986). Introduction to Quality Engineering. Tokyo: Asian Productivity Organization.

Taguchi G. (1987). System of Experimental Design. White Plains, NY: Unipub. Kraus International.

23

Joint Modeling of the Mean and Dispersion for the Analysis of Quality Improvement Experiments

Youngjo Lee
Seoul National University, Seoul, Korea

John A. Nelder
Imperial College, London, England

1. INTRODUCTION

The Taguchi method for analyzing quality improvement experiments has been much discussed. It first defines summarizing quantities called performance measures (PMs) and then analyzes them using analysis of variance. PMs are defined as functions of the response y; however, we believe that they should be regarded as quantities of interest derived after analysis of the basic response and defined as functions of the fitted values or parameter estimates. One of Taguchi's signal-to-noise ratios (SNRs) involves $\sum y^{-4}$. This is not a good estimate of $\sum \mu^{-4}$, though $\sum \hat{\mu}^{-4}$ might be acceptable. However, as Box (1988) showed, Taguchi's signal-to-noise ratios make sense only when the log of the response is normally distributed. The correct statistical procedure is (1) to analyze the basic responses using appropriate statistical models and then (2) to form quantities of interest and measures of their uncertainty. Taguchi's procedure inverts this established process of statistical analysis by forming the PMs first and then analyzing them. However, most writers concentrate on the analysis of PMs, though they may use other than signal-to-noise ratios. Miller and Wu (1996) refer to the Taguchi approach as performance measure modeling (PMM) and the established statistical approach as response function modeling (RFM). They, of course, recommend RFM. However, what they actually do seems to be closer to the PMM approach. The major difference is that they consider statistical models for responses before choosing PMs. Because of the initial data reduction to PMs, their primary tool for analysis is restricted to graphical tools such as the normal probability plot. Interpretation of such plots can be subjective. Because information on the adequacy of the model is in the residuals, analysis using PMs makes testing for lack of fit difficult or impossible.

In 1991, we (Nelder and Lee, 1991) published a paper giving a general method that allows analysis of data from Taguchi experiments in a statistically natural way, exploiting the merits of standard statistical methods. In this chapter, we provide a detailed exposition of our method and indicate how to extend the analysis to Taguchi experimental data for dynamic systems.

2. THE MODEL

Taguchi robust parametric design aims to find the optimal setting of control (i.e., controllable) factors that minimizes the deviation from the target value caused by uncontrollable noise factors. Robustness means that the resulting products are then less sensitive to the noise factors. Suppose a response variable y can be modeled by a GLM with $E(y_i) = \mu_i$ and $\text{var}\,(y_i) = \phi_i V(\mu_i)$, where ϕ_i are dispersion parameters and $V\,(\,)$ the variance function. The variance of y_i is thus the product of two components; $V(\mu_i)$ expresses the intrinsic variability due to the functional dependence of the variance on the mean μ_i, while ϕ_i expresses the extrinsic variability, which is independent of the range of means involved. Suppose we have control factors $C_1, ..., C_p$ and noise factors $N_1, ..., N_q$. In our 1991 paper we considered the following joint models for the mean and the dispersion

$$\eta = g(\mu) = f_1(C_1, ..., C_p, N_1, ..., N_q) \tag{1}$$

and

$$\zeta = \log(\phi) = f_2(C_1, ..., C_p, N_1, ..., N_q) \tag{2}$$

where $g\,(\,)$ is the link function for the mean, and $f_i(C_1, ..., C_p, N_1, ..., N_q)$ are linear models for experimental designs, e.g., the main effect of the model is $C_1 + ... + C_p + N_1 + ... + N_q$. The log link is assumed for the dispersion as a default; there are often insufficient data to discriminate between different link functions. We need to choose for each model a variance function, a link function, and terms in the linear predictor. By choosing an appropriate variance function for the mean, we aim to eliminate unnecessary complications in the model due to functional dependence between the mean and variance [the separation of Box (1988)]. It is useful if the final mean and dispersion models have as few common factors as possible. The link function for the mean should give the simplest additive model [the parsimony of Box (1988)].

Control factors occuring in $f_2(\ )$ only or in both $f_1(\ )$ and $f_2(\ )$ are used to minimize the extrinsic variance, and control factors occurring in $f_1(\ )$ only are then used to adjust the mean to a target without affecting the extrinsic variability.

If we analyze PMs such as SNRs, calculated over the noise factors for each combination of the control factors, it is then impossible to make inferences about the noise factors in the model for the mean. This reduction of data leads to the number of responses for the dispersion analysis being only a fraction of those available for the mean. We do not have such problems since we analyze the entire set of data; see Lee and Nelder (1998).

3. THE ALGORITHM

When a GLM family of distributions does not exist for a given $V(\mu_i)$, Wedderburn's (1974) quasi-likelihood (QL) is often used for inference from the mean model (1). However, it cannot be used for joint inference from both mean and dispersion models; for this we need Nelder and Pregibon's (1987) extended quasi-likelihood (EQL), defined by

$$-2Q^+ = \sum\left\{\frac{d_i}{\phi_i} + \log[2\pi\phi_i V(y_i)]\right\}$$

where $d_i = -2\int_{y_i}^{\mu_i} (y_i - u)/V(u)du$ denotes the GLM deviance component

For given ϕ_i, the EQL is, apart from a constant, the quasi-likelihood (QL) of Wedderburn (1974) for a GLM with variance function $V(\mu_i)$. Thus maximizing Q^+ with respect to β will give us the QL estimators with prior weights $1/\phi_i$, satisfying

$$\frac{\partial Q+}{\partial\beta} = \sum\left[\frac{y_i - \mu_i}{\phi_i V(\mu_i)}\right]\frac{\partial\mu_i}{\partial\beta} = 0$$

The EQL provides a scaled deviance with component d_i/ϕ_i, and this deviance may be used as a measure of discrepancy, so that we can create an analysis-of-deviance table for a nested set of models, as with GLMs. The differences of such deviances allow us to identify significant experimental factors on the same link scale and to compare different link functions for the mean model (1).

For given μ_i, the EQL gives a GLM with the gamma distribution for the deviance components d_i, and this forms the basis of the dispersion model. Thus, with the EQL we can identify significant experimental factors for both the mean and the dispersion models. However, when the number of mean parameters is relatively large compared with the sample size, dispersion estimators can be seriously biased without appropriate adjustment for the degrees of freedom. The REML technique removes this bias for mixed linear models (Patterson and Thompson, 1971). Cox and Reid (1987) extended the REML idea to a wider class of models that satisfy an orthogonality relation of the form $E(\partial^2 Q + \partial\beta\partial\gamma) = 0$. The Cox–Reid adjusted profile EQL becomes

$$-2Q_C^+ = \sum\left\{\frac{d_i}{\phi_i} + \log[2\pi\phi_i V(y_i)]\right\} + \log\det\left(\frac{X^t W^* X}{2\pi}\right)$$

where W^* is an $n \times n$ diagonal matrix with ith element $\{1/(\phi_i V(\mu_i))\}(\partial\mu_i/\partial\eta_i)$. Thus for inference from the dispersion model (2) we (Lee and Nelder, 1998) use Q_C^+; then $\partial Q_C^+/\partial\gamma = 0$ gives estimating equations for γ, $(-\partial^2 Q_C^+/\partial\gamma^2)^{-1}$ a variance estimate for $\hat{\gamma}$, and $-2Q_C^+$ the basis of a deviance test. To overcome the slow computation of REML estimation, we (Lee and Nelder, 1998) have developed an efficient approximation.

The EQL is the true likelihood for the normal and inverse Gaussian distributions, so our estimators (deviance tests) for β and γ are the ML and REML estimators (likelihood ratio and adjusted likelihood ratio tests), respectively. The EQL also gives good approximations for the remaining distributions of the GLM family. There are two approximations in the assumed model for the dispersion. The first lies in assuming that $E(d) = \phi$; in general the bias is small except in extreme cases, e.g., Poisson errors with small μ. Such biases enter the analysis for the mean only through the weight and do not much affect the estimates of β. The second approximation is the assumption of a gamma error for the dispersion analysis, regardless of the error chosen for the mean. The justification for this is the effectiveness of the deviance transform in inducing a good approximation to normality for all the GLM distributions (Pierce and Schafer, 1986), excluding extreme cases such as binary data; see also the simulation study of Nelder and Lee (1992).

In summary, our model consists of two interlinked GLMs, one for the mean and one for the dispersion as follows. The two connections, one in each direction, are marked. The deviance component from the model for the mean becomes the response for the dispersion model, and the inverse of the fitted values for the dispersion model give prior weights for the mean model. (See Table 1.) In consequence, we can use all the methods for GLMs for inferences from the joint models, including various model-checking procedures.

4. STATISTICAL MODELS FOR DYNAMIC SYSTEMS

Recently, there has been an emphasis on making the system robust over a range of input conditions, so the relationship between the input (signal factor) and output (response) is of interest. Following Lunani et al. (1997), we refer to this as a dynamic system. Miller and Wu (1996) and Lunani et al. (1997) have studied Taguchi's method for dynamic systems. Suppose we have a continuous signal factor M, measured at m values. These researchers consider models analogous to the mean and dispersion models

$$\eta = g(\mu) = I(M) * f_1(C_1, \dots, C_p, N_1, \dots, N_q) \tag{3}$$

and

$$\log(\phi) = f_2(C_1, \dots, C_p, N_1, \dots, N_q) \tag{4}$$

where $g(\ )$ is the link function for the mean and $I(\ )$ is the function describing the relationship between the input (signal factor) and output (response).

Table 1

GLM	Mean	Dispersion
Response	y	$\rightarrow d$
Mean	μ	ϕ
Variance	$\phi V(\mu)$	$2\phi^2$
Link	$\eta = g(\mu)$	$\zeta = \log \phi$
Deviance component	$d = 2\int_{\mu}^{y} \frac{y-u}{V(u)} du$	$2\left[-\log\left(\frac{d}{\phi}\right) + \frac{d-\phi}{\phi}\right]$
Prior weight	$1/\phi \leftarrow$	1

The function $I(\,)$ may be known a priori or may have to be identified. Lunani et al. assume that $I(M) = M\beta$, i.e., that it is a linear function without an intercept, and Miller and Wu select $I(M) = \beta_0 + \beta_1 M + \beta_2 M^2$. The $*$ operator in Eq. (3) represents the fact that parameters of $I(\,)$ are modeled as functions of C_i and N_j. In dynamic systems the signal factor is used to adjust the mean using the mean model (3), and control factors are set to optimize the sensitivity measure [see Miller and Wu (1996) and Lunani et al. (1997)].

The fitting of dynamic systems has so far been done in two stages; in stage I parameters in $I(\,)$ are estimated for each run, and in stage II models are fitted separately to each set of stage I parameter estimates. For example, with $I(M) = \beta_0 + \beta_1 M + \beta_2 M^2$ we fit $I(M)$ for each individual run, computing $\hat{\beta}_0$, $\hat{\beta}_1$, and $\hat{\beta}_2$, and then fit separate models for these as functions of C_i and N_j. However, the model chosen by this approach may not fit the data well because data reduction to PMs under the wrong model makes testing for lack of fit difficult. Our method analyzes the whole data set and does not require two stages of fitting. All that is necessary is that the software allow the specification of compound terms of the form $A.x$ in the linear predictor, denoting that the slope for x varies with the level of the factor A.

5. ADVANTAGES OF THE GLM APPROACH

The advantage of analyzing all the individual responses using two interlinked GLMs over the analysis of variance of PMs (with possible transformation of the data) are as follows:

1. Box's (1988) two criteria, separation and parsimony, cannot necessarily both be achieved by a single data transformation, while the GLM analysis achieves them separately by choosing appropriate variance and link functions for the two interlinked GLMs. Analysis is thus always carried out on the original data.
2. Any GLM can be used for modeling the means. Thus counts, proportions, and positive continuous quantities can be incorporated naturally into the model.
3. Our model uses all the information in the data. For example, the dispersion analysis has a response for each observation, just as with the mean. Compare this with the use of s_i^2 calculated over the noise factors for each combination of the control factors; this leads to the number of responses for the dispersion analysis being only a fraction of that available for the mean. Such s_i^2 do not use random variation but are functions of arbitrarily selected levels for the noise factors. Furthermore, when the s_i^2 $(\hat{\beta})$ are computed over

noise (signal) factors in static (dynamic) systems it is impossible to make inferences about those noise (signal) factors in the model for the dispersion (mean). With this approach, the signal factor cannot be included in the dispersion model (4) for dynamic systems.

4. The model is defined for any design. For example, our method can be used for dynamic systems as easily as for static systems, and we can also consider more general models such as $\log(\phi) = f_2(C_1, ..., C_p, N_1, ..., N_q, M)$ for (4).
5. The use of a GLM for fitting the dispersion model means that model-checking techniques, such as residual plots, developed for GLMs generally can be applied directly to both parts of the joint model.

6. CONCLUSION

Data from Taguchi experiments should be analyzed in a statistically natural way so that existing statistical methods can be used; this allows for statistically efficient likelihood inferences, such as the likelihood-ratio test, model-checking diagnostics to test the adequacy of the model, and maximum likelihood estimation or restricted maximum likelihood estimation to be used. Our method supports these desirable aims.

REFERENCES

Box GEP. (1988). Signal-to-noise ratios, performance criteria and transformations. Technometrics 30:1–17.

Cox DR, Reid N. (1987). Parameter orthogonality and approximate conditional inference. J Roy Stat Soc Ser B 49:1–39.

Lee Y, Nelder JA. (1998). Generalized linear models for the analysis of quality-improvement experiments. Can J Stat, 26: 95 105.

Lunani M, Nair VN, Wassweman GS. (1997). Graphical methods for robust design with dynamic characteristics. J Qual Technol 29: 327–338.

McCullagh P, Nelder JA. (1989). Generalized Linear Models. London: Chapman and Hall.

Miller A, Wu JCF. (1996). Improving a measurement system through designed experiments. Stat Sci 11: 122–136.

Nelder JA, Lee Y. (1991). Generalized linear models for the analysis of Taguchi-type experiments. Appl Stochast Models Data Anal 7: 107–120.

Nelder JA, Lee Y. (1992). Likelihood, quasi-likelihood and pseudo-likelihood: Some comparisons. J Roy Stat Soc Ser B 54:273–284.

Nelder JA, Pregibon D. (1987). An extended quasi-likelihood function. Biometrika 74: 221–232.

Patterson HD, Thompson R. (1971). Recovery of interblock information when block sizes are unequal. Biometrika 58: 545–554.

Pierce DA, Schafer DW. (1986). Residuals in generalized linear models. J Am Stat Assoc 81: 977-986.

Wedderburn RWM. (1974), Quasi-likelihood functions, generalized linear models and the Gauss–Newton method. Biometrika 61: 439–447.

24

Modeling and Analyzing the Generalized Interaction

Chihiro Hirotsu
University of Tokyo, Tokyo, Japan

1. INTRODUCTION

The analysis of interaction is the key in a wide variety of statistical problems including the analysis of two-way contingency tables, the comparison of multinomial distributions, and the usual two-way analysis of variance. It seems, however, that it has been paid much less attention than it deserves.

In the usual analysis of variance, both of the two-way factors have generally been assumed to be controllable, and the combination that gives the highest productivity has been searched for. We should, however, also consider the possibilities that the factors may be indicative or variational. By an indicative factor we mean a fixed but uncontrollable factor such as the region in the adaptability test of rice varieties where the problem is to choose the best level of the controllable factor (the variety of rice) for each level of the indicative factor (region) by considering the interaction between these two factors. Then a procedure is desired for grouping the levels of the indicative factor whose responses against the levels of the controllable factor are similar so that a common level of the controllable factor can be assigned to every level of the indicative factor within a group.

By a variational factor we mean a factor that is fixed and indicative within an experiment but acts as if it were a random noise when the result is extended to the real world. A typical example is the noise factor in Taguchi's parameter design, where the problem is to choose the level of the controllable factor to give not only the highest but also the most stable responses against the wide range of levels of the noise factor. For all these problems the usual omnibus F test for interaction is not very useful, and row-wise and/or columnwise multiple comparison procedures have been proposed (Hirotsu, 1973, 1983a, 1991a). Those procedures are also useful for modeling and analyzing contingency tables and multinomial distributions not restricted narrowly to the analysis of variance (Hirotsu, 1983b, 1993).

Another interesting problem is detecting a two-way changepoint for the departure from a simple additive or multiplicative model when there are intrinsic natural orderings among the levels of the two-way factors. Detecting a change in the sequence of events is an old problem in statistical process control, and there is a large body of literature dealing with this. These works, however, are mostly for univariate series of independent random variables such as normal, gamma, Poisson, or binomial [e.g., see, Hawkins (1977), Worsley (1986), and Siegmund (1986)]. Therefore in this chapter I discuss an approach to detecting a two-way changepoint.

2. MODELING THE INTERACTION IN THE ANALYSIS OF VARIANCE FRAMEWORK

Suppose that we are given two-way observations with replications and assume the model

$$y_{ijk} = \mu_{ij} + \varepsilon_{ijk}, \qquad i = 1, \ldots, a; j = 1, \ldots, b; k = 1, \ldots, r$$

where the ε_{ijk} are independently distributed as $N(0, \sigma^2)$. The μ_{ij} may be modeled simply by $\mu_{ij} = \mu + \alpha_i + \beta_j$ if the hypothesis of no interaction is accepted. When it is rejected, however, we are faced with a more complicated model, and it is desirable to have a simplified interaction model with fewer degrees of freedom. Several models have been proposed along this line, including those of Tukey (1949), Mandel (1967), and Johnson and Graybill (1972). The block interaction model obtained as a result of the row-wise and column-wise multiple comparisons is also a useful alternative (Hirotsu, 1973, 1983a, 1991a).

For row-wise multiple comparisons we define an interaction element between two rows, the mth and the nth, say, by

$$\mathbf{L}(m; n) = (1\sqrt{2})P_b'(\mu_m - \mu_n)$$

where $\mu_i = (\mu_{i1}, \ldots, \mu_{ib})'$ and P_b' is a $(b-1) \times b$ matrix satisfying $P_b'P_b = I_{b-1}$ and $P_bP_b' = I_b = b^{-1}j_bj_b'$ with I an identity matrix and $\mathbf{j}$ a vector of 1's. Then a multiple comparison procedure for testing $\mathbf{L}(m; n) = 0$ is given in Hirotsu (1983a) to obtain homogeneous subgroups of rows so that in each of them all interaction elements are zero. The columns can be dealt with similarly. Then the resulting model can be expressed as

$$\mu_{ij} = \bar{\mu}_{i.} + \bar{\mu}_{.j} - \bar{\mu}_{..} + (\alpha\beta)_{ij} \tag{1}$$

with $(\alpha\beta)_{i.} = 0, (\alpha\beta)_{.j} = 0$ and $(\alpha\beta)_{ij} = (\alpha\beta)_{i'j'}$ if $i,\ i' \in G_u$ and $j,\ j' \in J_v$, where $G_u, u = 1, ..., A$ and $J_v, v = 1, ..., B$ denote the homogeneous subgroups of rows and columns, respectively. We use the usual dot bar notation throughout the paper. Model (1) may be called the block interaction model with df$(A-1)(B-1)$ for interaction. The row-wise and/or columnwise multiple comparisons seem particularly useful for dealing with indicative or variational factors; see Hirotsu (1991a, 1991b, 1992) for details.

3. THE GENERALIZED INTERACTION

We encounter two-way table analysis even in the one-way analysis of variance framework if only we take the nonparametric approach.

The data in Table 1 are the half-life of the drug concentration in blood for low and high doses of an antibiotic. This is a simple two-sample problem. In the nonparametric approach, however, we change those data into rank data as given in Table 2. In Table 2 we are interested in whether the 1's are more likely to occur to the right than to the left for the high dose relative to the low dose since that would suggest that the high dose is more likely to prolong the half-life.

Table 3 is the result of a dose–response experiment and gives the same type of data with Table 2 where the ordered categories are thought to be tied ranks. Again the high categories seem to occur more frequently in the higher dose. It is the problem of analyzing interaction to confirm these observations statistically.

The outcome of a Bernoulli trial can also be expressed in a similar way to Table 2. We give an example in Table 4, where the probability of occurrence changes from 0.2 to 0.4 at the 11th trial.

The outcomes of an independent binomial sequence are also summarized similarly to Table 4. We give an example in Table 5, which is taken from a clinical trial for heart disease.

Table 1 Half-life of Antibiotic Drug

Dose (mg/(kg · day	Data				
25	1.55	1.63	1.49	1.53	2.14
200	1.78	1.93	1.80	2.07	1.70

Table 2 Rank Data Obtained from Table 1

	Rank									
Dose	1	2	3	4	5	6	7	8	9	10
25	1	1	1	1	0	0	0	0	0	1
200	0	0	0	0	1	1	1	1	1	0

Table 3 Usefulness in a Dose-Finding Experiment

Drug	1 Undesirable	2 Slightly undesirable	3 Not useful	4 Slightly useful	5 Useful	6 Excellent	Total
AF3mg	7	4	33	21	10	1	76
AF6mg	5	6	21	16	23	6	77

Table 4 Outcome of Bernoulli Trial with Probability Change at the 11th Trial

	Run																			
Outcome	1	2	3	4	5	6	7	8	9	10	11	12	13	14	15	16	17	18	19	20
Nonoccurrence	1	1	0	1	1	0	1	1	1	1	1	1	1	1	0	0	1	1	0	0
Occurrence	0	0	1	0	0	1	0	0	0	0	0	0	0	0	1	1	0	0	1	1
Total	1	1	1	1	1	1	1	1	1	1	1	1	1	1	1	1	1	1	1	1

Table 5 Independent Binomial Sequence

	Dose level (mg)				
Outcome	100	150	200	225	300
Failure	16	18	9	9	5
Success	20	23	27	26	9
Total	36	41	36	35	14

In any of Tables 2–5 we denote by p_{ij} the occurrence probability of the (i, j) cell, $i = 1, 2; j = 1, ..., k$. Then in Tables 2 and 3 we are interested in comparing two multinomials $(p_{i1}, ..., p_{ib} | p_{i.} = 1)$, $i = 1, 2$, and Tables 4 and 5 are concerned with comparisons of k binomials $(p_{1j}, p_{2j} | p_{.j} = 1)$, $j = 1, ..., k$. Regardless of the differences between the sampling schemes, however, we are interested in both cases in testing the null hypothesis,

$$\frac{p_{21}}{p_{11}} = \frac{p_{22}}{p_{12}} = \ldots = \frac{p_{2k}}{p_{1k}} \tag{2}$$

against the ordered alternative,

$$\frac{p_{21}}{p_{11}} \leq \frac{p_{22}}{p_{12}} \leq \ldots \leq \frac{p_{2k}}{p_{1k}} \tag{3}$$

taking into the account the natural ordering in columns. In (3) we assume that at least one inequality is strict. It then includes as its important special case a changepoint model,

$$\frac{p_{21}}{p_{11}} = \ldots = \frac{p_{2J}}{p_{1J}} < \frac{p_{2J+1}}{p_{1J+1}} = \ldots = \frac{P_{2k}}{p_{1k}} \tag{4}$$

where J is an unknown changepoint, the detection of which is an old problem in statistical process control

The hypotheses (2), (3), and (4) can be expressed in terms of the interaction parameters in the log-linear model

$$\log p_{ij} = \mu + \alpha_i + \beta_j + (\alpha\beta)_{ij}$$

The interaction term $(\alpha\beta)_{ij}$ can be interpreted as an odds ratio parameter in this context. Thus we can generalize the usual analysis of interaction into the analysis of odds ratio parameters in multinomials, where an ordered alternative hypothesis is often of particular interest.

Under the null hypothesis, Eq. (2), we base our statistical inference on the conditional distribution given sufficient statistics. Regardless of the sampling schemes, this leads to the hypergeometric distribution given all the row and column marginal totals [see Plackett (1981)]. This is why we need not distinguish Tables 2 and 3 from Tables 4 and 5.

Table 6 Rank Data for the Half-Life Data

Dose (mg/kg · day)	Rank 1	2	3	4	5	6	(7,8)	9	10	11	12	13	14	15	Total
25	0	0	1	0	1	1	1	0	0	0	0	0	0	1	5
50	1	1	0	1	0	0	1	0	0	0	1	0	0	0	5
200	0	0	0	0	0	0	0	1	1	1	0	1	1	0	5

4. A SAMPLE PROBLEM

Given half-life data (1.21, 1.63, 1.37, 1.50, 1.81) at a dose level of 50 mg/(kg · day) in addition to Table 1, we obtain Table 6. We also have placebo data in the dose–response experiment, with which we obtain Table 7.

Next suppose that the products from an industrial process are classified into three classes (1st, 2nd, 3rd) and their probabilities of occurrence are changed from (1/3, 1/3, 1/3) to (2/3, 1/6, 1/6) at the 11th trial. An example of the outcome is shown in Table 8. This is regarded as an independent sequence of trinomials.

It should be noted that in all three examples the row-wise and/or columnwise multiple comparisons are essential. Noting the existence of the natural orderings in both rows and columns, we are particularly interested in testing the null hypothesis

$$p_{i+11}/p_{i1} = p_{i+12}/p_{i2} = \dots = p_{i+1k}/p_{ik}, \qquad i = 1, \dots, a-1$$

against the ordered alternative

$$p_{i+11}/p_{i1} \le p_{i+12}/p_{i2} \le \dots \le p_{i+1k}/p_{ik}, \qquad i = 1, \dots, a-1$$

Table 7 Usefulness in a Dose-Finding Experiment

Drug	1 Undesirable	2 Slightly undesirable	3 Not useful	4 Slightly useful	5 Useful	6 Excellent	Total
Placebo	3	6	37	9	15	1	71
AF3mg	7	4	33	21	10	1	76
AF6mg	5	6	21	16	23	6	77

Table 8 Products Classified into Three Classes

	Run																			
Class	1	2	3	4	5	6	7	8	9	10	11	12	13	14	15	16	17	18	19	20
3rd	1	1	0	0	0	0	1	0	0	1	0	0	0	0	0	0	0	1	0	0
2nd	0	0	0	1	1	1	0	1	1	0	0	0	1	0	0	1	0	0	0	0
1st	0	0	1	0	0	0	0	0	0	0	1	1	0	1	1	0	1	0	1	1
Total	1	1	1	1	1	1	1	1	1	1	1	1	1	1	1	1	1	1	1	1

with at least one inequality strict. Again the alternative hypothesis includes as its special case a two-way changepoint model such that the inequality

$$p_{i'j}/p_{ij} < p_{i'j'}/p_{ij'} \tag{5}$$

holds only when $i \leq I, i' \geq I + 1$ and $j \leq J, j' \geq J + 1$, where (I, J) is the unknown changepoint. This is a natural extension of the one-way changepoint model (4).

5. TESTING THE ORDERED ALTERNATIVE FOR INTERACTION—TWO-SAMPLE CASE

The analyses of interaction in the analysis of variance model and in the log-linear model are parallel to some extent, at least for two-way tables [see Hirotsu (1983a, 1983b)], and here we give only the procedure for the latter for brevity.

5.1. Comparing Treatments

The most popular procedure for comparing treatments is Wilcoxon's rank sum test. In that procedure the jth category is given the score of the mid-rank,

$$\omega_j = y_{.1} + \cdots + y_{.j-1} + \frac{y_{.j} + 1}{2}$$

and the rank sum of each treatment is defined by

$$W_i = \sum_j \omega_j y_{ij}, \qquad i = 1, 2$$

where y_{ij} is the observed frequency in the (i,j)th cell. The standardized difference of the rank sums is then defined by

$$W(1;2) = \left(\frac{n-1}{\sigma^2 Iw}\right)^{1/2} \left(\frac{1}{y_{1.}} + \frac{1}{y_{2.}}\right)^{-1/2} \left(\frac{W_2}{y_{2.}} - \frac{W_1}{y_{1.}}\right)$$

where

$$\sigma_w^2 = \sum \omega_j^2 y_{.j} - \left(\frac{\sum \omega_j y_{.j}}{y_{..}}\right)^2$$

For evaluating the p-value of $W(1;2)$ we can use a normal approximation. The network algorithm of Mehta et al. (1989) can also be applied to give the exact p-value. As an example, for the data of Table 4 we obtain $W(1;2) = 1.320$ with the two-sided p-value 0.187 by the normal approximation.

Another possible approach is the cumulative chi-square method (Hirotsu, 1982). For this method we partition the original table at the jth column to obtain a 2×2 table by pooling columns as in Table 9 and calculate the goodness-of-fit chi-square statistic

$$\chi_j^2 = \frac{y_{..}(Y_{ij}\bar{Y}_{2j} - \bar{Y}_{1j}Y_{2j})^2}{y_{1.}y_{2.}Y_{.j}Y_{.j}}$$

$$= \frac{y_{..}^3(Y_{1j} - y_{1.}Y_{.j}/y_{..})^2}{y_{1.}y_{2.}Y_{.j}\bar{Y}_{.j}} \qquad j = 1, \ldots, k-1$$

Then the cumulative chi-square statistic is defined by

$$\chi^{*2} = \chi_1^2 + \ldots + \chi_{k-1}^2$$

The null distribution of χ^{*2} is well approximated by the distribution of the constant times the chi-square variable $d\chi_\nu^2$, where the constant d and the degrees of freedom ν are given by the formulas

Table 9 Calculating the Cumulative Chi-Square Statistic

Row	Column pooled		Total
	$(1, \ldots, j)$	$(j+1, \ldots, k)$	
1	Y_{1j}	$\bar{Y}_{1j}$	$y_{1.}$
2	Y_{2j}	$\bar{Y}_{2j}$	$y_{2.}$
Total	$Y_{.j}$	$\bar{Y}_{.j}$	$y_{..}$

$$d = 1 + \frac{2}{k-1}\left(\frac{\lambda_1}{\lambda_2} + \frac{\lambda_1 + \lambda_2}{\lambda_3} + \ldots + \frac{\lambda_1 + \ldots + \lambda_{k-2}}{\lambda_{k-1}}\right) \qquad \lambda_j = \frac{y_{.1} + \ldots + y_{.j}}{y_{.j+1} + \ldots + y_{.k}} \tag{6}$$

and

$$\nu = (k-1)/d$$

When the $y_{.j}$ are all equal as in Table 4, χ^{*2} is well characterized by the expansion

$$\chi^{*2} = \frac{k}{1 \times 2}\chi^2_{(1)} + \frac{k}{2 \times 3}\chi^2_{(2)} + \cdots + \frac{k}{(k-1)k}\chi^2_{k-1}$$

where $\chi^2_1, \chi^2_2, \ldots$ are the linear, quadratic, etc. chi-square components each with one degree of freedom (df) and are asymptotically mutually independent; see Hirotsu (1986) for details. More specifically, $\chi^2_{(1)}$ is just the square of the standardized Wilcoxon statistic. Thus the statistic χ^{*2} is used to test mainly but not exclusively the linear trend in p_{2j}/p_{1j} with respect to j. For the data of Table 4, $\chi^{*2} = 30.579$ and constants are obtained as $d = 6.102$ and $\nu = 3.114$. The approximated two-sided p-value is then obtained as 0.183.

5.2 Changepoint Analysis

The maximal component of the cumulative chi-square statistic

$$\chi^2_M = \max_j \chi^2_j$$

is known as the likelihood ratio test statistic for changepoint analysis and has been widely applied for the analysis of multinomials with ordered categorical responses since it is a very easy statistic to interpret. Some exact and efficient algorithms have been obtained for calculating its p-value, which is based on the Markov property of the sequence of the chi-square components, $\chi^2_1, \ldots, \chi^2_{k-1}$ [see Worsley (1986) and Hirotsu et al. (1992)]. Applying those algorithms to Table 4, we obtain the two sided p value 0.135 for $\chi^2_M = 5.488$, which gives moderate evidence for the change in the probability of occurrence.

In comparing the three statistics introduced above for testing the ordered alternatives (3), the Wilcoxon statistic tests exclusively a linear trend, max χ^2 is appropriate for testing the changepoint model (4), and χ^{*2} keeps a high power over a wide range of the ordered alternatives. As an example of comparing two multinomials with ordered categorical responses, the three methods are applied to the data of Table 3, and the results are summarized in Table 10. For reference, the usual goodness-of-fit

Table 10 Three Methods Applied to Table 3

Test statistic	Two-sided p-value
$W(1;2) = 2.488$	0.0128
$\chi^{*2} = 18.453$	0.0096
$\chi^2_M = 10.303$	0.0033
$\chi^2 = 12.762$	0.0257

chi-square value is shown at the bottom of the table; it does not take the natural ordering into account and as a consequence is not so efficient as the other three methods for the data.

6. TESTING THE ORDERED ALTERNATIVE FOR INTERACTION—GENERAL CASE

6.1. Comparing Treatments on the Whole

As an overall test for the association between ordered rows and columns, rank correlations such as Spearman's ρ or Kendall's τ and the Jonckheere test are well known. Here we introduce a doubly cumulative chi-square statistic defined by

$$\chi^{**2} = \sum_{i}^{a-1}\sum_{j}^{k-1}\chi^2_{ij}$$

$$\chi^2_{ij} = \frac{Y^3_{ab}(Y_{ij} - Y_{ib}Y_{aj}/Y_{ab})^2}{Y_{ib}(Y_{ab} - Y_{ib})Y_{aj}(Y_{ab} - Y_{aj})}$$

$$Y_{ij} = \sum_{l}^{i}\sum_{m}^{j} y_{lm}, \qquad i = 1, \ldots, a; j = 1, \ldots, k$$

so that $Y_{ab} = y_{..}$ is the grand total of observations. The (i, j)th component χ^2_{ij} is the goodness-of-fit chi-square value for the 2×2 table obtained in the same way as Table 9 by partitioning and pooling the original $a \times k$ data at the ith row and the jth column.

The statistic χ^{**2} is again well approximated by $d\chi^2_\nu$ with

$$d = d_1 d_2, \qquad \nu = (a-1)\frac{k-1}{d_1 d_2}$$

where d_1 is given by formula (6) and d_2 is calculated similarly from row margins,

$$d_2 = 1 + \frac{2}{a-1}\left(\frac{\gamma_1}{\gamma_2} + \frac{\gamma_1+\gamma_2}{\gamma_3} + \ldots + \frac{\gamma_1 + \ldots + \gamma_{a-2}}{\gamma_{a-1}}\right)$$

$$\gamma_i = (y_{1\cdot} + \cdots + y_{i\cdot})/(y_{i+1.} + \cdots + y_{a\cdot})$$

As an example, the doubly cumulative chi-square method is applied to Table 7. For calculating χ_{ij}^2 it is convenient to prepare Table 11.

The constants are obtained as

$$d = d_1 \times d_2 = 1.5125 \times 1.2431 = 1.8802, \qquad \nu = (3-1)\frac{6-1}{1.8802} = 5.319$$

Then the p-value of $\chi^{**2} = 0.00773 + \ldots + 1.41212 = 31.36087$ is evaluated as 0.0065 by the distribution 1.8802 $\chi^2_{5.319}$. This is highly significant, suggesting the dose dependence of responses.

6.2. Multiple Comparisons of Treatments

Although the doubly cumulative chi-square value generally behaves well in suggesting any relation between ordered rows and columns, it cannot point out the optimum level of treatment. For the dose–response experiment an interesting approach is to detect dose levels between which are observed the most significantly different responses or the steepest slope change. A possible approach to this is to partition rows between i and $i+1$, to obtain the appropriate statistic $S(1, \ldots, i; i+1, \ldots, k)$ to compare two groups of rows $(1, \ldots, i)$ and $(i+1, \ldots, k-1)$, and then to make multiple comparisons of $S(1, \ldots, i; i+1, \ldots, k)$ for $i = 1, \ldots, k-1$. For the rank-based approach, S can naturally be taken as the Wilcoxon statistic, which we denote by

Table 11 Calculating the Doubly Cumulative Chi-Square Statistic

	Ordered category										
Dose	(1)(2 ~ 6)		(1,2)(3 ~ 6)		(1 ~ 3)(4 ~ 6)		(1 ~ 4)(5, 6)		(1 ~ 5)(6)		Total
(1)	5	72	11	66	32	45	48	29	71	6	77
(2,3)	10	137	20	127	90	57	120	27	145	2	147
	$\chi^2_{11} = 0.00773$		$\chi^2_{12} = 0.01961$		$\chi^2_{13} = 7.88007$		$\chi^2_{14} = 10.03340$		$\chi^2_{15} = 6.06959$		
(1,2)	12	141	22	131	76	77	113	40	146	7	153
(3)	3	68	9	62	46	25	55	16	70	1	71
	$\chi^2_{21} = 1.01589$		$\chi^2_{22} = 0.11796$		$\chi^2_{23} = 4.46771$		$\chi^2_{24} = 0.33680$		$\chi^2_{25} = 1.41212$		
Total	15	209	31	193	122	102	168	56	216	8	224

$W(1, ..., i; i+1, ..., k)$. The statistic S can also be based on the cumulative chi-square statistic, which we denote by $\chi^{*2}(1, ..., i; i+1, \cdots, k)$. They are calculated as two-sample test statistics between the two subgroups of rows $(1, \cdots, i)$ and $(i+1, ..., k)$. The formula to obtain the asymptotic p-value of max $W(1, ..., i; i+1, ..., k)$ is given in Hirotsu et al. (1992), and the one for max $\chi^{*2}(1, ..., i; i+1, ..., k)$ in Hirotsu and Makita (1992), where the maximum is taken over $i = 1, ..., k-1$. The multiple comparison approaches applied to the data of Table 7 are summarized in Table 12.

6.3. Two-Way Changepoint Analysis

The maximal component of the doubly cumulative chi-square statistic, denoted by max max $\chi^2 ij$, can be useful for testing the two-way changepoint model Eq. (5). An efficient algorithm to obtain the exact p-value of max max χ^2_{ij} is proposed in Hirotsu (1994, 1997). Applying it to Table 8, the one-sided p-value of

$$\max_i \max_j \chi^2_{ij} = 7.500$$

is obtained as 0.0476, which suggests the increased probability of occurrence of the first class in later periods.

The max max chi-square value can also be used in the context of the dose–response experiment. When applied to the data of Table 7, the exact p-value of max max $\chi^2_{ij} = 10.033$ is evaluated as 0.014; see Hirotsu (1997) for details.

6.4. Modeling by the Generalized Linear Model

Another useful approach for modeling multinomials with ordered categories is to use a generalized linear model, such as proportional odds and proportional hazards models. The goodness-of-fit chi-square value of the block interaction model applied to the taste-testing data of five foods in five ordered categorical responses by Bradley et al. has been compared to fitting of the proportional odds model of Snell (1964) and its extension (McCullagh, 1980); see Hirotsu (1990, 1992) for details.

Table 12 Multiple Comparisons of Three Dose Levels

Test statistic	Two-sided p-value
max $W = W(1, 2; 3) = 2.7629$	0.011
max $\chi^{*2} = \chi^{*2}(1, 2; 3) = 24.010$	0.005

7. SOME EXTENSIONS

7.1. General Isotonic Inference

A monotonicity hypothesis in a dose–response relationship, say can be naturally extended to the convexity hypothesis (Hirotsu, 1986) and the downturn hypothesis (Simpson and Margolin, 1986), which are stated in the one-way analysis of variance setting as

$$H_c: \quad \mu_2 - \mu_1 \leq \mu_3 - \mu_2 \leq \ldots \leq \mu_a - \mu_{a-1}$$

and

$$H_D: \quad \mu_1 \leq \ldots \leq \mu_{\tau+1} \geq \mu_{\tau+2} \geq \ldots \geq \mu_a, \qquad \tau = 1, \ldots, \mu_{a-1}$$

respectively. In Hirotsu (1986) a statistic $\chi^{\dagger 2}$ is introduced for testing those hypotheses, and an application of its maximal component is also discussed in Hirotsu and Marumo (1995). These ideas can be extended to two-way tables, and a row-wise multiple comparisons procedure was introduced in Hirotsu et al. (1996) for classifying subjects based on the 24 h profile of their blood pressures, which returns to approximately its starting level after 24 h, where the cumulative chi-square and linear trend statistics are obviously inappropriate. For a more general discussion for the isotonic inference, one should refer to Hirotsu (1998).

7.2. Higher Way Layout

The ideas of the present chapter can be naturally extended to higher way layouts. As one of those examples, a three-way contingency table with age at four levels, existence of metastasis into a lymph node at two levels, and the soating grade at three levels, is analyzed in Hirotsu (1992). An example of highly fractional factorial experiments with ordered categorical responses is given in Hamada and Wu (1990); see also the discussion following that article.

8. CONCLUSION

The analysis of interaction seems to have been paid much less attention than it deserves. First, the character of the two-way factors should be taken into account in making statistical inference to answer actual problems most appropriately. Row-wise and/or columnwise multiple comparisons are particularly useful when one of the factors is indicative or variational. Second, analysis of the generalized interaction is required even in the one-way analysis of variance framework if the responses are ordered categorical, which includes rank data as an important special case. Then testing the ordered alternatives for interaction is of particular interest, and the cumulative chi-square statistic and its maximal component are introduced in addition to the well-known rank sum statistic. Based on these statistics, a method of multiple comparisons of ordered treatments is introduced as well as an overall homogeneity test. Third, the independent sequence of multinomials can be dealt with similarly to the multinomial data with ordered categories. For example, a sequence of Bernoulli trials can be dealt with as two multinomials with cell frequencies all zero or unity. In this context we are interested in changepoint analysis, for which the maximal component of the cumulative chi-square statistic is useful. When there are natural orderings in both rows and columns, the maximal component of the doubly cumulative chi-square statistic is introduced for detecting a two-way changepoint. Finally those row-wise and/or columnwise multiple comparisons are useful not only for comparing treatments but also for defining the block interaction model.

REFERENCES

Bradley RA, Katti SK, Coon TJ. (1962). Optimal scaling for ordered categories. Psychometrika 27: 355–374.

Hamada M, Wu CFJ. (1990). A critical look at accumulation analysis and related methods (with discussion). Technometrics 32: 119–130.

Hawkins DM. (1977). Testing a sequence of observations for a shift in location. J Am Stat Assoc 72: 180–186.

Hirotsu C. (1973). Multiple comparisons in a two-way layout. Rep Stat Appl Res JUSE 1–10.

Hirotsu C. (1982). Use of cumulative efficient scores for testing ordered alternatives in discrete models. Biometrika 69: 567–577.

Hirotsu C. (1983a). An approach to defining the pattern of interaction effects in a two-way layout. Ann Inst Stat Math A 35: 77–90.

Hirotsu C. (1983b). Defining the pattern of association in two-way contingency tables. Biometrika 579–589.

Hirotsu C. (1986). Cumulative chi-squared statistic as a tool for testing goodness of fit. Biometrika 73: 165–173.

Hirotsu C. (1990). Discussion on Hamada and Wu's paper. Technometrics 32: 133–136.

Hirotsu C. (1991a). Statistical methods for quality control—Beyond the analysis of variance. Proc 2nd HASA Workshop, St Kirik, pp. 213–227.

Hirotsu C. (1991b). An approach to comparing treatments based on repeated measures. Biometrika 75: 583–594.

Hirotsu C. (1992). Analysis of Experimental Data, Beyond Analysis of Variance (in Japanese). Tokyo: Kyoritsu-Shuppan.

Hirotsu C. (1993). Beyond analysis of variance techniques: Some applications in clinical trials. Int Stat Rev 61: 183–201.
Hirotsu C. (1994). Two-way changepoint analysis—The alternative distribution (in Japanese). Proc Annu Meeting Jpn Soc Math, Stat Math Branch, pp. 153–154.
Hirotsu C. (1997). Two-way change-point model and its application. Aust J Stat (to appear).
Hirotsu C. (1998). Isotonic inference. In: Encyclopedia of Biostatistics, New York: Wiley, to appear.
Hirotsu C, Makita S. (1992). Multiple comparison procedures based on the cumulative chi-squared statistic (in Japanese). Proc Annu Meeting Jpn Soc Appl Stat, pp. 13–17.
Hirotsu C, Kuriki S, Hayter AJ. (1992). Multiple comparison procedures based on the maximal component of the cumulative chi-squared statistic. Biometrika 78: 583–594.
Hirotsu C, Marumo K. (1995). changepoint analysis for subsequent mean differences and its application (in Japanese). Proc 63rd Annu Meeting Jpn Soc Stat, pp. 333–334.
Hirotsu C, Aono K, Adachi E. (1996). Profile analysis of the change pattern of the blood pressure within a day. Proc 64th Annu Meeting Jpn Soc Stat, pp. 50–51.
Johnson DE, Graybill FA. (1972). An analysis of a two-way model with interaction and no replicatiotn. J Am Stat Assoc 67: 309–328.
McCullagh P. (1980). Regression models for ordinal data. J Roy Stat Soc B 42: 109–142.
Mandel J. (1969). The partitioning of interaction in analysis of variance. J Res Natl Bur Stand B73: 309–328.
Mehta CR, Patel NR, Tsiatis TA. (1989). Exact significance testing to establish treatment equivalence with ordered categorical data. Biometrics 40: 819–825.
Plackett RL (1981). The Analysis of Categorical Data, 2nd ed. London: Griffin.
Siegmund D. (1986). Boundary crossing probabilities and statistical applications. Ann Stat 14: 361–404.
Simpson and Margolin. (1986).
Snell EJ. (1964). A scaling procedure for ordered categorical data. Biometrics 20: 592–607.
Tukey JW. (1949). One degree of freedom for non-additivity. Biometrics 5: 232–242.
Worsley KJ. (1986). Confidence regions and tests for a change point in a sequence of exponential family random variables. Biometrika 27: 103–117.

25

Optimization Methods in Multiresponse Surface Methodology

André I. Khuri
University of Florida, Gainesville, Florida

Elsie S. Valeroso
Montana State University, Bozeman, Montana

1. INTRODUCTION

One of the primary objectives in a response surface investigation is the determination of the optimum of a response of interest. Such an undertaking may also be carried out when several responses are under consideration. For example, in a particular chemical experiment, a resin is required to have a certain minimum viscosity, high softpoint temperature, and high percentage yield (see Chitra, 1990, p. 107). The actual realization of the optimum depends on the nature of the response(s) and the form of the hypothesized (empirical) model(s) being fitted to the data at hand.

Optimization in response surface methodology (RSM) has received a great deal of attention, particularly from experimental researchers. This is evidenced by the numerous articles on optimization that have appeared in a variety of professional journals. See, for example, Fichtali et al. (1990), Floros (1992), Floros and Chinnan (1988a, 1988b), Guillou and Floros (1993), Mouquet et al. (1992), and the two review articles by Khuri (1996) and Myers et al. (1989), to name just a few.

For the most part, current optimization techniques in RSM apply mainly to single-response models. There are, however, many experimental situations where several response variables are of interest and can subsequently be measured for each setting of a group of control variables. Such experiments are referred to as *multiresponse experiments*. For example, the quality of a product may depend on several measurable characteristics (responses). Hill and Hunter (1966) were perhaps the first authors to make reference to multiresponse applications in chemistry and chemical engineering. A review of RSM techniques applicable to multiresponse experiments is given by Khuri (1996). See also Khuri and Cornell (1996, Chapter 7).

The optimization problem in a multiresponse setting is not as well defined as in the single-response case. In particular, when two or more responses are considered simultaneously, their data are multivariately distributed. In this case, the meaning of "optimum" is unclear, because there is no unique way to order such data. Obviously, the univariate approach of optimizing the responses individually and independently of one another is not recommended. Conditions that are optimal for one response may be far from optimal or even physically impractical for the other responses from the experimental point of view.

The purpose of this chapter is to provide a comprehensive survey of the various methods of multiresponse optimization currently in use in RSM. A comparison of some of these methods is made in Section 3 using two numerical examples from the semiconductor and food science industries.

2. METHODS OF MULTIRESPONSE OPTIMIZATION

Multiresponse optimization requires finding the settings of the control variables that yield optimal, or near optimal, values for the responses under consideration. Here, "optimal" is used with reference to conditions deemed more acceptable, or more desirable, than others with respect to a certain criterion. Multiresponse optimization techniques can be graphical or analytical.

2.1 Graphical Techniques

In the graphical approach to optimization, response models are fitted individually to their respective data. Contour plots are generated and then superimposed to locate one or more regions in the factor space where all the predicted responses attain a certain degree of "acceptability." There can be several candidate points from which the experimenter may choose. Note that these plots limit consideration of the control variables to only two. If there are more, then the remaining variables are assigned fixed values. In this case, a large number of plots will have to be generated.

Contour plotting was initially used in the early development of RSM. For example, it was described by Hill and Hunter (1966) in reference to an article by Lind et al. (1960). More recently, an improved graphical technique was deployed using computer-generated contour surfaces, with three control variables, instead of two, represented on the same diagram. This technique was discussed, for example, by Floros and Chinnan (1988b), who credited Box (1954) and Box and Youle (1955) for being the originators of this idea.

It is worth noting here that renewed interest in the graphical approach has evolved in recent years due to advances in computer technology. This approach is simple and easily adaptable to most commonly used computer software packages. However, it has several disadvantages. For example, its capability is limited in large systems involving several control variables and responses. Also, since only two or three control variables can be represented in the same plot, the number of generated plots can be quite large, as was mentioned earlier. This makes it difficult to identify one set of conditions as being optimal. Furthermore, the graphical approach does not account for the possibility of having correlated responses, which may also be heteroscedastic. Obviously, graphs based on such responses are not very reliable and may adversely affect the finding of optimum conditions. In particular, failure to recognize multi-collinearities among the responses can lead to meaningless results in the fitting of the response models (see Box et al., 1973) and hence in the determination of optimum conditions.

2.2 Analytical Techniques

Analytical techniques apply mainly to linear multiresponse models. Let r denote the number of response variables, and let $\boldsymbol{x} = (x_1, x_2, ..., x_k)'$ be a vector of k related control variables. The model for the ith response is of the form

$$y_i = \boldsymbol{f}_i'(\boldsymbol{x})\boldsymbol{\beta}_i + \varepsilon_i, \qquad i = 1, 2, ..., r \tag{1}$$

where $\boldsymbol{f}_i(\boldsymbol{x})$ is a vector of order $p_i \times 1$ whose elements consist of powers and products of powers of the elements of $\boldsymbol{x}$ up to degree $d_i(\geq 1)$, $\boldsymbol{\beta}_i$ is a vector of p_i unknown constant coefficients, and ε_i is random experimental error. Suppose that there are n sets of observations on $y_1, y_2, ..., y_r$. The corresponding design settings of $\boldsymbol{x}$ are denoted by $\boldsymbol{x}_1, \boldsymbol{x}_2, ..., \boldsymbol{x}_n$. From (1) we have

$$y_{ui} = \boldsymbol{f}_i'(\boldsymbol{x}_u)\boldsymbol{\beta}_i + \varepsilon_{ui}, \qquad i = 1, 2, ..., r; \ u = 1, 2, ..., n \tag{2}$$

where y_{ui} is the uth observation on y_i. Model (2) can be written in vector form as

$$y_i = X_i \beta_i + \varepsilon_i, \qquad i = 1, 2, ..., r \tag{3}$$

where y_i and ε_i are the vectors of y_{ui}'s and ε_{ui}'s, respectively, and X_i is a matrix of order $n \times p_i$. It is assumed that X_i is of full column rank and that $E(\varepsilon_i) = \mathbf{0}$ and $\mathrm{Var}(\varepsilon_i) = \sigma_i^2 I_n$, where I_n is the identity matrix ($i = 1, 2, ..., r$). Furthermore, we assume that $\mathrm{Cov}(\varepsilon_i, \varepsilon_j) = \sigma_{ij} I_n$, $i \neq j$. Let $\Sigma = (\sigma_{ij})$. The models in Eq. (3) can be combined into a single linear multiresponse model of the form

$$y = X\beta + \varepsilon \tag{4}$$

where X is a block-diagonal matrix, diag $(X_1, X_2, ..., X_r)$, $\beta = [\beta_1' : \beta_2' : ... : \beta_r']'$, and $\varepsilon = [\varepsilon_1' : \varepsilon_2' : ... : \varepsilon_r']'$. Hence, $\mathrm{Var}(\varepsilon) = \Sigma \otimes I_n$, where $\otimes$ denotes the direct product of matrices. The best linear unbiased estimator (BLUE) of β is given by (see Khuri and Cornell, 1996, Chapter 7)

$$\hat{\beta} = [X'(\Sigma^{-1} \otimes I_n)X]^{-1} X'(\Sigma^{-1} \otimes I_n)y \tag{5}$$

In general, $\hat{\beta}$ depends on the variance-covariance matrix Σ, which is unknown and must therefore be estimated. Zellner (1962) proposed the estimate $\hat{\Sigma} = (\hat{\sigma}_{ij})$, where

$$\hat{\sigma}_{ij} = \frac{1}{n} y_i'[I_n - X_i(X_i'X_i)^{-1}X_i'][I_n - X_j(X_j'X_j)^{-1}X_j']y_j, \qquad i, j = 1, 2, ..., r \tag{6}$$

Srivastava and Giles (1987, p. 16) showed that $\hat{\Sigma}$ is singular if $r > n$. They demonstrated that $r \leq n$ is a necessary, but not sufficient, condition for the nonsingularity of $\hat{\Sigma}$. Using $\hat{\Sigma}$ in place of Σ in Eq. (5) produces the following estimate of β.

$$\hat{\beta}_e = [X'(\hat{\Sigma}^{-1} \otimes I_n)X]^{-1} X^{-1}(\hat{\Sigma}^{-1} \otimes I_n)y \tag{7}$$

This is known as Zellner's seemingly unrelated regression (SUR) estimate of β. It is also referred to as an estimated generalized least squares (EGLS) estimate of β. It can be computed using PROC SYSLIN (SAS, 1990a). In particular, if $X_i = X_0$ ($i = 1, 2, ..., r$), then it is easy to show that (5) reduces to

$$\hat{\beta} = [I_r \otimes (X_0'X_0)^{-1} X_0']y \tag{8}$$

In this case, the BLUE of $\boldsymbol{\beta}_i$ coincides with its ordinary least squares (OLS) estimate, which does not depend on $\boldsymbol{\Sigma}$, that is,

$$\hat{\boldsymbol{\beta}}_i = (\boldsymbol{X}_0'\boldsymbol{X}_0)^{-1}\boldsymbol{X}_0'\boldsymbol{y}_i, \qquad i = 1, 2, \ldots, r \tag{9}$$

This special case occurs when the response models in (1) are of the same degree and form and are fitted using the same design.

From Eqs. (1) and (7), the ith predicted response, $\hat{y}_{ei}(\boldsymbol{x})$, at a point $\boldsymbol{x}$ in a region R is given by

$$\hat{y}_{ei}(\boldsymbol{x}) = \boldsymbol{f}_i'(\boldsymbol{x})\hat{\boldsymbol{\beta}}_{ei}, \qquad i = 1, 2, \ldots, r \tag{10}$$

where $\hat{\boldsymbol{\beta}}_{ei}$ is the portion of $\hat{\boldsymbol{\beta}}_e$ in Eq. (7) that corresponds to $\boldsymbol{\beta}_i$.

Now by a multiresponse optimization of the responses we mean finding an $\boldsymbol{x}$ in R at which $\hat{y}_{ei}(\boldsymbol{x})$, $i = 1, 2, \ldots, r$, attain certain optimal values. The term "optimal" is defined acccoding to some criterion. In the next two sections, two optimality criteria are defined and discussed.

The Desirability Function Approach

The desirability function approach (DFA) was introduced by Harrington (1965). The response models in (1) are first fitted individually using OLS estimates of the $\boldsymbol{\beta}_i$'s, namely,

$$\hat{\boldsymbol{\beta}}_i^* = (\boldsymbol{X}_i'\boldsymbol{X}_i)^{-1}\boldsymbol{X}_i'\boldsymbol{y}_i, \qquad i = 1, 2, \ldots, r$$

The corresponding predicted responses are

$$\hat{y}_i^*(\boldsymbol{x}) = \boldsymbol{f}_i'(\boldsymbol{x})\hat{\boldsymbol{\beta}}_i^*, \qquad i = 1, 2, \ldots, r \tag{11}$$

The $\hat{y}_i^*(\boldsymbol{x})$'s are then transformed into desirability functions denoted by $d_i(\boldsymbol{x})$, where $0 \leq d_i(\boldsymbol{x}) \leq 1$, $i = 1, 2, \ldots, r$. The value of $d_i(\boldsymbol{x})$ increases as the "desirability" of the corresponding response increases. In a production process, the responses $y_1, y_2, \ldots, y_r$ usually measure particular characteristics of a product.

The choice of the desirability function is subjective and depends on how the user assesses the desirability of a given product characteristic. Harrington (1965) used exponential-type desirability transformations. Later, Derringer and Suich (1980) introduced more general transformations that offer the user greater flexibility in setting up desirability values. Derringer and Suich considered one-sided and two-sided desirability transformations. The former are employed when the $\hat{y}_i^*(\boldsymbol{x})$'s are to be maximized. In this case, $d_i(\boldsymbol{x})$ is defined by

$$d_i(\boldsymbol{x}) = \begin{cases} 0 & \text{if } \hat{y}_i^* \leq u_i \\ \left[\dfrac{\hat{y}_i^*(\boldsymbol{x}) - u_i}{v_i - u_i}\right]^s & \text{if } u_i < \hat{y}_i^* < v_i \\ 1 & \text{otherwise} \end{cases} \tag{12}$$

where u_i is the minimum acceptable value of $\hat{y}_i^*$ and v_i is such that higher values of $\hat{y}_i^*$ would not lead to further increase in the desirability of the ith response ($i = 1, 2, ..., r$). The value s is specified by the user. Note that if the minimization of $\hat{y}_i^*(\boldsymbol{x})$ is desired, then $d_i(\boldsymbol{x})$ is chosen as

$$d_i(\boldsymbol{x}) = \begin{cases} 0 & \text{if } \hat{y}_i^* \geq \tilde{v}_i \\ \left[\dfrac{\hat{y}_i^*(\boldsymbol{x}) - \tilde{v}_i}{\tilde{u}_i - \tilde{v}_i}\right]^s & \text{if } \tilde{u}_i < \hat{y}_i^* < \tilde{v}_i \\ 1 & \text{otherwise} \end{cases} \tag{13}$$

where $\tilde{u}_i$ and $\tilde{v}_i$ are specified values ($i = 1, 2, 3, ..., r$). Two-sided desirability transformations are used when y_i has both minimum and maximum constraints. The corresponding $d_i(\boldsymbol{x})$ is given by

$$d_i(\boldsymbol{x}) = \begin{cases} \left[\dfrac{\hat{y}_i^*(\boldsymbol{x}) - u_i}{c_i - u_i}\right]^s & \text{if } u_i \leq \hat{y}_i^* \leq c_i \\ \left[\dfrac{\hat{y}_i^*(\boldsymbol{x}) - v_i}{c_i - v_i}\right]^t & \text{if } c_i < \hat{y}_i^* \leq v_i \\ 0 & \text{otherwise} \end{cases}$$

where here u_i and v_i are, respectively, minimum acceptable and maximum acceptable values of $\hat{y}_i^*$, c_i is that value of $\hat{y}_i^*$ considered "most desirable" (target value), and s and t are specified by the user.

Once the desirability functions for all the responses have been chosen, the $d_i(\boldsymbol{x})$'s are then combined into a single function, denoted by $d(\boldsymbol{x})$, which measures the overall desirability of the responses. Derringer and Suich (1980) adopted the geometric mean of the $d_i(\boldsymbol{x})$'s as such a function, that is,

$$d(\boldsymbol{x}) = \left[\prod_{i=1}^{r} d_i(\boldsymbol{x})\right]^{1/r}$$

We note that $0 \leq d(\boldsymbol{x}) \leq 1$ and that $d(\boldsymbol{x}) = 0$ if any of the $d_i(\boldsymbol{x})$'s is equal to zero. Thus if a product does not meet a specified characteristic, it is deemed unacceptable. Large values of d correspond to a highly desirable product. Hence, optimum conditions are found by maximizing $d(\boldsymbol{x})$ over the experimental region. The multiresponse optimization problem has therefore been reduced to the maximization of the single function $d(\boldsymbol{x})$.

More recently, Derringer (1994) referred to the desirability function approach as the desirability optimization methodology. He also provided information concerning software availability for its computer implementation. Note that the actual maximization of $d(\boldsymbol{x})$ can be carried out only by using search methods, as opposed to gradient-based methods, because $d(\boldsymbol{x})$ is not differentiable at certain points. Del Castillo et al. (1996) proposed modified desirability functions that are everywhere differentiable so that more efficient gradient-based optimization procedures can be used.

The Generalized Distance Approach (GDA) was introduced by Khuri and Conlon (1981). The responses are assumed to be adequately represented by polynomial models of the same degree and form within the experimental region R. In this case, the $\boldsymbol{X}_i$'s in models (3) are equal to a common matrix $\boldsymbol{X}_0$. The estimates of $\boldsymbol{\beta}_i$, $i = 1, 2, \ldots, r$, and the corresponding expressions for the predicted responses are given by Eqs. (9) and (11), respectively.

If the assumptions made earlier in Section 2.2 concerning the distributions of the responses are valid, then

$$\text{Var}[\hat{y}_i(\boldsymbol{x})] = \boldsymbol{f}'(\boldsymbol{x})(\boldsymbol{X}_0'\boldsymbol{X}_0)^{-1}\boldsymbol{f}(\boldsymbol{x})\sigma_{ii}$$

and

$$\text{Cov}[\hat{y}_i(\boldsymbol{x}), \hat{y}_j(\boldsymbol{x})] = \boldsymbol{f}'(\boldsymbol{x})(\boldsymbol{X}_0'\boldsymbol{X}_0)^{-1}\boldsymbol{f}(\boldsymbol{x})\sigma_{ij}, \qquad i \neq j$$

where $\boldsymbol{f}(\boldsymbol{x})$ is the common form of $\boldsymbol{f}_i(\boldsymbol{x})$, $i = 1, 2, \ldots, r$, and σ_{ij} is the (i, j)th element of $\boldsymbol{\Sigma}$, the variance-covariance of the responses. Hence, if $\hat{\boldsymbol{y}}(\boldsymbol{x}) = [\hat{y}_1(\boldsymbol{x}) : \hat{y}_2(\boldsymbol{x}) : \ldots : \hat{y}_r(\boldsymbol{x})]'$ is the vector of predicted responses, then its variance-covariance matrix is given by

$$\text{Var}[\hat{\boldsymbol{y}}(\boldsymbol{x})] = \boldsymbol{f}'(\boldsymbol{x})(\boldsymbol{X}_0'\boldsymbol{X}_0)^{-1}\boldsymbol{f}(\boldsymbol{x})\boldsymbol{\Sigma} \tag{14}$$

Since $\boldsymbol{X}_i = \boldsymbol{X}_0$ for $i = 1, 2, \ldots, r$, an unbiased estimator of $\boldsymbol{\Sigma}$ is given by

$$\hat{\boldsymbol{\Sigma}}_0 = \frac{1}{n - p_0}\boldsymbol{Y}'[\boldsymbol{I}_n - \boldsymbol{X}_0(\boldsymbol{X}_0'\boldsymbol{X}_0)^{-1}\boldsymbol{X}_0']\boldsymbol{Y} \tag{15}$$

where $\boldsymbol{Y} = [\boldsymbol{y}_1 : \boldsymbol{y}_2 : \dots : \boldsymbol{y}_r]$ is the $n \times r$ matrix of multiresponse data and p_0 is the number of columns of $\boldsymbol{X}_0$ [see Khuri and Conlon (1981), formula 2.3]. If $r \leq n - p_0$, then $\hat{\boldsymbol{\Sigma}}_0$ will be nonsingular provided that $\boldsymbol{Y}$ is of rank r. Using $\hat{\boldsymbol{\Sigma}}_0$ in place of $\boldsymbol{\Sigma}$ in (14), an unbiased estimator of $\text{Var}[\hat{\boldsymbol{y}}(\boldsymbol{x})]$ is obtained, namely,

$$\widehat{\text{Var}}[\hat{\boldsymbol{y}}(\boldsymbol{x})] = \boldsymbol{f}'(\boldsymbol{x})(\boldsymbol{X}_0'\boldsymbol{X}_0)^{-1}\boldsymbol{f}(\boldsymbol{x})\hat{\boldsymbol{\Sigma}}_0$$

The main idea behind the generalized distance approach is based on measuring the distance of $\hat{\boldsymbol{y}}(\boldsymbol{x})$ from the so-called ideal optimum, which is defined as follows: Let $\hat{\phi}_i$ denote the optimum value if $\hat{y}_i(\boldsymbol{x})$ obtained individually over a region R, $i = 1, 2, \dots, r$. Let $\hat{\boldsymbol{\phi}} = (\hat{\phi}_1, \hat{\phi}_2, \dots, \hat{\phi}_r)'$. If these individual optima are attained at the same point in R, then an ideal optimum is said to be achieved. In general, the occurrence of such an optimum is very rare, since the $\hat{\phi}_i$'s attain their individual optima at different locations in R. In this case, we search for the location of a near ideal optimum, a point $\boldsymbol{x}_0$ in R at which $\hat{\boldsymbol{y}}(\boldsymbol{x})$ is "closest" to $\hat{\boldsymbol{\phi}}$. Here, "closeness" is determined by a metric $\rho[\hat{\boldsymbol{y}}(\boldsymbol{x}), \hat{\boldsymbol{\phi}}]$ defined as follows:

$$\begin{aligned} \rho[\hat{\boldsymbol{y}}(\boldsymbol{x}), \hat{\boldsymbol{\phi}}] &= [(\hat{\boldsymbol{y}}(\boldsymbol{x}) - \hat{\boldsymbol{\phi}})'\{\widehat{\text{Var}}[\hat{\boldsymbol{y}}(\boldsymbol{x})]\}^{-1}(\hat{\boldsymbol{y}}(\boldsymbol{x}) - \hat{\boldsymbol{\phi}})]^{1/2} \\ &= \left[\frac{(\hat{\boldsymbol{y}}(\boldsymbol{x}) - \hat{\boldsymbol{\phi}})'\hat{\boldsymbol{\Sigma}}_0^{-1}(\hat{\boldsymbol{y}}(\boldsymbol{x}) - \hat{\boldsymbol{\phi}})}{\boldsymbol{f}'(\boldsymbol{x})(\boldsymbol{X}_0'\boldsymbol{X}_0)^{-1}\boldsymbol{f}(\boldsymbol{x})}\right]^{1/2} \end{aligned} \tag{16}$$

Thus the multiresponse optimization problem in this approach has been reduced to the minimization of $\rho[\hat{\boldsymbol{y}}(\boldsymbol{x}), \hat{\boldsymbol{\phi}}]$ with respect to $\boldsymbol{x}$ over R. Optimum conditions found in this manner result in a so-called compromise ideal optimum.

Several other metrics were proposed in Khuri and Conlon (1981), for example,

$$\rho_1[\hat{\boldsymbol{y}}(\boldsymbol{x}), \hat{\boldsymbol{\phi}}] = \left[\sum_{i=1}^{r} \frac{[\hat{y}_i(\boldsymbol{x}) - \hat{\phi}_i]^2}{\hat{\sigma}_{0ii}\boldsymbol{f}'(\boldsymbol{x})(\boldsymbol{X}_0'\boldsymbol{X}_0)^{-1}\boldsymbol{f}(\boldsymbol{x})}\right]^{1/2}$$

$$\rho_2[\hat{\boldsymbol{y}}(\boldsymbol{x}), \hat{\boldsymbol{\phi}}] = \left[\sum_{i=1}^{r} \frac{[\hat{y}_i(\boldsymbol{x}) - \hat{\phi}_i]^2}{\hat{\phi}_i^2}\right]^{1/2}$$

where $\hat{\sigma}_{0ii}$ is the ith diagonal element of $\hat{\boldsymbol{\Sigma}}_0 (i = 1, 2, \dots, r)$. The metric ρ_1 is appropriate whenever the responses are statistically independent. The metric ρ_2 measures the total relative deviation of $\hat{\boldsymbol{y}}(\boldsymbol{x})$ from $\hat{\boldsymbol{\phi}}$. It can be used when $\hat{\boldsymbol{\Sigma}}_0$ is ill-conditioned.

Remark 1. It should be noted that in the generalized distance approach, $\hat{\phi}_i$ is treated as a fixed quantity, when in fact it is random ($i = 1, 2, ..., r$). To account for the randomness in the elements of $\hat{\boldsymbol{\phi}}$, Khuri and Conlon (1981) developed a rectangular confidence region, C_{ϕ}, on $\boldsymbol{\phi}$, the vector of true individual optima over the region R. For a fixed $\boldsymbol{x}$ in R, the maximum of $\rho[\hat{\boldsymbol{y}}(\boldsymbol{x}), \boldsymbol{\eta}]$ is obtained with respect to $\boldsymbol{\eta}$ in C_{ϕ}. This maximum provides a conservative estimate of $\rho[\hat{\boldsymbol{y}}(\boldsymbol{x}), \boldsymbol{\phi}]$, the metric that should be minimized with respect to $\boldsymbol{x}$ instead of $\rho[\hat{\boldsymbol{y}}(\boldsymbol{x}), \boldsymbol{\phi}]$. The maximum so obtained, which is a function of $\boldsymbol{x}$, is minimized with respect to $\boldsymbol{x}$ over R. A more detailed discussion concerning this max-min approach is also given in Khuri and Cornell (1996, Chapter 7).

The computer implementation of Khuri and Conlon's (1981) generalized distance approach, including the use of the confidence region C_{ϕ}, is available through the MR (for multiple responses) software written by Conlon (1988). A copy of the MR code along with the accompanying technical report and examples can be downloaded from the Internet at ftp://ftp.stat.ufl.edu/pub/mr.tar.Z. Note that the mr.tar.Z file is compressed. It should be uncompressed and then compiled. Furthermore, MR fits a second-degree polynomial model to each response.

An Extension of Khuri and Conlon's (1981) GDA. The generalized distance approach (GDA) described earlier requires that all fitted response models be of the same form and degree and depend on all the control variables under consideration. Valeroso (1996) extended the GDA by making it applicable to models that are not necessarily of the same degree or form. The following is a summary of Valeroso's extension.

The models considered are of the form given in (1). The SUR (or EGLS) estimates of $\boldsymbol{\beta}_i$ are obtained from formula (7). The expressions for the predicted responses are given by formula (10). Let $\hat{\boldsymbol{y}}_e(\boldsymbol{x}) = [\hat{y}_{e1}(\boldsymbol{x}), \hat{y}_{e2}(\boldsymbol{x}), ..., \hat{y}_{er}(\boldsymbol{x})]'$. Then,

$$\hat{\boldsymbol{y}}_e(\boldsymbol{x}) = \boldsymbol{\Lambda}'(\boldsymbol{x})\hat{\boldsymbol{\beta}}_e \tag{17}$$

where $\boldsymbol{\Lambda}'(\boldsymbol{x}) = \text{diag}[\boldsymbol{f}_1'(\boldsymbol{x}), \boldsymbol{f}_2'(\boldsymbol{x}), ..., \boldsymbol{f}_r'(\boldsymbol{x})]$. An estimate of the variance-covariance matrix of $\hat{\boldsymbol{y}}_e(\boldsymbol{x})$ is approximately of the form

$$\widehat{\text{Var}}[\hat{\boldsymbol{y}}_e(\boldsymbol{x})] = \boldsymbol{\Lambda}'(\boldsymbol{x})[\boldsymbol{X}'(\hat{\boldsymbol{\Sigma}}^{-1} \otimes \boldsymbol{I}_n)\boldsymbol{X}]^{-1}\boldsymbol{\Lambda}(\boldsymbol{x})$$

where the elements of $\hat{\boldsymbol{\Sigma}}$ are given in (6). The metric ρ defined in (16) is now replaced by

$$\rho_e[\hat{y}_e(x), \hat{\phi}_e] = [(\hat{y}_e(x) - \hat{\phi}_e)'\{\Lambda'(x)[X'(\hat{\Sigma}^{-1} \otimes I_n)X]^{-1}\Lambda(x)\}^{-1}(\hat{y}_e(x) - \hat{\phi}_e)]^{1/2} \tag{18}$$

where $\hat{\phi}_e = (\hat{\phi}_{e1}, \hat{\phi}_{e2}, ..., \hat{\phi}_{er})'$ and $\hat{\phi}_{ei}$ is the individual optimum of $\hat{y}_{ei}(x)$ over the region R. Minimizing the metric ρ_e over R results in a simultaneous optimization of the r predicted responses.

Valeroso's (1996) extension also includes an accountability of the randomness of $\hat{\phi}_e$ by applying a max-min approach similar to the one described in Remark 1.

2.3. Other Optimization Procedures

There are other optimization procedures that involve more than one response. Some of these procedures, however, are not truly multivariate in nature since they do not seek simultaneous optima in the same fashion as in Section 2.2.

The Dual Response Approach

The dual response approach (DRA) was introduced by Myers and Carter (1973). It concerns the optimization of a single response, identified as the primary response, subject to equality constraints on another response labeled the secondary response. Both responses are fitted to second-degree models. Biles (1975) extended this idea by considering more than one secondary response.

Del Castillo and Montgomery (1993) presented an alternative way to solve the DRA problem by using a nonlinear optimization procedure called the generalized reduced gradient (GRG) algorithm. They demonstrated the advantages of this algorithm and made a reference to software packages for its computer implementation.

The DRA can be used in experimental situations where both the mean and variance of a process are of interest. One is considered the primary response and the other the secondary response [see Vining and Myers (1990) and Myers et al. (1992)]. Previously, the DRA was used by Khuri and Myers (1979) to provide an improvement to the method of ridge analysis, which is an optimization procedure for a single response represented by a second-degree model within a spherical region [see Draper (1963)]. The modification imposed certain quadratic constraints for the purpose of limiting the size of the prediction variance. More recently, several authors elaborated further on the use of the DRA in conjunction with the modeling of both the mean and variance. For example, Lin and Tu (1995) suggested using the mean squared error (MSE) as a new objective function to be minimized. This MSE is the sum of the estimated process variance and the square of the differnce between the estimated process mean and some target value. Copeland and Nelson (1996) proposed using direct function minimization based on Nelder and Mead's (1965) simplex method. Lin and Tu (1995, p. 39) made an interesting comment by stating that the use of the DRA for solving the mean-variance problem can work well only when the mean and variance are independent.

Optimization via Constrained Confidence Regions

Optimization via constrained confidence regions (Del Castillo, 1996) is somewhat related to the DRA. The responses are fitted individually using either first-degree or second-degree models. Confidence regions on the locations of the constrained stationary points for the individual responses are obtained if their corresponding models are of the second degree. If some of the models are of the first degree, then confidence cones on the directions of steepest ascent (or descent) are used. These regions (or cones) are then treated as constraints in a nonlinear programming problem where one response is defined as a primary response. The next step requires finding a solution that lies inside all the confidence regions and/or cones.

A Fuzzy Modeling Approach

The fuzzy modeling approach of Kim and Lin (1998) is based on the so-called fuzzy multiobjective optimization methodology. It is assumed that the degree of satisfaction of the experimenter with respect to the ith response is maximized when $\hat{y}_i^*(\boldsymbol{x})$ [see formula (11)] is equal to its target value T_i and decreases as $\hat{y}_i^*(\boldsymbol{x})$ moves away from T_i, $i = 1, 2, \ldots, r$. If $y_i^{\min}$ and $y_i^{\max}$ denote lower and upper bounds on the ith response, respectively, then the degree of satisfaction with respect to the ith response is defined by a function called the membership function, which we denote by $m_i[\hat{y}_i^*(\boldsymbol{x})]$, $i = 1, 2, \ldots, r$, and is given by

$$m_i[\hat{y}_i^*(\boldsymbol{x})] = \begin{cases} 0 & \text{if } \hat{y}_i^* \leq y_i^{\min} \text{ or } \hat{y}_i^* \geq y_i^{\max} \\ 1 - \dfrac{T_i - \hat{y}_i^*(\boldsymbol{x})}{T_i - y_i^{\min}} & \text{if } y_i^{\min} < \hat{y}_i^* \leq T_i \\ 1 - \dfrac{\hat{y}_i^*(\boldsymbol{x}) - T_i}{y_i^{\max} - T_i} & \text{if } T_i < \hat{y}_i^* < y_i^{\max} \end{cases}$$

The values of $y_i^{\min}$ and $y_i^{\max}$ can be chosen as the individual optima of $\hat{y}_i^*(\boldsymbol{x})$ over a region R. We note that the definition of this function is similar to that of the desirability function. Simultaneous optimization of the responses is achieved by maximizing the minimum degree of satisfaction, that is, $\min_i\{m_i[\hat{y}_i^*(\boldsymbol{x})], i = 1, 2, ..., r\}$. Additional constraints may be added to this formulation as appropriate.

The Procedure of Chitra (1990)

The procedure of Chitra (1990) is similar to the generalized distance approach. Chitra defined different types of objective functions to be minimized. These functions measure deviations of the responses from their target values. The procedure allows the inclusion of several constraints on the responses and control variables.

Remark 2. The generalized distance approach is the only multiresponse optimization procedure that takes into account the variance-covariance structure of the responses. We recall that this structure affects the fit of the models. It should therefore be taken into consideration in any simultaneous optimization. Also, in order to avoid any difficulties caused by multicollinearities among the responses, the multiresponse data should first be checked for linear dependences among the columns of $\boldsymbol{Y}$ [see formula (15)]. Khuri and Cornell (1996, pp. 255–265) provide more details about this and show how to drop responses considered to be linearly dependent on other responses.

The extension of the generalized distance approach in Section 2.2 makes it now possible to apply this procedure to models that are not of the same form or dependent on the same control variables. On the other hand, the desirability function approach, although simple to apply, is subjective, as it depends on how the user interprets desirabilities of the various responses. The user should be very familiar with the product whose characteristics are measured by the responses under consideration. Derringer (1994, p. 57) provided some insight into the choice of desirability values. He stated that "the process of assigning desirability curves and their weights is best done by consensus in the early stages of product conception. The consensus meeting should include an expert facilitator and representatives from all functional areas involved with the product." Care should therefore be exercised in setting up desirability functions. Improperly assessed desirabilities can lead to inaccurate optimization results.

It should be recalled that in Derringer and Suich (1980), no account was given of the variance-covariance matrix of the responses, not even at the modelling stage. Del Castillo et al. (1996, p. 338), however, recommended using Zellner's (1962) SUR estimates to fit the models in (1) [see formula (7)]. Furthermore, the desirability function approach has no built-in procedure for detecting those responses, if any, that are either linearly dependent or highly multicollinear. Ignoring such dependences can affect the overall desirability and hence the determination of optimum conditions.

3. EXAMPLES

In this section, we illustrate the application of the extended generalized distance approach (GDA) and the desirability function approach (DFA) of Section 2.2 and the dual response approach (DRA) using the GRG algorithm of Section 2.3. We present two examples, one from the semiconductor industry and the other from the food industry.

3.1 A Semiconductor Example

An experiment was conducted to determine the performance of a tool used to polish computer wafers. Three control variables were studied: x_1 = down force, x_2 = table speed, and x_3 = slurry concentration. The measured responses were removal rate of metal (RR), oxide removal rate (OXRATE), and within-wafer standard deviation (WIWSD). The objective of the experiment was to maximize y_1 = selectivity and minimize y_2 = nonuniformity, where

$$y_1 = \frac{\text{RR}}{\text{OXRATE}} \quad \text{and} \quad y_2 = \frac{\text{WIWSD}}{\text{RR}}$$

A Box–Behnken design with eight replications at the center and two replications at each noncentral point was used. Each treatment run required two wafers. The first wafer was used to measure RR and WIWSD. The second wafer was used to measure OXRATE. The design points and corresponding values of y_1 and y_2 are given in Table 1.

Before determining the optima associated with y_1 and y_2, we need to select models that provide good fits to these responses. Since the models are fitted using Zellner's (1962) seemingly unrelated regression (SUR) parameter estimation [see formula (7)], measures of the goodness of fit for SUR models should be utilized. These include Sparks' (1987) PRESS statistic and McElroy's (1977) R^2 statistic. The latter is interpreted the same way as the univariate R^2 in that it represents the proportion of the total variation explained by the SUR multiresponse model. These measures provide the user with multivariate variable selection techniques, which, in general, require screening a large number of subset models. To reduce the number of models considered, Sparks (1987) recommends using the univariate R^2, adjusted R^2, and Mallows' C_p statistics to identify

Table 1 Experimental Design and Response Values (Semiconductor Example)

Coded control variables			Responses	
x_1	x_2	x_3	y_1	y_2
0	0	0	0.49074	0.18751
0	0	0	0.39208	0.19720
1	0	1	0.85866	0.12090
1	0	1	0.74129	0.16544
−1	0	1	0.33484	0.65322
−1	0	1	0.29645	0.75198
1	−1	0	0.57887	0.15566
1	−1	0	0.62203	0.10841
1	1	0	0.70656	0.14648
1	1	0	0.88189	0.09600
0	0	0	0.43939	0.24803
0	0	0	0.46587	0.23759
−1	1	0	0.30218	0.55831
−1	1	0	0.36169	0.71183
0	1	1	0.60465	0.23622
0	1	1	0.53486	0.26489
0	−1	−1	0.48908	0.24406
0	−1	−1	0.43681	0.38756
−1	0	−1	0.25005	0.63051
−1	0	−1	0.19546	0.72421
0	−1	1	0.52298	0.25327
0	−1	1	0.42990	0.25019
0	0	0	0.45782	0.32923
0	0	0	0.46910	0.29522
1	0	−1	0.63714	0.12583
1	0	−1	0.79454	0.19912
0	1	−1	0.88856	0.27198
0	1	−1	0.84218	0.29578
−1	−1	0	0.13258	0.62442
−1	−1	0	0.13665	0.53618
0	0	0	0.49810	0.29392
0	0	0	0.46321	0.37023

"good" subset models. For each combination of such models, Sparks' PRESS and McElroy's R^2 statistics are computed. The "best" multiresponse model is the one with the smallest PRESS statistic value and a value of McElroy's R^2 close to 1. On this basis, the following models were selected for y_1 and y_2:

$$\hat{y}_{e1}(\boldsymbol{x}) = 0.441 + 0.2382x_1 + 0.1109x_2 - 0.0131x_3 + 0.0429x_2^2 + 0.0912x_3^2 - 0.0773x_2x_3 \quad (19)$$

$$\hat{y}_{e2}(x) = 0.2727 - 0.2546x_1 + 0.0014x_2 - 0.0114x_3 + 0.1216x_1^2 \quad (20)$$

The SUR parameter estimates, their estimated standard errors, the values of the univariate R^2, adjusted R^2, and C_p statistics and values of McElroy's R^2 and Sparks' PRESS statistics are given in Table 2. Note that the SUR parameters estimates were obtained using PROC SYSLIN in SAS (1990a), and the univariate R^2, adjusted R^2, and C_p statistics were computed using PROC REG in SAS (1989). From Table 2 it can be seen that models (19) and (20) provide good fits to the two responses.

On the basis of models (19) and (20), the individual optima of $\hat{y}_{e1}(\boldsymbol{x})$ and $\hat{y}_{e2}(\boldsymbol{x})$ over the region $R = \{(x_1, x_2, x_3) | \sum_{i=1}^{3} x_i^2 \leq 2\}$ are given in Table 3. These values were computed using a Fortran program written by Conlon (1992), which is based on Price's (1977) optimization procedure. The simultaneous optima of $\hat{y}_{e1}(\boldsymbol{x})$ and $\hat{y}_{e2}(\boldsymbol{x})$ over R were determined by using the extension of the GDA (see Section 2.2). The minimization of ρ_e in (18) was

Table 2 SUR Parameter Estimates and Values of C_p, R^2, and Adjusted R^2 (Semiconductor Example)

	Responses[a]	
Parameter	$\hat{y}_{e1}$	y_{e2}
Intercept	0.4410(0.0190)	0.2727(0.0135)
x_1	0.2382(0.0155)	−0.2546(0.0135)
x_2	0.1109(0.0155)	0.0014(0.0135)
x_3	−0.0131(0.0155)	−0.0114(0.0135)
x_1x_2		
x_1x_3		
x_2x_3	−0.0773(0.0219)	
x_1^2		0.1216(0.0191)
x_2^2	0.0429(0.0219)	
x_3^2	0.0912(0.0219)	
C_p	6.17	4.39
R^2	0.91	0.93
adj.R^2	0.89	0.91

[a]The number in parentheses is the standard error.
Note: McElroy's $R^2 = 0.9212$; Sparks' PRESS statistic = 103.9.

carried out using a program written in PROC IML of SAS (1990b). The results are shown in Table 3.

To apply the DFA, we use formulas (12) and (13) for $d_1(\boldsymbol{x})$ and $d_2(\boldsymbol{x})$, respectively, where

$$d_1(\boldsymbol{x}) = \begin{cases} 0 & \text{if } \hat{y}_{e1} \leq 0 \\ \dfrac{\hat{y}_{e1}(\boldsymbol{x})}{0.95} & \text{if } 0 < \hat{y}_{e1} < 0.95 \\ 1 & \text{otherwise} \end{cases}$$

and

$$d_2(\boldsymbol{x}) = \begin{cases} 0 & \text{if } \hat{y}_{e2} \geq 1.0 \\ \dfrac{\hat{y}_{e2}(\boldsymbol{x}) - 1}{0.20 - 1} & \text{if } 0.20 < \hat{y}_{e2} < 1.0 \\ 1 & \text{otherwise} \end{cases}$$

Note that the values 0.95 and 0.20 in d_1 and d_2, respectively, are of the same order of magnitude as the individual maxima and minima of $\hat{y}_{e1}$ and $\hat{y}_{e2}$, respectively. Note also that, on the basis of a recommendation by Del Castillo et al. (1996, p. 338), we have used the SUR predicted responses, $\hat{y}_{e1}(\boldsymbol{x})$ and $\hat{y}_{e2}(\boldsymbol{x})$, instead of $\hat{y}_1^*(\boldsymbol{x})$ and $\hat{y}_2^*(\boldsymbol{x})$. The latter two are the ones normally used in the DFA and are obtained by fitting the models individually [see formula (11)]. The overall desirability function $d(\boldsymbol{x}) = [d_1(\boldsymbol{x})d_2(\boldsymbol{x})]^{1/2}$ was maximized over R using the Fortran program written by Conlon (1992). Alternatively, Design-Expert (Stat-Ease, 1993) software can also be used to maximize $d(\boldsymbol{x})$. The DFA results are given in Table 4.

The results for the DRA are given in Table 5. In applying this procedure to the present example, each of the two responses was considered as the

Table 3 Individual and GDA Simultaneous Optima for the Semiconductor Example

Response	Optimum	Location
Individual optima		
$\hat{y}_{e1}(\boldsymbol{x})$	Max = 0.8776	(0.7888, 0.9031, −0.7479)
$\hat{y}_{e2}(\boldsymbol{x})$	Min = 0.1302	(0.9443, −0.0468, 0.9689)
Simultaneous optima (GDA)		
$\hat{y}_{e1}(\boldsymbol{x})$	Max = 0.8641	(0.9976, 0.9127, −0.3961)
$\hat{y}_{e2}(\boldsymbol{x})$	Min = 0.1463	(0.9976, 0.9127, −0.3961)

Note: Minimum value of ρ_e in Eq.18 is 0.8610.

Table 4 DFA Simultaneous Optima for the Semiconductor Example

Response	Optimum	Location
$\hat{y}_{e1}(\boldsymbol{x})$	Max = 0.8772	(0.8351, 0.7951, −0.8172)
$\hat{y}_{e2}(\boldsymbol{x})$	Min = 0.1556	(0.8351, 0.7951, −0.8172)

Note: The maximum of $d(\boldsymbol{x})$ over R is 0.9609.

primary response. Its optimum value was then obtained over R using the constraint that the other response is equal to its individual optimum from Table 3. Values of the DRA optima in Table 5 were computed on the basis of the GRG algorithm using the "solver" tool, which is available in the Microsoft Excel (Microsoft, 1993) spreadsheet program. For more details on how to use this tool, see Dodge et al. (1995).

The results of applying GDA, DFA, and DRA are summarized in Table 6. We note that the results are similar to one another. The maxima of $\hat{y}_{e1}(\boldsymbol{x})$ under GDA and DFA are close, and both are higher than the maximum under DRA. Their overall desirability values are also higher.

3.2. A Food Industry Example

Tseo et al. (1983) investigated the effects of x_1 = washing temperature, x_2 = washing time, and x_3 = washing ratio on springiness (y_1), thiobarbituric acid number (y_2), and percent cooking loss (y_3) for minced mullet flesh. It is of interest to simultaneously maximize y_1 and minimize y_2 and y_3. The design settings in the original and coded variables and the corresponding multiresponse data are given in Table 7. Note that the design used is a central composite design with three center point replications and an axial parameter equal to 1.682. The same data set was reproduced in Khuri and Cornell (1996, pp. 295–296).

The multivariate variable selection techniques [Sparks' (1987) PRESS statistic and McElroy's (1977) R^2 statistic] mentioned in the previous section were used, and the following SUR models were obtained:

Table 5 DRA Optima for the Semiconductor Example

Response	Optimum	Location
$\hat{y}_{e1}(\boldsymbol{x})$	0.7639	(1.0, 0.3299, 0.9440)
$\hat{y}_{e2}(\boldsymbol{x})$	0.1488	(1.0, 0.7751, −0.6319)

Table 6 Comparison of GDA, DFA, and DRA Results for the Semiconductor Example

	Method		
	GDA	DFA	DRA
Optimal response value	(0.86,0.15)	(0.88,0.16)	(0.76,0.15)
Optimal settings	(1.0,0.91,-0.40)	(0.84,0.80,-0.82)	See Table 5.
Minimum metric (ρ_e)	0.8610	1.1768	Not applicable.
Overall desirability	0.9537	0.9609	0.8967

$$\hat{y}_{e1}(\boldsymbol{x}) = 1.8807 - 0.0974x_1 - 0.0009x_2 + 0.0091x_3 - 0.1030x_1^2 + 0.0013x_2^2 + 0.0028x_3^2$$

$$\hat{y}_{e2}(\boldsymbol{x}) = 22.5313 + 5.6609x_1 - 0.1719x_2 - 1.2268x_3 + 7.8739x_1^2 + 0.1489x_2^2 + 2.6920x_1x_2 + 0.1752x_2x_3$$

$$\hat{y}_{e3}(\boldsymbol{x}) = 17.8118 + 0.7442x_1 - 0.0120x_2 - 1.0710x_3 + 3.4798x_1^2 + 0.8288x_2^2 + 1.6731x_3^2 + 1.3020x_1x_2 + 1.9716x_1x_3$$

Table 7 Experimental Design and Response Values (Food Industry Example)

Original control variables			Coded control variables			Responses		
X_1	X_2	X_3	x_1	x_2	x_3	y_1	y_2	y_3
26.0	2.8	18.0	–1.000	–1.000	–1.000	1.83	29.31	29.50
40.0	2.8	18.0	1.000	–1.000	–1.000	1.73	39.32	19.40
26.0	8.2	18.0	–1.000	1.000	–1.000	1.85	25.16	25.70
40.0	8.2	18.0	1.000	1.000	–1.000	1.67	40.81	27.10
26.0	2.8	27.0	–1.000	–1.000	1.000	1.86	29.82	21.40
40.0	2.8	27.0	1.000	–1.000	1.000	1.77	32.20	24.00
26.0	8.2	27.0	–1.000	1.000	1.000	1.88	22.01	19.60
40.0	8.2	27.0	1.000	1.000	1.000	1.66	40.02	25.10
21.2	5.5	22.5	–1.682	0.000	0.000	1.81	33.00	24.20
44.8	5.5	22.5	1.682	0.000	0.000	1.37	51.59	30.60
33.0	1.0	22.5	0.000	–1.682	0.000	1.85	20.35	20.90
33.0	10.0	22.5	0.000	1.682	0.000	1.92	20.53	18.90
33.0	5.5	14.9	0.000	0.000	–1.682	1.88	23.85	23.00
33.0	5.5	30.1	0.000	0.000	1.682	1.90	20.16	21.20
33.0	5.5	22.5	0.000	0.000	0.000	1.89	21.72	18.50
33.0	5.5	22.5	0.000	0.000	0.000	1.88	21.21	18.60
33.0	5.5	22.5	0.000	0.000	0.000	1.87	21.55	16.80

The estimated standard errors for the parameter estimates, the values of the univariate R^2, adjusted R^2, and C_p statistics, and values of McElroy's R^2 Sparks' PRESS statistics are given in Table 8. We can see that the fits of the three models are quite good.

The individual optima and the GDA simultaneous optima over the region $R = \{(x_1, x_2, x_3) | \sum_{i=1}^{3} x_i^2 \leq 3\}$ are given in Table 9.

The results of the DFA are presented in Table 10. Here, the desirability values were computed using the functions

$$d_1(\boldsymbol{x}) = \begin{cases} 0 & \text{if } \hat{y}_{e1} \leq 1.3 \\ \dfrac{\hat{y}_{e1}(\boldsymbol{x}) - 1.3}{2.5 - 1.3} & \text{if } 1.3 < \hat{y}_{e1} < 2.5 \\ 1 & \text{otherwise} \end{cases}$$

$$d_2(x) = \begin{cases} 0 & \text{if } \hat{y}_{e2} \geq 51 \\ \dfrac{\hat{y}_{e2}(\boldsymbol{x}) - 51}{17 - 51} & \text{if } 17 < \hat{y}_{e2} < 51 \\ 1 & \text{otherwise} \end{cases}$$

Table 8 SUR Parameter Estimates and Values of C_p, R^2, and Adjusted R^2 (Food Industry Example)

Parameter	Responses[a]		
	$\hat{y}_{e1}$	$\hat{y}_{e2}$	$\hat{y}_{e3}$
Intercept	1.8807(0.0207)	22.5313(0.8854)	17.8118(0.8097)
x_1	−0.0974(0.0097)	5.6609(0.5538)	0.7442(0.3818)
x_2	−0.0009(0.0097)	−0.1719(0.5538)	−0.0120(0.3818)
x_3	0.0091(0.0097)	−1.2268(0.5538)	−1.0710(0.3818)
x_1x_2		2.6920(0.7234)	1.3020(0.4370)
x_1x_3			1.9716(0.4323)
x_2x_3		0.1752(0.7158)	
x_1^2	−0.1030(0.0107)	7.8739(0.5823)	3.4798(0.4198)
x_2^2	0.0013(0.0107)	0.1489(0.5823)	0.8288(0.4198)
x_3^2	0.0028(0.0107)		1.6731(0.4162)
C_p	6.61	7.82	8.59
R^2	0.93	0.95	0.87
Adj. R^2	0.88	0.91	0.74

[a]The number inside parentheses is the standard error.
Note: McElroy's $R^2 = 0.9271$; Sparks' PRESS statistic = 230.21.

Table 9 Individual and GDA Simultaneous Optima for the Food Industry Example

Response	Optimum	Location
Individual optima		
$\hat{y}_{e1}(\boldsymbol{x})$	Max = 1.9263	(−0.4661, −0.3418, 1.6276)
$\hat{y}_{e2}(\boldsymbol{x})$	Min = 18.8897	(−0.5347, 1.1871, 1.1415)
$\hat{y}_{e3}(\boldsymbol{x})$	Min = 17.4398	(−0.2869, 0.2365, 0.4970)
Simultaneous optima (GDA)		
$\hat{y}_{e1}(\boldsymbol{x})$	Max = 1.9136	(−0.5379, 1.0435, 0.8622)
$\hat{y}_{e2}(\boldsymbol{x})$	Min = 19.3361	(−0.5379, 1.0435, 0.8622)
$\hat{y}_{e3}(\boldsymbol{x})$	Min = 17.9834	(−0.5379, 1.0435, 0.8622)

Note: Minimum value of ρ_e in Eq. (18) is 0.9517.

$$d_3(\boldsymbol{x}) = \begin{cases} 0 & \text{if } \hat{y}_{e3} \geq 30 \\ \dfrac{\hat{y}_{e3}(\boldsymbol{x}) - 30}{14 - 30} & \text{if } 14 < \hat{y}_{e3} < 30 \\ 1 & \text{otherwise} \end{cases}$$

In setting up these functions, we assumed that the ranges of acceptable values for the three responses are $1.3 < y_1 < 2.5$, $17 < y_2 < 51$, and $14 < y_3 < 30$.

Finally, for the DRA, each of the three responses was considered to be the primary response, and its optimum value over R was obtained under the constraints that the other two responses are equal to their respective individual optima from Table 9. The results are shown in Table 11.

A summary of the optimization results of applying GDA, DFA, and DRA to this example is given in Table 12. Here also the results are similar, with the GDA and DFA providing slightly smaller minima for y_2 and y_3 than the DRA.

Table 10 DFA Simultaneous Optima for the Food Industry Example

Response	Optimum	Location
$\hat{y}_{e1}(\boldsymbol{x})$	Max = 1.9127	(−0.4504, 0.6176, 0.8081)
$\hat{y}_{e2}(\boldsymbol{x})$	Min = 19.8768	(−0.4504, 0.6176, 0.8081)
$\hat{y}_{e3}(\boldsymbol{x})$	Min = 17.6386	(−0.4504, 0.6176, 0.8081)

Note: The maximum of $d(\boldsymbol{x})$ over R is 0.7121.

Table 11 DRA Optima for the Food Industry Example

Response	Optimum	Location
$\hat{y}_{e1}(\boldsymbol{x})$	1.9145	(−0.5617, 1.1228, 0.9415)
$\hat{y}_{e2}(\boldsymbol{x})$	20.5507	(−0.3514, 0.2824, 0.5605)
$\hat{y}_{e3}(\boldsymbol{x})$	18.6124	(−0.5077, 1.0716, 1.2625)

Table 12 Comparison of GDA, DFA, and DRA Results for the Food Industry Example

	GDA	DFA	DRA
Optimal response values	(1.91, 19.34, 17.98)	(1.91, 19.88, 17.64)	(1.91, 20.55, 18.61)
Optimal settings	(−0.54, 1.04, 0.86)	(−0.45, 0.62, 0.81)	See Table 11
Minimum metric (ρ_e)	0.9517	1.2832	Not applicable
Overall desirability	0.7098	0.7121	0.6885

ACKNOWLEDGEMENT

We acknowledge the help of Ms. Terri L. Moore in providing the technical background for the example in Section 3.1.

REFERENCES

Biles WE. (1975). A response surface method for experimental optimization of multi-response processes. Ind Eng Chem, Process Des Dev 14:152–158.

Box GEP. (1954). The exploration and exploitation of response surfaces: Some general considerations and examples. Biometrics 10:16–60.

Box GEP, Youle PV. (1955). The exploration and exploitation of response surfaces. An example of the link between the fitted surface and the basic mechanism of the system. Biometrics 11: 287–323.

Box GEP, Hunter WG, MacGregor JF, Erjavec J. (1973). Some problems associated with the analysis of multiresponse data. Technometrics 15:33–51.

Chitra SP. (1990). Multi-response optimization for designed experiment. Am Stat Assoc Proc Stat Comput Sect, pp. 107–112.

Conlon M. (1988). MR: Multiple response optimization. Tech Rep No. 322, Department of Statistics, University of Florida, Gainesville, FL.

Conlon M. (1992). The controlled random search procedure for function optimization. Commun Stat Simul Comput B21: 919–923.

Copeland KAF, Nelson PR. (1996). Dual response optimization via direct function minimization. J Qual Technol 28: 331–336.

Del Castillo E. (1996). Multiresponse process optimization via constrained confidence regions. J Qual Technol 28: 61–70.

Del Castillo E, Montgomery DC. (1993). A nonlinear programming solution to the dual response problem. J Qual Technol 25:199–204.

Del Castillo E, Montgomery DC, McCarville DR. (1996). Modified desirability functions for multiple response optimization. J Qual Technol 28: 337–345.

Derringer GC. (1994). A balancing act: Optimizing a product's properties. Qual Prog 27: 51–58.

Derringer GC, Suich R. (1980). Simultaneous optimization of several response variables. J Qual Technol 12:214–219.

Dodge M, Kinata C, Stinson C. (1995). Running Microsoft Excel for Windows 95. Washington, DC: Microsoft Press.

Draper NR. (1963). "Ridge analysis" of response surfaces. Technometrics 5: 469–479.

Fichtali J, Van de Voort FR, Khuri AI. (1990). Multiresponse optimization of acid casein production. J Food Process Eng 12: 247–258.

Floros JD. (1992). Optimization methods in food processing and engineering. In: Hui YH, ed. Encyclopedia of Food Science and Technology, Vol. 3. New York: Wiley, pp 1952–1965.

Floros JD, Chinnan MS. (1988a). Seven factor response surface optimization of a double-stage lye (NaOH) peeling process for pimiento peppers. J Food Sci 53:631–638.

Floros JD, Chinnan MS. (1988b). Computer graphics-assisted optimization for product and process development. Food Technol 42:72–78.

Guillou AA, Floros JD. (1993). Multiresponse optimization minimizes salt in natural cucumber fermentation and storage. J Food Sci 58:1381–1389.

Harrington EC. (1965). The desirability function. Ind Qual Control 21:494–498.

Hill WJ, Hunter WG. (1966). A review of response surface methodology: A literature survey. Technometrics 8: 571–590.

Khuri AI. (1996). Multiresponse surface methodology. In: Ghosh S, Rao CR, eds. Handbook of Statistics, Vol. 13. Amsterdam: Elsevier Science, pp 377–406.

Khuri AI, Conlon M. (1981). Simultaneous optimization of multiple responses represented by polynomial regression functions. Technometrics 23:363–375.

Khuri AI, Cornell JA. (1996). Response Surfaces. 2nd ed. New York: Marcel Dekker.

Khuri AI, Myers RH. (1979). Modified ridge analysis. Technometrics 21: 467–473.

Kim KJ, Lin DKJ. (1998). Dual response surface optimization: A fuzzy modeling approach. J Qual Technol 30:1–10.

Lin DKJ, Tu W. (1995). Dual response surface optimization. J Qual Technol 27: 34–39.

Lind EE, Goldin J, Hickman JB. (1960). Fitting yield and cost response surfaces. Chem Eng Prog 56: 62–68.

McElroy MB. (1977). Goodness of fit for seemingly unrelated regressions: Glahn's $R^2_{y.x}$ and Hooper's $\bar{r}^2$. J Econometrics 6:381–387.
Microsoft (1993). Microsoft Excel User's Guide, Version 4.0. Redmond, WA: Microsoft Corporation.
Mouquet C, Dumas JC, Guilbert S. (1992). Texturization of sweetened mango pulp: optimization using response surface methodology. J Food Sci 57:1395–1400.
Myers RH, Carter WH. (1973). Response surface techniques for dual response systems. Technometrics 15:301–317.
Myers RH, Khuri AI, Carter WH. (1989). Response surface methodology: 1966–1988. Technometrics 31:137–157.
Myers RH, Khuri AI, Vining G. (1992). Response surface alternatives to the Taguchi robust parameter design approach. Am Stat 46:131–139.
Nelder JA, Mead R. (1965). A simplex method for function minimization. Comput J. 7:308–313.
Price WL. (1977). A controlled random search procedure for global optimization. Comput J. 20:367–370.
SAS (1989). SAS/STAT User's Guide, Vol. 2, Version 6. 4th ed. Cary, NC: SAS Institute, Inc.
SAS (1990a). SAS/ETS, Version 6. Cary, NC: SAS Institute, Inc.
SAS (1990b). SAS/IML Software, Version 6. Cary, NC: SAS Institute, Inc.
Sparks RS. (1987). Selecting estimators and variables in the seemingly unrelated regression model. Commun Stat Simul Comput B16:99–127.
Srivastava VK, Giles DEA. (1987). Seemingly Unrelated Regression Equations Models. New York: Marcel Dekker.
Stat-Ease (1993). Design-Expert User's Guide, Version 4.0. Minneapolis, MN: Stat-East, Inc.
Tseo CL, Deng JC, Cornell JA, Khuri AI, Schmidt RH. (1983). Effect of washing treatment on quality of minced mullet flesh. J Food Sci 48. 163–167.
Valeroso ES. (1996). Topics in multiresponse analysis and optimization. Unpublished PhD Thesis, Department of Statistics, University of Florida, Gainesville, FL.
Vining GG, Myers RH. (1990). Combining Taguchi and response surface philosophies: A dual response approach. J Qual Technol 22:38–45.
Zellner A. (1962). An efficient method of estimating seemingly unrelated regressions and tests for aggregation bias. J Am Stat Assoc 57:348–368.

26
Stochastic Modeling for Quality Improvement in Processes

M. F. Ramalhoto
Technical University of Lisbon, Lisbon, Portugal

1. INTRODUCTION

In any service industry there are essentially two types of products to be considered, product service and product supply. *Product service* can be defined as how the service has been provided, and *product supply* is what has been provided (this is, in many cases, what is commonly called product). The product service is usually provided through a service delivery process of a queuing system. The service delivery process is essentially described by a queuing model. This paper deals only with the product service.

To develop policies to provide consistently high product service for a wide range of customer types and arrival and service rates at "reasonable" cost is one of the ultimate targets of most queuing system managers. Usually, those are not easy targets. The present chapter presents a methodology to address them.

In Section 2 the differences between product service and product supply are discussed. In Section 3 a way is provided of quantifying delay and discomfort in the queuing system of the service industry in order to achieve a product service of high quality. Six external queuing system quality dimensions and four internal queuing system quality dimensions are defined to address delay and discomfort. The external quality dimensions—performance, flexibility, serviceability (responsiveness), reliability, courtesy (empathy), and appearance (tangibles)—provide a way to establish a kind of channel of communication between the queuing system managers and operators and their customers (they allow the managers to understand their customers' expectations and perceptions of the queuing system). The first three internal quality dimensions—timeliness, integrity, and predictability—provide a way to establish a kind of channel of communication between the managers and the actual physics of the queuing system (they allow the managers to identify and understand the limitations of the production process). The fourth internal quality dimension—customer satisfaction—provides a way to establish a kind of channel of communication between the managers and their market competitors. Once we have established the channels of communication we have to learn how to use them to communicate efficiently and to find the solution or the way of coping with the identified problems. Most of those problems have to do with the design of the service delivery process.

Behind a service delivery process there is usually a queuing model responsible for its failure or its success. In Section 4 the most relevant queuing models addressing the reduction of delay and discomfort and their functional relationship with the basic queuing model parameters are presented and discussed (two analytical queuing models that consider the quality dimension flexibility, one queuing model that considers the customers' perceptions of waiting and service, and a brief reference to approximations and bounds for queuing models with time-dependent arrival rates and to retrial queuing models). Usually, there are more than one queuing model able to respond to the needs of a particular service delivery process. Each queuing model option might lead to different levels of delay and discomfort reduction, impact on customer satisfaction, and costs. The aim is to find the "optimal" choice that balances it all. In Section 5 a simulation-decision framework, called total quality queue management, is described that explicitly considers and evaluates alternative queuing model options and makes the necessary decisions by selecting those particular options that provide the best projected performance scores, in terms of specified scoring criteria, based on measures linked to the quality dimensions selected. Section 6 consists of conclusions and further remarks.

2. PRODUCT SERVICE AND PRODUCT SUPPLY

2.1. Distinguishing Product Service and Product Supply

There might be situations where a clear cutoff between the product service and the product supply is too difficult to achieve. However, usually the product supply is an object and the product service is not. Also, in most cases, a poor product service might ruin an excellent quality product supply and vice versa. Therefore, both the quality improvement of product service and that of product supply have to be looked for and considered equally important.

Quality improvement of the product supply is linked to stochastic maintenance, reliability, quality control, and experimental design techniques. Furthermore, an important problem is how to achieve a high-quality product supply without increasing cost. In many situations the study of interactions among maintenance, reliability, and control charts, through a total quality management (TQM) approach, might help to reach that goal. However, that is not the concern of this chapter, which deals only with the product service.

It has been recognized by several authors including Deming (see, e.g., Ref. 1) that people who work in queuing systems are usually not aware that they too have a product to sell and that this product is the service they are providing. The product service is frequently invisible to the operators. They have difficulties in seeing the impact of their performance on the success or failure of the organization that employs them, on the security of their jobs, and on their wages. Perhaps it would make sense to propose a quality index (based on some of the quality dimensions to be defined next) for most of the relevant queuing systems of common citizens' everyday life (that would also help their operators to understand better the importance of their mission). Just imagine all the queuing systems relevant to our everyday life operating under the customer satisfaction criterion efficiently, adequately, and at controlled costs.

2.2. Identification of Differences

Product service cannot be stored, so apparently at least some measurements must be almost immediate. In fact, product service is intangible and ephemeral or perishable. It cannot be stockpiled and must be produced on demand (it should be noted that similar constraints now exist on the production of at least some product supply, owing to the new requirements in manufacturing production, such as just-in-time or zero inventory). Frequently, the delivery of the product service involves the customer and begins a very time-sensitive relationship with the customer. The involvement of the customer also makes the definition of quality of the product service vary over time much more quickly than that of the product supply. Customers also add uncertainties to the process, because it is often difficult to determine their exact requirements and what they regard as an acceptable standard for the product service. This problem is magnified by the fact that standards are very often subjective, based on personal preferences or moods rather than on technical performance that can be easily measured [2]. Whereas a product service may have completely satisfied a customer yesterday, exactly the same product service may not do so today because of the customer's mood. On the other hand, with the same equipment and for the same required service, because of the mood of the operator (if the operator is a human), the product service might be of poor quality today even if usually it is not. Queuing and waiting in general are at the same time personal and emotional. Qualitative and quantitative aspects of human behavior toward waiting have to be addressed. In most cases if customers are pleasantly occupied while waiting (entertainment, socially relevant information, opportunity to make interesting contacts, job opportunities, extra information about the queuing system itself, etc.), their perception of the length of the waiting time and of whether it is "reasonable" may differ substantially. Unlike the product supply, which can usually be sampled and tested for quality, the product service cannot, at least not easily. The record of an inspection of the product service cannot be assumed to be a "true" reflection of its quality. For instance, during inspection the operator (if a human) might be quicker, more courteous, and more responsive to customers than if left alone. (However, if the operator feels pleasure in providing a high quality product service and is proud of contributing to the higher standards of the queuing system, he or she works well even without any kind of inspection.) Moreover, unlike the control of quality in the product supply [1], the quality of the product-service depends both on the operator and on the customer. Also, product service can be classified as poor by some and good by others. Indeed, its qualification, good or faulty, need not be consistent.

On assessing the effectiveness of a product service, quantitative and qualitative factors have to be taken into consideration. It is also expected that different individuals will have different judgments and different opinions about many factual issues. Nevertheless, if the process continues long enough, the observers are expected to independently arrive at very similar interpretations. That, obviously, encourages the development of mechanisms of communicatiotn between the system's management and their customers. Moreover, product service is delivered at the moment it is produced. Any quantification or measurement taken is thus too late to avoid a failure or defect with that particular customer. However, that situation might be alleviated if a communication mechanism is already in operation (for instance, at the exit the customer could be asked, or given a short and clear questionnaire, to quantify the product service just received according to the quality dimensions to be defined in the next section and to briefly state what he or she would like to see improved in it; means of contacting the customer for mutually relevant communication in the future should also be recorded if the customer is interested). The success of the communication mechanism depends heavily on showing customers that they have been heard by the system managers and that their relevant opinions really make a difference.

Nevertheless, product service quality must always be balanced between customer expectations and their perception of the product service received. A higher quality product service is one with which the customers' perceptions meet or exceed their expectations. It is obvious that it is much more difficult to define quantitative terms for the features that contribute to the quality of product service than to quantify the quality of the product supply. Therefore, the primary area of difficulty is that of identifying appropriate quality "measures" (quantities resulting from measurements or quantification) that we call here quality dimensions. These quality dimensions also serve as a common language among the customers, operators, and managers.

3. QUALITY DIMENSIONS

I shall classify the quality dimensions into external and internal.

3.1. External Quality Dimensions

The quality dimensions—performance, flexibility, serviceability (responsiveness), reliability, courtesy (empathy), and appearance (tangibles)—are here called external quality dimensions and defined as follows, in a slightly different way than in Refs 3 and 4. Note that all external quality dimensions are defined from the customer's viewpoint.

Performance is the primary operating characteristics of the queuing system. It can be "measured" by, for instance, the "absence or perceived absence of waiting time," "total sojourn time in the system not exceeding X units of time." "competitive price," etc.

Flexibility is the queuing system's built-in ability to quickly respond to changes in demand. It can be "measured" by, for instance, the duration of a traffic peak (how quickly the peak is gotten rid of).

Serviceability (responsiveness) is the ability of the queuing system to respond to the individual needs of a particular customer. It can be measured by, for instance, the time to respond to those individual needs, including length of time to answer enquiries or to answer complaints.

Reliability is the ability to always perform the product service dependably, knowledgeably and accurately, and as expected by the customer.

Courtesy (empathy) is the caring, individualized attention provided to the customer, the effort to understand the customer's needs, the ability to convey trust and confidence. Those are factors more linked to standards of preferential human behavior, which are most subjective and difficult to control and evaluate. They need separate attention and joint research work with other specialists in order to set up adequate ways of quantifying them.

Appearance (tangibles) is the quality appearance of the physical environment and materials, facilities, equipment, personnel, and communications used to produce the product service. To quantify this quality dimension, joint research work with other specialists is also required to set up the right questions to lead to an adequate way of quantifying them.

The first four dimensions are mainly concerned with the cost–benefit characteristics of the particular queuing system under study. In fact, in many situations once they reach reasonably high ranks it is easier to improve the last two dimensions. Otherwise, a very kind operator who does not know the job well will very soon be considered to be of little use to the customer. An office full of well-dressed operators and sophisticated equipment is not necessarily the most important factor for the customer, particularly if the first four dimensions are not ranked high. They might even represent an insult for the customer who knows that, directly or indirectly, he or she is paying for that luxury.

Those quality dimensions are of great value as facilitators of system improvement but not in the ongoing business of monitoring and improving product service quality and cost reduction. They can be obtained only after the product service is delivered. Also, they reflect the views of the customer and not necessarily the real state of the system. They indicate the targets, from the point of view of the customers, that must be aimed for. However, a lot more might be learned by comparing the ranking of those quality dimensions with the "real" state of the system (for instance, by establishing priority targets and identifying the need to add more relevant quality dimensions). In fact, other external quality dimensions could be envisaged, such as managers', operators', and, when applicable, customers' commitment to quality. That is, of course, another external quality dimension that is difficult to measure but not so difficult to quantify.

3.2. Internal Quality Dimensions

We need "measures" that will help us to deliver what the customer expects or to improve the queuing system beyond customers' expectations at reasonable prices. For that, the quality dimensions timeliness, integrity, predictability, and "customer satisfaction," called here internal quality dimensions, are adopted. The quality dimension timeliness has been referred to, by several authors, as one of the most influential components in the quality of a product service, because the product service has to be produced on demand.

Timeliness is formed by the *access time*, which is the time taken to gain attention from the system; *time queuing*, which is the time spent waiting for service (and which can be influenced by the length of the queue and/or its integrity); and *action time*, which is the time taken to provide the required product service.

Integrity deals with the completeness of service and must set out what elements are to be included in order for the customer to regard the service as satisfactory. This quality dimension will set out precisely what features are essential to the product service.

Predictability refers to the consistency of the service and also the persistence or frequency of the demand. Standards for predictability identify the proper processes and procedures that need to be followed. They may include standards for the availability of people, materials, and equipment and schedules of operation.

Customer satisfaction is defined here as the way to provide the targets of success, which may be based on relative market position for the provision of a specific queuing system.

So far, we have established external and internal channels of communication and "measures" that tie together, in equal terms though with different roles, the managers, their operators (as part of the production process), customers, and market competitors. The aim is to build up a fair partnership of system managers, operators, market competitors, and customers, all able to communicate among themselves and committed to quality improvement and cost reduction of the system. Let me call this the *manager tetrahedron concept* (see Fig. 1). This concept allows a TQM

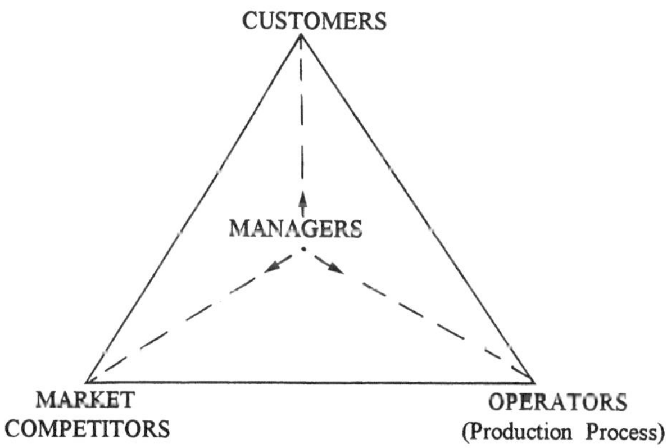

Figure 1 Manager tetrahedron.

approach to the quality of queuing systems in the way discussed, for instance, in Refs. 4–6.

Furthermore, the first internal quality dimension is clearly part of the theory of queues. Namely, access time has to do with the theory of retrial queues, and queuing time and action time are waiting time and service time, respectively. Unlike manufacturing, the production process in queuing systems of the service industry is usually quite visible to customers, since they are often part of this process. Therefore, it is crucial to place some quality improvement efforts on improving the production process. The service delivery process might be seen as the process of producing the product-service. Parasuraman et al [7], through external quality dimensions, have also identified the service delivery process as the key to improving product service quality and building customer loyalty. To improve the service delivery process essentially means to improve the queuing model behind it. Timeliness provides basic measures of its performance.

Let me now give examples of queuing model studies relevant to the quality improvement of the service delivery process.

4. SOME EXAMPLES OF IMPORTANT QUEUING MODELS IN QUALITY SERVICE

Some "product service failures or defects are very often linked to "unacceptable access time," "unacceptable queuing time," "unacceptable action time," and "unacceptable sojourn time in the system." All are clearly measured in queuing theory terms. Those failures or defects, as already mentioned, might ruin the ranking of most of the other quality dimensions. The way to prevent those failures or defects rests in the quality of the design of the process delivery of the queuing system. Often, if nothing is done to spread out the arrival pattern or to change the service rate or to modify the service discipline, the queuing system experiences very uneven traffic flows and serious failures or defects occur in the product service. All of those possible failures or defects have costs. Very often the cost of delay is to lose customers.

4.1. Two Queuing Models that Consider the Quality Dimension Flexibility

Queuing models that address the queuing system quality have to be able to efficiently deal with the peak duration that might occur in those systems. Very often, the rate of arrival to the system is very uneven, subject to random fluctuation, or periodically time-dependent. Designing such a queuing system specially to meet the peak demands is not always the best action to take, because it can be costly and the excess capacity can have negative psychological effects on the customer.

On the other hand, a poor rank in flexibility might lead to poor ranks in almost all the other quality dimensions. Most traditional queuing models are unable to respond quickly to changes in their environment. (The basic queuing parameter, namely, the number of operators, is usually assumed to be unchanged no matter what is happening in the queuing system.) The result is unacceptable queue sizes and waiting times. Long queues are, with few exceptions (e.g., the restaurant with excellent food, product supply at a good price), always considered an indication of poor product service.

Ramalhoto and Syski [8] show how quality management concepts of satisfying the customer can be incorporated into the design of queuing models. They propose and study a queuing model that aims to provide managers with a way of dealing with some temporary peak situations, that is to say, to have high ranking in the flexibility quality dimension. The model is essentially a G/G/c/FCFS (or a G/G/c/c+d/FCFS, i.e. first come first served queuing model with c, c operators and d waiting position; d is omitted when equal to zero or infinite) queuing model under the following additional decision rule, called here rule 1.

Rule 1. If the queue size exceeds b (the action line), introduce another server (or k servers, $k \geq 1$); when it falls below a (the prevention or alarm line), withdraw one server (or k servers, $k \geq 1$), $b > a$.

For the M/M/ queuing model c, (i.e., first come first served queuing model with Poisson arrival process and exponential service times distribution with $c$operators and infinite waiting positions) under rule 1, the equilibrium distribution of the state of the two-dimensional Markov process that characterizes the queuing model is derived. Some first-passage-time problems useful in the quality design of the queuing system are solved. Several extensions of these analytical results to more general settings, including nonhomogeneous Poisson arrivals, are discussed.

For the M/M/c queue under rule 1, where the arrival rate is denoted by λ and the service rate by μ, $\rho = \lambda/[(c+k)\mu]$, $\tau = \lambda/(c\mu)$, $\rho < \tau$, $\rho < 1$, and $e_{(i,n)}$ for $i = 0, 1, 2, \ldots; n = c, c+k$, denote the steady-state probability of having i customers in the queuing system and n operators serving, Ramalhoto and Suski prove, among other results, that [Ref. 8, p. 163, Eqs. (9) and (10)]

$$
\begin{array}{lll}
 & e_{(i,c)} = (c!/i!)c^{i-c}(b-a)e_{(b-1,c)}, & 0 \le i \le c-1 \\
 & e_{(i,c)} = (b-a)e_{(b-1,c)}, & c \le i \le a \\
\text{up,} & e_{(i,c)} = (b-i)e_{(b-1,c)}, & a+1 \le i \le b-1 \\
\text{down,} & e_{(i,c+k)} = \rho(1-\rho^{i-a})(1-\rho)^{-1}e_{(b-1,c)}, & a+1 \le i \le b-1 \\
 & e_{i,c+k)} = \rho^{i-b+1}(1-\rho^{b-a})(1-\rho)^{-1}e_{(b-1,c)}, & i \ge b
\end{array}
$$

$$
(b-a)\left[\left(\frac{c!}{c^c}\right)\sum_{i=0}^{c-1}\left(\frac{c^i}{i!}\right) + a - c + 1 + \frac{b-a-1}{2}\right] + \rho(1-\rho)^{-1}
$$
$$
\left[b-a-1-\frac{\rho(1-\rho^{b-a-1})}{1-\rho}+\frac{1-\rho^{b-a}}{1-\rho}\right] = \frac{1}{e_{(b-1,c)}}
$$

A measure of preference to use $c+k$ operators for a short period of time (Ref. 8, p. 164, Eqs. (18) and (19)], is given by $D_{(b,c+k)}$, the entrance probability to the set of states (i, c) for $i = 0, \ldots, a-1$, before entering the set of states $(i, c+k)$ for $i = b+1, b+2, \ldots$, when starting from the boundary state $(b, c+k)$. The value of $D_{(b,c+k)}$ gives an indication of the tendency toward c operators, when starting with $c+k$ operators.

$$
\frac{1}{D_{(b,c+k)}} = 1 + \zeta(a+1,\rho)\frac{\rho(1-\tau^{b-a+1})}{1-\tau^{b-a}}
$$

By letting $\tau \to 1$, Ramalhoto and Syski [8] obtained

$$
D_{(b,c+k)} = \left[1 + \frac{\rho(1-\rho^{b-a})}{1-\rho}\,\frac{b-a+1}{b-a}\right]^{-1}
$$

with $\rho = c/(c+k)$.

Other rules could be considered as alternatives to rule 1; for instance:

Rule 2. When the queue size exceeds b (the action line), shorten the service time (for instance, by deferring some tasks to be worked out later, by dividing and scheduling when the service can be provided in multiple separate segments, or by reducing the quality of service).

Rule 3. Identify classes of service needed by customers (each class requiring a different service time and being of different "value"), and treat the customers in separate queues, when the total queue length exceeds b (the action line).

Which rule is preferable? Section 5 addresses this question.

Affinity Operators

There are several important examples of queuing systems in the service industry, where it is "more efficient" to have a customer serviced by one operator than by any other. Thus the system schedules customers on the queue of their affinity operator. To address the inevitable imbalance in the number of customers assigned to each operator, there are several policies that can be considered. Any conventional queuing model under rule 3 might also be seen as a related model. Nelson and Squillante [9] consider a general threshold policy that allows overloaded operators to transfer some of their customers to underloaded operators. They vary four policy control parameters. Decomposition and matrix-geometric techniques yield closed-form solutions. They illustrate the potential sojourn time benefits even when the costs of violating affinities are large and experimentally determine optimal threshold values. One of the important applications of those models is in maintenance after sales, which has become a significant portion of manufacturing quality.

4.2. An Analytical Queuing Model that Considers the Customers' Perceptions of Waiting and Service

Conventional queuing control theory considers the costs of waiting in terms of time and money. For instance, Kitaev and Rykov [10] collect the newest results of the theory of Markov (semi-Markov and semi-regenerative) decision processes related to queuing models and show its applications to the control of arrivals, service mechanism, and service discipline. The theory of Markov decision processes claims that under certain conditions there exists an optimal Markov stationary strategy that can be constructed according to an optimal principle based on an optimality equation. Usually this approach does not account for customers' perceptions of waiting time and service.

Carmon et al. [11] examine how the service should be divided and scheduled when it can be provided in multiple separate segments. They analyze variants of this problem by using a model with a conventional function describing the waiting cost, which is modified to account for some aspects of the psychological cost of waiting in line. They analytically show, in some particular cases, that considerations of the psychological cost can result in prescriptions that are inconsistent with those dictated by conventional queuing control. From these results and the comments in the previous sections, it is obvious that psychologically based queuing research has a very important role to play in quality improvement in service industries.

4.3. Numerical Approximations for Queuing Models with Time-Dependent Arrival Processes

In any real-life queuing system of a service industry, there is seasonal (daily, weekly, and so on) patterns of traffic, rush hours and slack times. Queuing models with nonhomogeneous stochastic process arrivals better reflect these time-dependent traffic situations. However, the analysis of time-dependent behavior is very difficult and very often impossible, even for the simplest conventional queuing models. Nevertheless, the infinite server queue with nonhomogeneous Poisson arrivals and general service time distribution is one of the very rare exceptions, where time-dependent analysis is completely known and useful in practice.

In Ref. 12, it is shown that in the ergodic M/M/r/r + d queuing model, on the one hand, the distribution of almost any relevant queuing characteristic can be rewritten in terms of the third Erlang formula (the probability of nonimmediate service), which depends only on r and $r\rho$, where ρ is the traffic intensity. On the other hand, the number of waiting customers, number of servers occupied, number of customers in the system, waiting time in the queue, and total sojourn time in the system, in the stationary state, are sums of the corresponding random variables of the M/M/r/r loss queuing model (well approximated by the infinite-server queue for almost any value of the basic parameters involved, and even for the time-dependent case) and of the M/M/1/1/ + (d–1) queuing model, respectively, weighted by the third Erlang formula. The third Erlang formula value also indicates a "heavy/low" traffic situation. An extension of some of those results to the M/M/r/r + d queuing model with constant retrial rate is presented in Ref. 13, where the probability of not avoiding the orbit parallels the role of the third Erlang formula. In both models the decomposition's physical properties seem to be robust to several generalizations, including the time-dependent (transient) case. However, in many situations, namely, the time-dependent ones, there is no closed formula for most of the probability distributions. Therefore, exact comparisons are not possible. Approximations and bounds might be obtained through this decomposition approach.

It is well known that many queuing system practitioners empirically approximate the M_t/M/r/r + d queuing model with nonhomogeneous Poissosn arrivals by the infinite-server queuing model with nonhomogeneous Poisson arrivals. Based on this practice and on the results presented in Ref. 12, Ref. 14 provides a simple-to-use empirical approximation method to obtain bounds and approximations for the M_t/G/r/r + d queuing model. Other authors, such as Whitt and his coauthors at the AT&T Laboratories, have developed other, more sophisticated approaches to tackle the problem of obtaining approximations and bounds for the $M_t/G/r/r+d$ queuing model. A lot of research work is still needed on this queuing model. Its great importance in the service industry has already been shown, for instance, in Ref. 15.

4.4. The Retrial Queuing Model

As shown in the previous section, the access time to the queuing system is one of the components of the internal quality dimension timeliness. In fact, usually, a customer whose first call for access to the queuing system is unsuccessful will repeat the call, once or several times, in quick succession, thus giving rise to the phenomenon of repeated attempts. The retrial queuing model studies this phenomenon. The effect of repeated attempts is to lead to additional theoretical difficulties, even for the M/M/1/1 queuing model with constant retrials. The study of the M/M/r/r queuing model with retrials involves multidimensional random walks. Approximations and numerical methods for this queuing model date back to 1947 [16], but Ref. 17 is the first book completely dedicated to retrial queues.

When a queuing system is very successful it is usually because more customers are seeking access. If not properly controlled, the number of customers seeking access might eventually ruin the queuing system's quality reputation. Therefore, it is crucial to understand the interplay of the basic parameters—λ, arrival rate; ν, service rate; and α, access rate—and their influence on the most relevant quality dimensions. Figures 2–4 illustrate the kind of three-dimensional surfaces that represent the mean and the

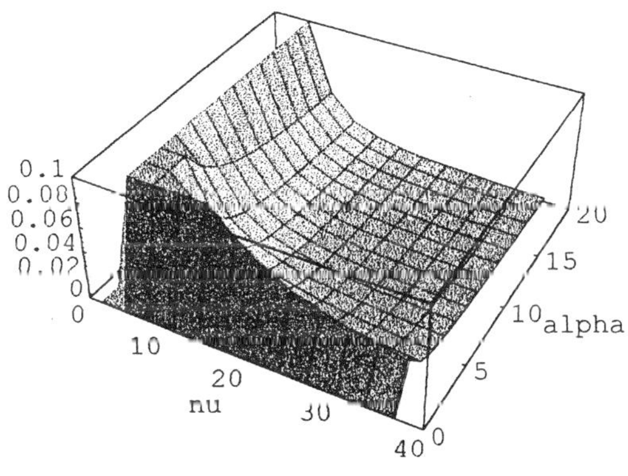

$\nu>(1+\lambda\alpha^{-1}$

Figure 2 Mean value of the waiting time in the M/M/1/1 queuing model with constant retrial rate and $\lambda = 1.5$.

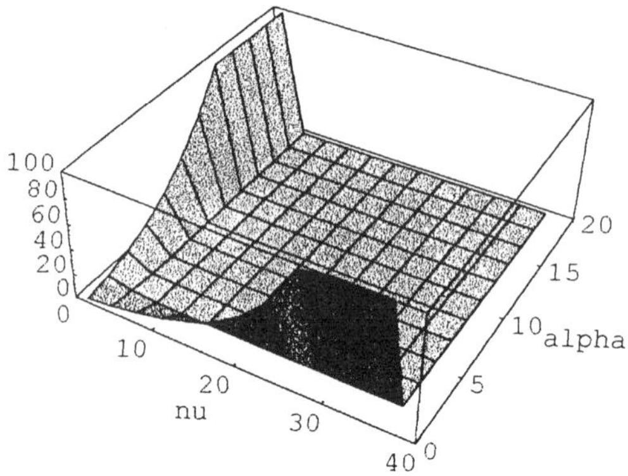

Figure 3 Variance of the waiting time in the M/M/1/1 queuing model with constant retrial rate and $\lambda = 1.5$.

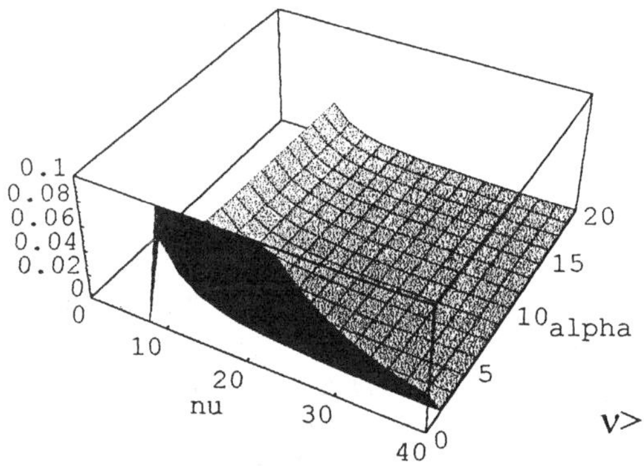

$\nu > (\lambda+\alpha)(2\alpha)^{-1}\,[(\alpha(\alpha+4\lambda))^{-1} - \alpha]$

Figure 4 Mean value of the waiting time in the M/M/1/1 + 1 queuing model with constant retrial rate and $\lambda = 1.5$.

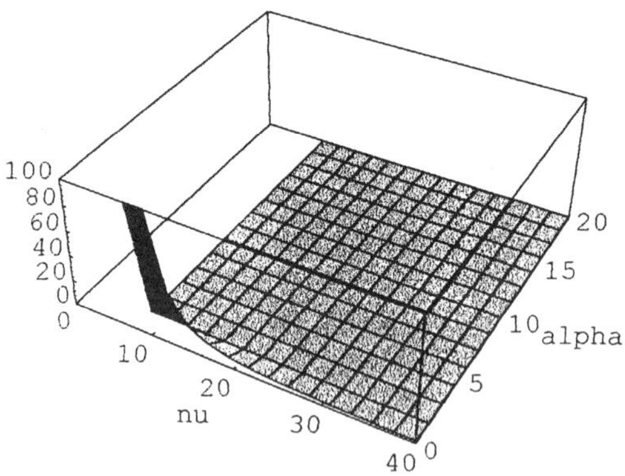

Figure 5 Variance of the waiting time in the M/M/1/1 + 1 queuing model with constant retrial rate and $\lambda = 1.5$.

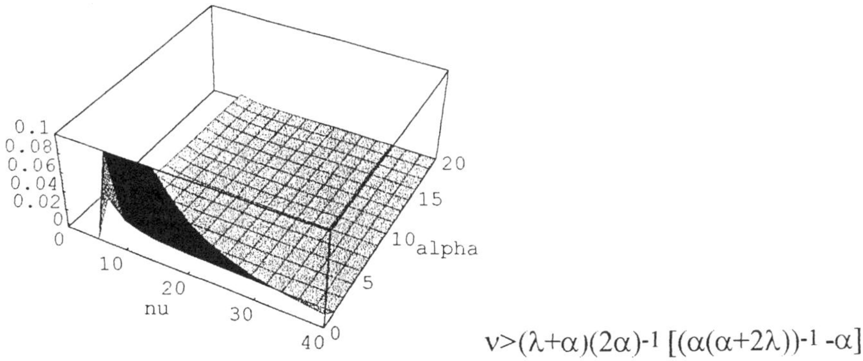

Figure 6 Mean value of the waiting time in the M/M/2/2 queuing model with constant retrial rate and $\lambda = 1.5$.

variance of the waiting time, as functions of α and ν, for the M/M/1/1 (one server and no waiting position), M/M/1/1 + 1 (one server and one waiting position), and M/M/2/2 retrial queuing models with constant retrial rate α and for different ergodicity intensities. Results of this kind help to evaluate the range of arrival, retrial, and service rates that provide consistently high product service quality in an increasingly successful queuing system. (Also, for example, providing k extra servers, as in Section 4.1, when λ or/and α increase beyond a certain threshold might be an adequate short-term policy to maintain the high quality of the product service in an increasing successful queuing system).

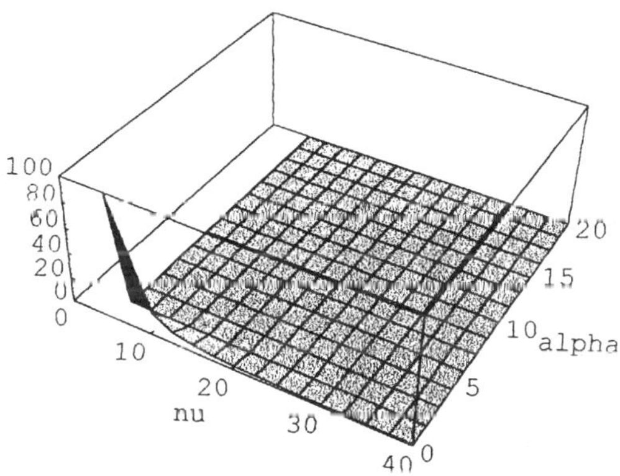

Figure 7 Variance of the waiting time in the M/M/2/2 queuing model with constant retrial rate and $\lambda = 1.5$.

Remark 1. Perhaps it should be noted that if the design of the delivery process is no longer fit for the purposes required, it will cause a kind of "common cause variation." The temporary changes in the arrival or service mechanism will cause a kind of "special cause variation." Both causes of variation have to be addressed.

Remark 2. The following types of robustness are desirable: (1) The queuing model behind the delivery process is robust if its expected performance is not too much affected by "reasonable" changes in the arrival and departure rates. (2) The operator is robust if its performance is not too much affected by "reasonable" product service changes required by the customer.

Remark 3. Instead of setting up direct inspection of the operators, promote channels of communication among customers and operators to build a joint commitment to improving the quality of the product service.

Remark 4. Whenever possible, eliminate or substantially reduce waiting time and queue size. Managers and operators should network with customers through, for instance, new technologies in order to have customers' arrivals as close as possible to the instant they begin service.

Remark 5. Specific goals should be set for certain quality dimensions, such as access time not greater than x, duration of peaks not greater than y, queuing time (waiting time) not greater than z, action time (service time) not greater than h, and delivery process idle time not greater than u. A cost–benefit analysis should be established for queuing systems in monopolistic or ugently needed service industries.

Remark 6. An efficiently run queuing system should inform its customers at arrival that (1) on average, the waiting time to initiate service is shorter than a certain value and (2) its queuing size (if not visible) is shorter than a certain value. Whenever needed, it should spread out arrivals, for instance, by (3) setting up appointment schemes, (4) pricing at peak load intervals, when applicable, and (5) establishing priority schemes for special classes of customers.

Remark 7. Build on the ISO 9000 gains by introducing a request for a good understanding of customers' needs as well as operators' limitations (by the manager tetrahedron concept) and the use of an adequate delivery process design.

Queuing theory certainly has a role to play in the search for the better adjusted models to the needs of quality management of service industry queuing systems. However, the probabilistic results needed to understand and control the stochastic behavior of those queuing systems cannot all be determined analytically and need an interdisciplinary approach. They have to be obtained by a mixture of educated intuition (based on some of the queuing analytical and algorithmic results available), heuristics, simulation, and decision making guided by research findings on the psychology of waiting.

In fact, what seems to be required here is the creation of a framework with the ability to jointly consider data management (from the internal and external quality dimensions selected), process delivery design (robust queuing models, including psychologically based queuing models), and decision making also based on cost–benefit analysis. As already stressed, in most situations the service delivery process is the one that, more often than not, needs special attention.

5. EMPIRICAL MODEL BUILDING FOR THE QUALITY IMPROVEMENT OF QUEUING SYSTEMS

Usually more than one queuing model is capable of responding to the need to improve or redesign a particular service delivery process. Each queuing model option might lead to different levels of reduction of delay and discomfort, impact on customer satisfaction, and costs. The aim, in most cases, is to find the "optimal" solution that balances the customer delay and discomfort against operator idleness at the same cost.

Ramalhoto [18] formulated a practical simulation decision framework that considers and evaluates alternative queuing model options and makes the necessary decisions by selecting those particular options that provide the best projected performance scores, in terms of specified scoring criteria, based on measures linked to the quality dimensions selected. The queuing model options are defined as "control parameters" in this framework. For instance, the queuing models corresponding to rules 1, 2, and 3, respectively, defined in Section 4.1, can be represented quantitatively by the following three basic control parameters: X_1, the regular size of the service staff; X_2, the percentage by which the service times for each customer are to be reduced or expedited (as a function of queue length or any other relevant quantity); X_3, the amount by which the regular service staff is augmented by other personnel (such as secretarial or clerical staff to meet periods of heavy demand); X_4, the number of different classes of service needed by customers; and X_5, the percentage of the regular service staff to allocate to each of those different classes of service. This framework is called total quality queue management.

5.1. The Total Quality Queue Management Framework

Basically, the total quality queue management framework consists of four components: a stochastic demand model, a decision system, an outcome calculator, and a scoring system. The stochastic demand model represents our projection (and the uncertainties in our projection) of the rates of arrival and service requirements of the customers. The decision system searches systematically over the multidimensional space defined by the control parameters $X_1, \ldots, X_5$ to find an optimal combination of values, $X_1^*, \ldots, X_5^*$, for these control parameters that will yield the "best" system performance given the stochastic demand that has been specified for the particular problem.

To enable the decision system to compute and evaluate the consequences of any specific set of control parameter values, it has to use the results of the outcome calculator and the scoring system. The outcome calculator and the scoring system have to be constructed as entirely separate and independent systems.

The outcome calculator calculates (or projects) the specific outcome(s) that will result from any specific assumptions concering customer demand and any specific decisions concerning the values of the control parameters. In particular, for any such combination of assumptions, the outcome calculator must be able to compute the pertinent outcome parameters (which are defined in terms of objective physical quantities such as queue length, customer waiting time, service cost, and other pertinent descriptors of the outcomes) that may be needed to evaluate the queuing system performance in terms of the selected quality dimensions. Clearly, the outcome calculator is concerned with the objective physical outcomes of the queuing system (in principle, it has nothing to do with the customers' goals, objectives, priorities, or expectations). It should be able to provide the real ranking value of the quality dimensions selected.

The scoring system has to be concerned with the subjective desirability of the outcomes in terms of customers' expectations, perceptions of waiting and service, and current goals and objectives. That should be done through, for instance, a careful analysis of complaints, behavioral queuing research, and relevant customer questionnaires and surveys addressing the quality dimensions selected. The purpose of the scoring system is to assign to each outcome a ranking of the quality dimensions selected that corresponds, as accurately as possible, to the customers' real objectives and expectations for that particular queuing system.

The actual implementation of the total quality queue management framework to a specific queuing system might be done, for instance, following a "value-driven" approach. However, other approaches might be envisaged.

One of the interesting features of this framework is that we have two ranking schemes for the quality dimensions selected. The first is an inevitable consequence of the structure of the queuing system and its relevant physical law (it reflects the voice of the real system), and the second reflects the customers' perceptions and expectations of the queuing system (it reflects the voice of the customer). So the comparison of the two rankings might be very important to the queuing system's learning process.

The total quality queue management framework is expected to help managers gain insight into the main factors that influence product service quality and identify process changes that will improve it.

6. CONCLUSIONS AND FURTHER REMARKS

Studies have shown that indicators often distort a program from the beginning by forcing a focus on the indicators rather than on the true underlying goals. The result is generally a lack of sustained success. And in many cases there is no success at all save in the artificial indicators, which can often be manipulated with little effect on the underlying process. Unfortunately, in several situations the harm caused by those artificial indicators is very painful. That is indeed a serious risk to be avoided. Therefore, an effective process of judging the costs and consequences of the choices necessarily incorporates a learning process. An important result of such learning is a shared vision with the managers, operators (many operators know a lot about their jobs and about the queuing system they are working with and also have the capability of taking direct action), and customers about how the process works and how it should work in order to confront the challenges it faces.

In this chapter, product service is treated as "manufacturing in the field." It is advocated that it should be carefully planned, audited for quality control, and regularly reviewed for performance improvement and customer reaction. The methodology presented is an attempt to construct a learning queuing system that is able to assess (internally and externally) its own actions and judge and adjust the process through which it acts. It relies on teamwork among customers, operators, and managers to unify some goals, on a scientific approach, and on decision making based on reliable data. In fact, it is based on analysis, simulation, data, policies, and options. The idea is also to question policies whenever appropriate. Adequate data have to be collected and studied statistically, and options have to be analyzed, including the option to change policies.

In real life, changes are very often costly in terms of money, time, psychological tensions, and so on, for many reasons (e.g., the new changes in practice do not perform as well as expected), and many things can go irreversibly wrong. Therefore, whenever possible, the total quality queue management framework, or any other adequate system's thinking framework, should be used (and its solutions tested, including the budgeted costs) in connection with virtual reality experimentation technology.

Remark 8. There is also a need to systematically judge all the other aspects of the queuing system, namely, the product supply and information technology involved. Key benchmark measures and standards are also needed.

Remark 9. When an appointment scheme for the arrivals is being used, most likely the manager will prefer a tight schedule that limits idle time, while a customer may prefer to arrive late to avoid waiting. When commitment on both sides is lacking, cost penalties on both sides often lead to more successful appointment schemes with "reasonable" average idle time and waiting time.

Remark 10. In a queuing system there are essentially two main reasons for customer dissatisfaction: (1) a waiting time that exceeds a threshold level and (2) dissatisfaction with the service received. The latter can be caused by the poor quality of either the product service or the product supply or an excessively high cost.

Remark 11. Interdisciplinary advanced studies in the fields of data analysis, decision analysis, queuing theory, quality management, and the psychology of individuals, time, and change are needed to create more successful queuing systems for the service industries. Communication, information, and commitment are also important tools. Queuing system studies could also incorporate the latest behavioral queuing research (accumulated across the fields of psychology, marketing, economics, and sociology) to alleviate the human tensions and humiliation of waiting.

Remark 12. Future customer visits to any queuing system when alternatives exist (nonmonopolistic or non-urgently needed service) heavily depend on the price and quality of the service product quality and the product supply provided by this queuing system. It is well accepted by all that having to wait beyond certaitn limits is one of the crucial factors in customer satisfaction. However, as shown in Ref. 19, customers' final dissatisfaction with waiting for service is also very highly correlated with their global retrospective (dis)satisfaction judgments, which affect their future actions.

Remark 13. Little has been written on how queuing system quality is related to conventional productivity, profitability, and sales performance measures. The direct effect of queuing system redesign initiatives on those measures needs deeper investigation.

Remark 14. It should be noted that the service industry is spreading over to manufacturing. In the "service factory" concept, the service is identified as the "fifth competitive priority" (as opposed to the traditional four competitive priorities—cost, quality, delivery, and flexibility) in manufacturing strategy. The idea is that the manufacturing organization can become more competitive by employing a broader range of services provided by the factory personnel and facilities (for instance, maintenance after sales of their own goods).

Remark 15. It should be emphasized that a high quality queuing system is only the result of constant effort to control and to improve each single aspect of the queuing system as well as not to disregard synergies, integral (total) management, and strong market awareness.

ACKNOWLEDGEMENT

The Author thanks Professor Gomez-Corral for the elaboration of the figures 2 to 7.

The author has also benefited from discussions with colleagues from Princeton University, Maryland University, and Rutgers University during her half-year sabbatical in 1998 at Princeton University.

This work was carried out with the support from the "Fundaçao para a Ciência e a Tecnologia" under the contract number 134–94 of the Marine Technology and Engineering Research Unit-Research Group on Queueing Systems and Quality Management, and the project INTAS 96–0828, 1997–2000, on "Advances in Retrial Queueing Theory".

REFERENCES

1. WE Deming. Out of the Crisis, MIT-CAES. Cambridge, MA: Cambridge University Press, 1986.
2. CA King. Service quality assurance is different for advanced engineering studies. Qual Prog June: 14–18, 1985.
3. MF Ramalhoto. Queueing system of the service industry—A TQM approach. In: GK Kanji, ed. Total Quality Management. Proceedings of the First World Congress. London: Chapman & Hall, 1995, pp 407–411.
4. B Bergman, B. Klefsjo. Quality from customer needs to customer satisfaction. McGraw-Hill, New York, 1994.

5. GK Kanji. Quality and statistical concepts. In: GK Kanji, ed. Total Quality Management. Proceedings of the First World Congress. London: Chapman & Hall, 1995, pp. 3–10.
6. K Dahlgaard, K Kristensen, GK Kanji. TQM-leadership. In: GK Kanji, ed. Total Quality Management. Proceedings of the First World Congress. London: Chapman & Hall, 1995, pp 73–84.
7. A Parasuraman, LL Berry, VA Zeithaml. Understanding customer expectations of service. Sloan Manage Rev Spring: 39–48, 1991.
8. MF Ramalhoto, R. Syski. Queueing and quality service. Invest Oper 16(2):155–172, 1995.
9. RD Nelson, MS Squillante. Stochastic analysis of affinity scheduling and load balancing in parallel queues. (Submitted for publication)
10. MY Kitaev, VV Rykov. Controlled Queueing Systems. Boca Raton, FL: CRC Press, 1995.
11. Z Carmon, JG Shanthikumar, TF Carmon. A psychological perspective on service segmentation models: The significance of accounting for consumers' perceptions of waiting and service. Manage Sci 41(11): 1806–1815, 1995.
12. MF Ramalhoto. Generalizations of Erlang formulae and some of their 2nd order properties. In: A Bachem, V Derigs, M Junger, R Schrader, eds. Operations Research '93, GMOOR. Physica-Verlag, Heidelberg, 1993, pp 412–417.
13. MF Ramalhoto, A Gomez-Corral. Some decomposition formulae for the M/M//r/r+d queue with constant retrial rate. Communications in Statistics–Stochastic Models: 14(1 & 2), 1998, 123–145.
14. MF Ramalhoto. The state of the $M_t/G/\infty$ queue and its importance to the study of the M/G/r/r+d queue. (Submitted for publication).
15. L Green, P Kolesar, A Svoronos. Some effects of nonstationarity on multi-server Markovian queueing system. Oper Res 39:502–511, 1991.
16. I Kosten. On the influence of repeated calls in the theory of probabilities of blocking. Ingenieur 59:1–25, 1947.
17. GI Falin, JGC Templeton. Retrial Queues. London: Chapman & Hall, 1997.
18. MF Ramalhoto. Stochastic modelling in the quality improvement of service industries—Some new approaches. proceedings of International Conference on Statistical Methods and Statistical Computing for Quality and Productivity Improvement, ICSQP'95, Seoul, 1995, pp 27–35.
19. Z Carmon, D Kahneman. The experienced utility of queueing: Experience profiles and retrospective evaluation of simulated queues. Working paper, Fuqua School, Duke University, 1993.

27

Recent Developments in Response Surface Methodology and Its Applications in Industry

Angela R. Neff
General Electric, Schenectady, New York

Raymond H. Myers
Virginia Polytechnic Institute and State University, Blacksburg, Virginia

1. INTRODUCTION

It is interesting to note that there are a limited number of areas of statistics that are almost entirely motivated by and dependent on real problems. They do not progress merely because of innovative mathematical rigor, but rather their development is a function of the increased complexity of problems faced by practitioners. Such is the case with response surface methodology (RSM). The fundamental goal remains the same as it was in the late 1940s and early 1950s: to find optimum process conditions through experimental design and statistical analysis. While the term "quality improvement" became a classic and overused term in the 1980s and 1990s, RSM dealt with quality improvement problems 30 years earlier.

There is no question that RSM has received unprecedented attention in recent years and has been the beneficiary of Genichi Taguchi and the quality era. It has been put forth as a serious alternative to specific Taguchi methodology. In fact, the RSM approach has been suggested as a collection of tools that will allow for the adoption of Taguchi principles while providing a more rigorous approach to statistical analysis. Much progress continues to be made as RSM benefits from mathematical optimization methods, statistical graphics, robust fitting, new design ideas, Bayesian statistics, optimal design theory, generalized linear models, and many other advances. Researchers in all fields are able to focus on applications of RSM because of the substantial improvement in the software that is used for RSM. There is no doubt that high quality software is one of the better communication links between the statistics researcher and the user.

In this chapter we discuss and review some of the recent developments in RSM and how they are having and will continue to have an impact on applications in industry.

2. MEAN AND VARIANCE MODELING AND ROBUST PARAMETER DESIGN

Along with the realization that product quality depends on understanding process variation as well as targeting of the mean came the concept of response surface modeling for both the process mean and variance. Taguchi's clever consideration and use of noise models allowed this area to advance. Robust parameter design (RPD) is a principle that emphasizes proper choice of levels of controllable process variables (*parameters* in Taguchi's terminology) in order to manufacture a product with minimal variation around a predetermined target. These controllable process variables (controlled in experiments as well as in product and process design) are referred to as *control factors*. It is assumed that most of the variation around the target is due to the inability to control a second set of variables called *noise factors*. Some examples of noise factors are environmental conditions, raw material properties, variables related to how the consumer handles or uses the product, and even the tolerances around control factors. [The reader is referred to Myers and Montgomery (1995) for illustrations of control and noise variables for various applications.] The objective in RPD is to design the process by selecting levels of the control factors in order to achieve *robustness* (insensitivity) to the inevitable changes in the noise factors. This can be achieved through the appropriate design and analysis of experiments that include noise as well as control factors, since even the noise factors are often within our control for purposes of experimentation. This philosophy is perhaps Taguchi's greatest contribution to the quality movement.

Compared to the design and analysis techniques utilized by Taguchi (Taguchi and Wu (1980)), response surface methods can accomplish RPD through more rigorous analysis and efficient experimentation. For more on the RSM approach compared to Taguchi's methods, read Vining and Myers (1990), Myers et al. (1992a), Khattree (1996), and Lucas (1994).

Independent of the approach taken, however, the ability to incorporate robustness to noise factors into a process design depends on the existence and detection of at least one control × nose interaction. It is the structure of these interactions that determines the nature of nonhomogeneity of process variance that characterizes the parameter design problem. For illustration, consider a problem involving one control factor, x, and one noise factor, z. Figure 1 shows two potential outcomes of the relationship between factors x and z and their effects on the response, y. In Figure 1a, it can be seen that the response y is robust to variability in the noise factor z when the variable x is controlled at its low level. When x is at its high level, however, the change in z has an effect of 15 units on the response. In other words, the presence of the xz interaction indicates that there is an opportunity to reduce the response variability through proper choice of the level of the control factor. In contrast, Figure 1b shows that when there is no control × noise interaction, the variability in y induced by the noise factor cannot be "designed out" of the system, since the variability is the same (i.e., homogeneous) at both levels of the control factor.

While the estimation of control × noise interactions is important for understanding how best to control process variance, the control factor main effects, as well as interactions among control factors, are equally important for understanding how to drive the response mean to its target. The dual response surface approach, which addresses both process mean and variance, begins with the response model,

$$y(\mathbf{x}, \mathbf{z}) = \beta_0 + \mathbf{x}'\boldsymbol{\beta} + \mathbf{x}'\mathbf{B}\mathbf{x} + \mathbf{z}'\gamma + \mathbf{x}'\Delta\mathbf{z} + \varepsilon \tag{1}$$

In the response model, $\mathbf{x}$ and $\mathbf{z}$ represent the $r_x \times 1$ and $r_z \times 1$ vectors of control and noise factors, respectively. The $r_x \times r_x$ matrix $\mathbf{B}$ contains coeffi-

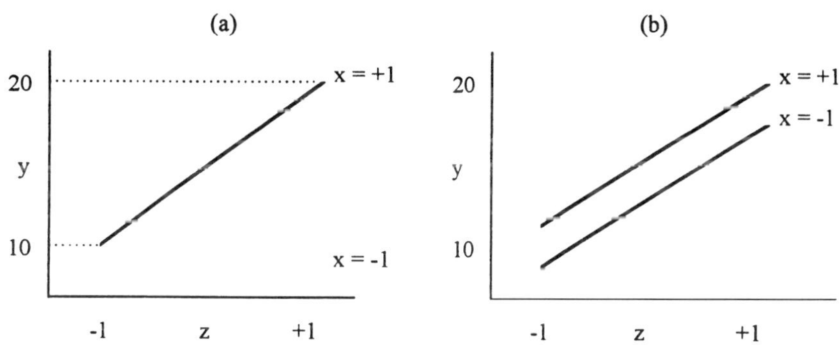

Figure 1 (a) Control by noise interaction. (b) No control by noise interaction.

cients of control × control interactions (which includes quadratics), while the matrix Δ is an $r_x \times r_z$ matrix of control × noise interactions. While it is possible to have interactions or even quadratics among noise factors, the previously defined response model will accommodate many real-life applications. It is assumed that $\varepsilon \sim N(\mathbf{0}, \sigma^2\mathbf{I})$, implying that any nonconstancy of variance of the process is due to an inability to control the noise variables. The assumption on the noise variables is such that the experimental levels of each z_j is centered at some mean μ_j with the ± 1 levels set at $\mu_{z_j} \pm c\sigma_{z_j}$, where c is a constant, 1, $\frac{1}{2}$, etc. As a result, it is assumed that

$$E(\mathbf{z}) = \mathbf{0}, \qquad \mathrm{Var}(\mathbf{z}) = \sigma_z^2\mathbf{I}_{r_z}$$

thus implying that noise variables are uncorrelated with known variance.

Taking expectation and variance operators on the response model in (1), we can obtain estimates of the mean and variance response surfaces as

$$\hat{E}_z[y(\mathbf{x})] = b_0\mathbf{x}'\mathbf{b} + \mathbf{x}'\hat{\mathbf{B}}\mathbf{x}$$

and

$$\mathrm{Var}[y(\mathbf{x}) = (\hat{\gamma} + \hat{\Delta}'\mathbf{x})'\mathbf{V}(\hat{\gamma} + \hat{\Delta}'\mathbf{x}) + \hat{\sigma}_\varepsilon^2$$

An equivalent form of the variance model, under the assumption that $V = \sigma^2\mathbf{I}$, is given by

$$\mathrm{Var}_z[y(\mathbf{x})] = \hat{\sigma}_z^2\mathbf{I}'(\mathbf{x})\mathbf{I}(\mathbf{x}) + \hat{\sigma}_\varepsilon^2$$

where $\mathbf{l}(\mathbf{x}) = (\hat{\gamma} + \hat{\Delta}'\mathbf{x})$, which is the vector of partial derivatives of $y(\mathbf{x}, \mathbf{z})$ with respect to $\mathbf{z}$. In these equations, $\mathbf{b}$, $\hat{\gamma}$, $\hat{\mathbf{B}}$, and $\hat{\Delta}$ contain regression coefficients from the fitted model of Eq. (1), with $\hat{\sigma}_\varepsilon^2$ representing the error mean square from this model fit. Notice the role that $\hat{\Delta}$ plays in the variance model, recalling that it contains the coefficients of the important control × noise interactions. Running the process at the levels of $\mathbf{x}$ that minimize $\|\mathbf{l}(\mathbf{x})\|$ will in turn minimize the process variance. If however, $\hat{\Delta} = \mathbf{0}$, the process variance does not depend on $\mathbf{x}$, and hence one cannot create a robust process by choice of settings of the control factors (illustrated previously with the simple example in Figure 1).

Various analytical techniques have been developed for the purpose of process understanding and optimization based on the dual response surface models. Vining and Myers (1990) proposed finding conditions in $\mathbf{x}$ that minimize $\mathrm{Var}_z[y(\mathbf{x})]$ subject to $\hat{E}_z[y(\mathbf{x})]$ being held at some acceptable level. Lin and Tu (1996) consider a mean squared error approach for the "target is best" case. Other methods, given in Myers et al. (1997), focus on the distribution of response values in the process. These include the development of prediction intervals for future response values as well as the development of tolerance intervals to include at least 100% of the process values with some specified probability. An example taken from Myers et al. (1997) will be used to graphically illustrate the dual response surface approach and the usefulness of the analytical measures previously mentioned.

The data for this example, taken from Montgomery (1997), comes from a factorial experiment conducted in a U.S. pilot plant to study the factors thought to influence the filtration rate of a chemical bonding substance. Four factors were varied in this experiment: pressure (x_1), formaldehyde concentration (x_2), stirring rate (x_3, and temperature (z). There is interest in maximizing filtration rate while also dealing with the variation transmitted by fluctuations of temperature in the process. For this reason, temperature is treated as a noise variable. All four factors are varied at the ± 1 levels in a 2^4 factorial arrangement, with the ± 1 levels of temperature assumed to be at $\pm\sigma_z$, representing temperature variability in the process. The fitted response model is given by (Montgomery, 1997).

$$\hat{y} = 70.025 + 10.8125z + 4.9375x_2 + 7.3125x_3 - 9.0625x_2z + 8.3125x_3z - 0.5625x_2x_3$$

with $R^2 = 0.9668$ and $\hat{\sigma}_\varepsilon = 4.5954$. Note that there are two control $\times$ noisc interactions present in the model, indicating that the variability transmitted from temperature fluctuations can be reduced through proper choice of formaldehyde concentration (x_2) and stirring rate (x_3). Pressure (x_1) was found to have no significant effect on filtration rate (y). The estimated mean and variance models are therefore given by

$$E[y(x_2, x_3)] = 70.02 + 4.9375x_2 + 7.3125x_3 - 0.5625x_2x_3$$

and

$$\text{Var}_z[y(x_2, x_3)] = (10.8125 - 9.0625x_2 + 8.3125x_3)^2 + (4.5954)^2$$

Figure 2 shows the overlaid contour plots for the response surface models of the process mean and standard deviation. The trade-off between maximizing filtration rate while attempting to minimize variance is evident. Figure 3 contains a contour plot of mean filtration rate along with the locus of points $1(x_2, x_3) = 0$, defining a line of minimum estimated process variance. The shaded region represents a 95% confidence interval around this linc of minimum variance. From Figure 3, the mean–variance trade-off becomes even more clear, since we can achieve barely more than 73 gal/hr for the estimated process mean while minimizing the process variance (with coordinates $x_2 = 1, x_3 = 0.2$).

In Figure 4 we see lower 95% one-sided prediction limits, while Figure 5 depicts lower 95% tolerance limits on filtration rates with probability 0.95. Both of these illustrations indicate that the process should be operated at a

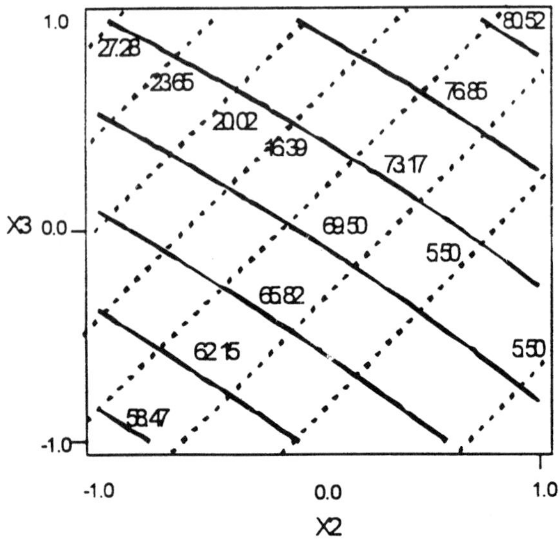

Figure 2 Contour plot of both the mean filtration rate and the process standard deviation.

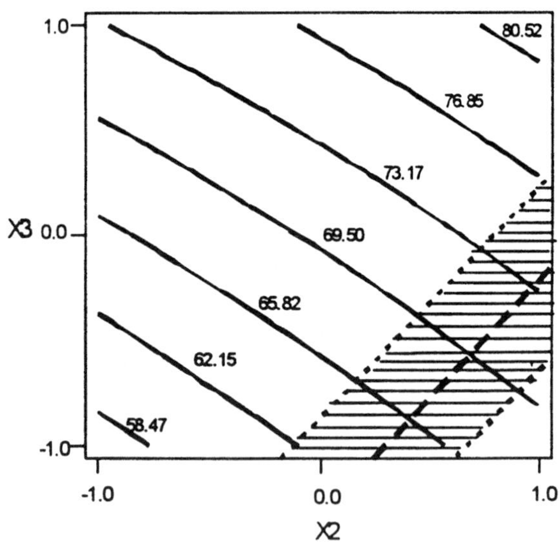

Figure 3 Contour plot of mean filtration rate and the line of minimum process variance with its 95% confidence region.

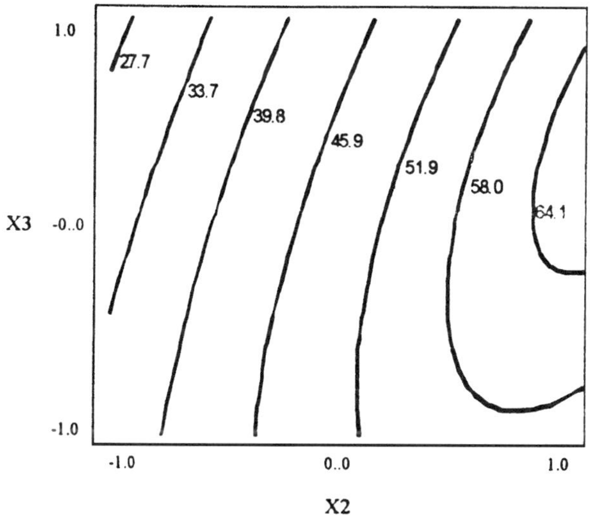

Figure 4 Contour plot of lower 95% one-sided prediction limits.

high concentration of formaldehyde (x_2) with reasonable flexibility in the operating level of stirring rate (x_2).

Combining the information from the four plots provides powerful insights into the process, namely, that operating at the ($x_2 =$

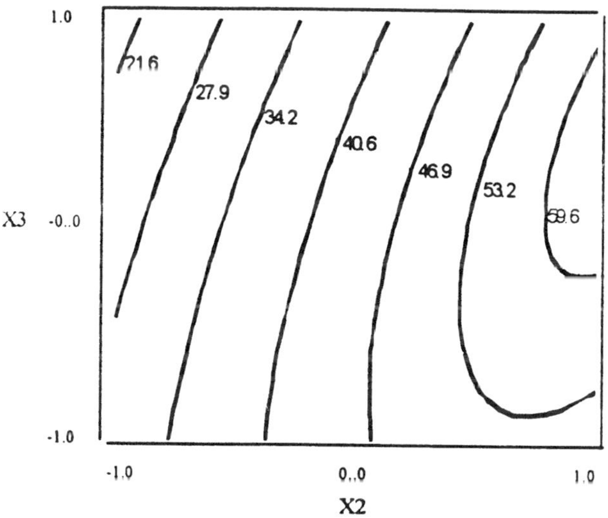

Figure 5 Contour plot of 0.95 content lower 95% one-sided tolerance limits.

$1.0, x_3 = -0.21$) condition will minimize the process variance with promising results indicated by the prediction and tolerance limits.

Taguchi's parameter design has had a profound effect on the rise in interest and use of RSM in industry. There are developments in other areas of interest, however, that should and likely will enhance its role, not only in traditional quality improvement but also in biostatistics and biomedical applications.

3. NEW DEVELOPMENTS IN RSM

3.1. Role of Computer-Generated Designs

The computer has been an important tool in the construction of experimental designs since the early 1980s. However, the focus has been almost entirely on criteria that have their underpinnings steeped in normal theory linear models. In this situation, of course, the alphabetic optimality criteria developed by Kiefer (1959) and others can be applied without knowledge of the parameters. However, as we emphasize in what follows, many of the response surface applications in the present and the future involve nonlinear and/or non-normal theory applications in which optimal designs depend on knowledge of the parameters. Uncertainty about model parameters in these cases as well as uncertainties in more standard cases about model assumptions, goals, the presence of outliers, or missing design points result in the need for considerations of design robustness as a serious alternative to optimal design. Almost without exception, commercial computer software deals with design optimality and does not address robustness. It is clear that computer-generated design cannot reach its full potential without considering these matters as well as dealing with various kinds of graphical methodology that allow the practitioner to compare and evaluate experimental designs. In what follows we discuss computer graphics that relate to RSM designs and provide some insight into new developments. These new developments necessitate design robustness as a companion to the RSM analysis tools that are currently finding use in industry.

3.2. Role of Creative Computer Graphics

Practitioners of RSM are undoubtedly familiar with the use of three-dimensional and contour plots for visualizing a predicted response. In a multiresponse optimization problem, the practice of overlaying multiple contour plots is extremely helpful for visualizing any potential compromises that must be made in order to determine the process optimum. Statistical software packages such as Design-Expert and Minitab (version 12) have built-in features for generating these overlaid plots.

There are also graphical techniques that are extremely useful for evaluating the prediction capability of experimental designs. Two such graphical methods that are discussed here are variance dispersion graphs and prediction variance contour plots. Both of these graphical techniques enable the user to visualize the stability of prediction variance throughout the design space, thus providing a mechanism for comparing competing designs.

The graphical technique referred to as the variance dispersion graph (VDG) was developed by Giovannitti-Jensen and Myers (1989) and Myers et al. (1992b). A variance dispersion graph for an RSM design displays a "snapshot" of the stability of the scaled prediction variance, $v(\mathbf{x}) = N \text{ Var } \hat{y}(\mathbf{x})/\sigma^2$, and how the design compares to an "ideal." For a spherical design [see Rozum (1990) and Rozum and Myers (1991) for extensions to cuboidal designs], the VDG contains four graphical components:

1. A plot of the spherical variance V^r against the radius r. The spherical variance is essentially $v(\mathbf{x})$ averaged (via integration) over the surface of a sphere of radius r.
2. A plot of the maximum $v(\mathbf{x})$ on a radius r against r.
3. A plot of the minimum $v(\mathbf{x})$ on a radius r against r.
4. A horizontal line at $v(\mathbf{x}) = p$, to represent the "ideal" case.

Figure 6 illustrates the utility of VDGs for comparison of two spherical designs for $k = 3$ variables, the CCD with $\alpha = \sqrt{3}$ and three center points and the Box–Behnken design, also with three center points. Both designs have been scaled so that points are at a radius $\sqrt{3}$ from the design center. The following represent obvious conclusions from the two VDGs in Figure 6 (Myers and Montgomery, 1995):

1. Note that there is very little difference between the minimum, average, and maximum of $v(\mathbf{x})$ for the CCD, indicating that it is nearly rotatable. This should not be surprising since $\alpha = 1.682$ results in exact rotatability.
2. The values of $v(x)$ are very comparable for the two designs near the design center. Any difference is accounted for by the difference in sample sizes ($N = 17$ for the CCD, $N = 15$ for the BBD).
3. The CCD appears to be the better design for prediction from radius 1.0 to $\sqrt{3}$, based on greater stability in $v(\mathbf{x})$ and a max $v(\mathbf{x})$ that is smaller than that of the BBD.
4. The comparison with the ideal design [$v(\mathbf{x}) = 10.0$] is readily seen for both designs.

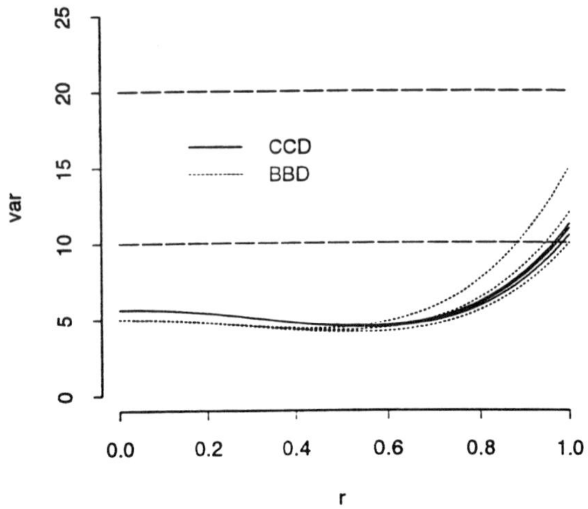

Figure 6 Variance dispersion graphs for CCD and Box–Behnken designs for $k = 3$ design variables.

Another graphical method of displaying the stability of the prediction variance is a display of contours of constant prediction variance. Like the VDG, this technique enables the user to visualize the behavior of $v(x)$ over the design space. Unlike the VDG, a contour plot of $v(x)$ allows one to determine the *direction* in which $v(x)$ is most unstable. This technique is now illustrated through a comparison of two competing designs of equal size, the $3 - 11A$ *hybrid* with one additional center point and a $D-$optimal design. The 12-run D-optimal design was generated using SAS Proc Optex, assuming the three-factor full quadratic model. The candidate list from which the design was selected was structured to be similar to the spherical space encompassed by the hybrid design. Contour plots of the unscaled prediction standard error $(v(x)/N)$ were generated for each design, under the assumption of a full quadratic model. Figures 7a and 7b contain these contour plots for the hybrid and D-optimal designs, respectively. Note that each contour plot represents a slice of the design space where factor C is fixed at its midpoint condition, and therefore the center contour represents the standard error of prediction at the center of the design space. Studying these plots provides information about two key aspects of the designs: (1) nearness to rotatability and (2) stability/consistency of prediction variance throughout the space.

The hybrid design, known to be nearly rotatable, also has very stable and consistent prediction variance throughout the space. The D-optimal

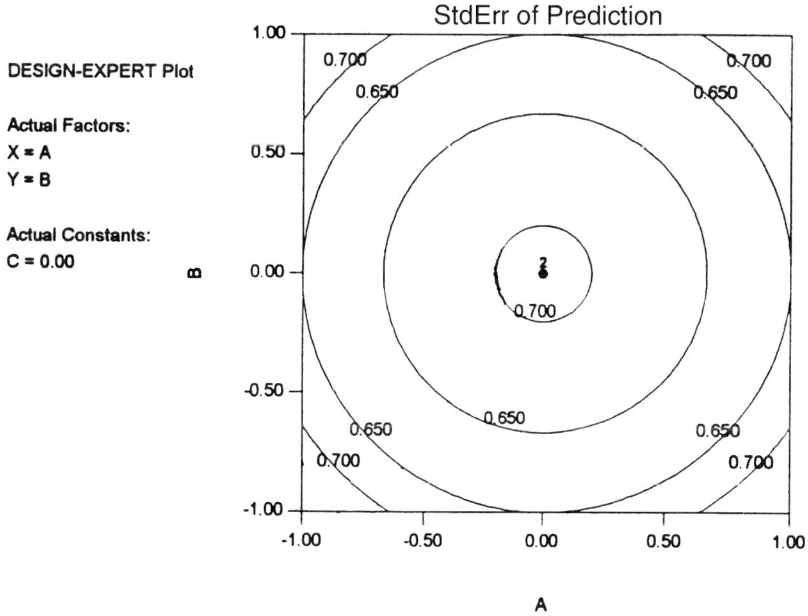

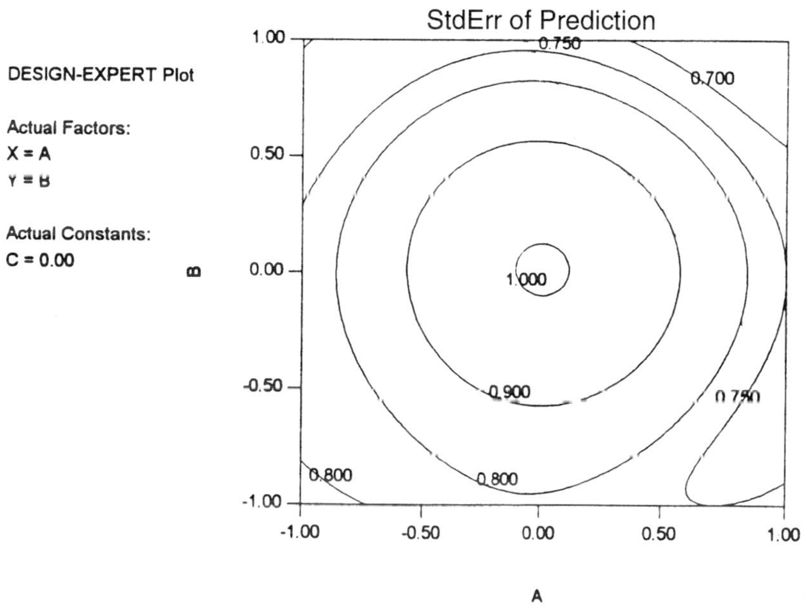

Figure 7 Contours of standard error of prediction for (a) hybrid 311A and (b) *D*-optimal design.

design, in contrast, is not rotatable, which can be seen by the inconsistency of the prediction variance in the corners of the plotted space. In addition to the D-optimal design being unstable in the center of the design space, we also observe an overall higher degree of prediction variability throughout relative to that of the hybrid design.

Independent of the designs studied, however, the power of the graphical techniques is evident. Graphical tools such as those presented in this section allow the researcher to quickly gain information about design performance and characteristics of the response surface.

3.3. Bayesian or Two-Stage Design

In more and more applications, the ability to design an experiment depends on a priori knowledge of the response surface model. For example, when designing experiments for nonlinear models, the parameters of the nonnormal error models must be known. Even for the case of the linear model, identification of "optimal" designs depends on knowledge of the model regressors. In fact, we can say that it is rare when we truly know enough to design the experiment effectively without invoking prior information or conducting a preliminary experiment.

Consider the following logistic regression model, used frequently in biomedical applications:

$$y_i = \frac{1}{1 + \exp[-(\beta_0 + x_i\beta_1)]} + \varepsilon_i, \qquad i = 1, 2, ..., N$$

where $y_i \in \{0, 1\}$ indicates whether the ith subject responded to dose x_i of a given drug. It is therefore assumed that ε_i is approximately Bernoulli (0, $p_i(1 - p_i)$), where

$$p_i = \frac{1}{1 + \exp[-(\beta_0 + x_i\beta_1)]}, \qquad i = 1, 2, ..., N$$

The corresponding Fisher information matrix is given by

$$\mathbf{I}(\beta) = \begin{bmatrix} \sum p_i, q_i & \sum p_i q_i x_i \\ \sum p_i q_i x_i & \sum p_i q_i x_i^2 \end{bmatrix}$$

Note that the information matrix is a function of the unknown β's. This makes it impossible to directly use traditional design optimality criteria for generating an efficient design, since they depend on being able to optimize some norm on the Fisher information matrix. For example, construction of the D-optimal design for the above model would require that the doses x_1, x_2, ..., x_n be chosen such that $\mathrm{Det}[N^{-1}\mathbf{I}(\beta)]$ is maximized. In order to do this, the scientist would be forced to make his or her best guess at the values of β_0 and β_1. The resulting design will be D-optimal for the specified values, which unfortunately may be very different from the truth, thus resulting in an inefficient design.

Chaloner and Verdinelli (1995) review a Bayesian approach to design-optimality that incorporates prior information about the unknown parameters in the form of a probability distribution. This provides a mechanism for building in robustness to parameter misspecification, since a distribution of the parameter is specified, not merely a point estimate. The resulting Bayesian design optimality criterion is a function of the Fisher information matrix, integrated over the prior distribution on the parameters. For example, the Bayesian D-optimal design for the previously defined logistic model is found by choosing the levels of x that will maximize the expression

$$\int \log \mathrm{Det}[N^{-1}\mathbf{I}(\beta)]\pi(\beta)d\beta$$

where $\pi(\beta)$ is the prior probability distribution of $\beta = [\beta_0, \beta_1]$. Other creative approaches have been taken that provide a robustness to parameter misspecification. For example, a minimax approach is provided by Sitter (1992).

A two-stage design is another method used to achieve robustness to parameter misspecification. The strategy behind designing in two stages is to generate parameter information from data in the first stage that can then be used to select the remaining experimental runs with maximum efficiency. A two stage procedure may implement any pair of design criteria that meet the first-stage objective as well as the objective of the combined design. For example, Abdelbasit and Plackett (1983) and Minkin (1987) studied the efficiency of two-stage D-optimal designs for binary responses, thus applying D optimality to obth stages. Myers et al. (1996) developed a two-stage procedure for the logistic regression model that uses D optimality in the first stage followed by Q optimality in the second.

To illustrate the two-stage method, a brief description of the two-stage D-optimal design (D-D optimality) procedure for the logistic model is now given. The first step in the D-D (and also D-Q) procedure is the selection of a first-stage D-optimal design. In order to implement D optimality in the first stage, the experimenter must estimate the unknown β with a best guess, $\mathbf{b}_0$. The N_1 runs for the first-stage design are then chosen to satisfy the first-stage D-optimality criterion, given by

$$\underset{D}{\mathrm{Max}}\, \mathrm{Det}[(N_1^{-1}\mathbf{I}(\beta)]_{\beta=\mathbf{b}_0}$$

with β replaced by $\mathbf{b}_0$ and D representing all possible designs of size N_1. After design and execution of the first-stage experiment, N_1 observations are available to estimate β. The "best guess" of β is updated by replacing $\mathbf{b}_0$ with the MLE of β. The second stage of the two-stage process uses $\mathbf{b}$, thus making it conditional on the results of the first stage. To complete the $D-D$ procedure, it is necessary to choose a set of N_2 second-stage design points that will create a combined design that is conditionally D optimal. The N_2 points are chosen to satisfy

$$\underset{D}{\text{Max}}\, \text{Det}[(N_1 + N_2)^{-1}[\mathbf{I}_1(\beta) + \mathbf{I}_{2|1}(\beta)]\}_{\beta=\mathbf{b}}$$

where D is now the set of all possible designs of size N_2 and $\mathbf{I}_1(\beta)$ is fixed after the first stage.

Letsinger (1995) and Myers et al. (1996) evaluated the efficiency of two-stage procedures relative to their single-stage competitors. In doing so, they showed that the best performance of the two-stage designs was achieved when the first-stage design contained only 30% of the combined design size, thus reserving 70% of the observations for the second stage, when more parameter information is present.

Even for the normal linear model, successful implementation of design optitmality criteria is often difficult in practice. This is due to the fact that the model content must be known a priori. In other words, the experimenter must be able to specify which regressors are needed to model the response, in order to generate the most efficient design for constructing the specified model. If too many regressors are specified, some design points (and consequently valuable resources) may be wasted on estimatiton of unimportant terms. If too few regressors are specified, then some terms that are needed in the model may not even be estimatable.

Suppose an experimenter identifies a set of regressors, x, containing all $p + q$ regressors he or she believes might be needed in modeling the behavior of a response y. The linear model is written as $\mathbf{y} = \mathbf{X}\beta + \varepsilon$, with $\mathbf{y}$ denoting the n observations to be collected in an experiment, under the assumption that $\mathbf{y}|\beta, \sigma^2 \sim N(\mathbf{X}\beta, \sigma^2\mathbf{I})$. The model matrix, $\mathbf{X}$, has dimensions $n \times (p + q)$, with the $p + q$ columns defined by the set of regressors, x. Quite often, the experimenter has knowledge of the process or system that allows him or her to identify p of the regressors as *primary terms*. These are the terms that the experimenter strongly believes are needed in modeling the response. The remaining terms are the *potential terms*, i.e., those terms about which the experimenter has uncertainty. For example, the experimenter may know from past experience that certain process variables must be included in the model as main effects (i.e., linear terms) but is uncertain if higher order interactions (such as quadratics) are needed. The key is to incorporate this information into the experimental design, so that limited resources are first focused on estimation o the primary terms (in this case the main effects), while also using some resources for estimation of the potential terms.

DuMouchel and Jones (1994) proposed a Bayesian *D*-optimality criterion for the efficient estimation of both primary and potential terms. Let β_{pri} and β_{pot} represent the parameters corresponding to primary and potential terms, respectively. The approach taken by DuMouchel and Jones is to assume a diffuse prior distribution (arbitrary prior mean with infinite prior variance) for β_{pri}. This is reasonable because these parameters are expected to be significantly different from zero, but no assumption of direction of effect is made. The potential terms, however, are perceived to have smaller coefficients than the coefficients of primary terms. For this reason, β_{pot} is assigned an $N(\mathbf{0}, \sigma^2\tau^2\mathbf{I})$, with σ^2 and τ^2 known. Fortunately, the design can be constructed independently of σ^2. The value of τ^2, however, affects the choice of the design, since it reflects the degree of uncertainty associated with the potential terms relative to σ^2. Under the assumption that primary and potential terms are uncorrelated (achieved through proper scaling of the x's), the joint prior distribution assigned to β_{pri} and β_{pot} is the $N(\mathbf{0}, \sigma^2\tau^2\mathbf{K}^{-1})$, where $\mathbf{K}$ is a $(p+q) \times (p+q)$ diagonal matrix whose first p diagonal elements equal 0 and whose remaining q diagonal elements equal 1. Under the assumption that $\mathbf{y}|\beta, \sigma^2 \sim N(\mathbf{X}\beta, \sigma^2\mathbf{I})$, the resulting posterior distribution of $\beta = [\beta_{pri}, \beta_{pot}]'$ is also normal, with mean $\mathbf{b} = (\mathbf{X}'\mathbf{X} + \mathbf{K}\tau^2)^{-1}\mathbf{X}'\mathbf{y}$ and variance $\mathbf{V} = \sigma^2(\mathbf{X}'\mathbf{X} + \mathbf{K}/\tau^2)^{-1}$. The Bayesian *D*-optimal design is that which minimizes the Bayes risk, proportional to

$$\log \mathrm{Det}[\mathbf{V}] = \log \mathrm{Det}[\sigma^2(\mathbf{X}'\mathbf{X} + \mathbf{K}/\tau^2)^{-1}]$$

In practice, the appropriate design may be found by selecting the rows of $\mathbf{X}$ from a predefined candidate list, so that $|\mathbf{V}|$ is minimized. Note that the diagonals of $\mathbf{V}$ associated with β_{pot} are somewhat stabilized through prior information (given through τ), identical to the technique used in ridge regression. The other diagonals of $\mathbf{V}$ associated with β_{pri}, however, are more dependent on design. As a result, the Bayesian *D*-optimal design will support estimation of both β_{pri} and β_{pot}, but with higher priority given to β_{pri}.

Consider an application in which three factors are to be studied, with emphasis being placed on the estimation of main effects and interactions (β_{pri} =[intercept, $\beta_1, \beta_2, \beta_3, \beta_{12}, \beta_{13}, \beta_{23}]'$) while there is still some interest in the estimation of quadratics ($\beta_{pot} = [\beta_{11}, \beta_{22}, \beta_{33}]$). The performance of the Bayesian *D*-optimal designs versus the familiar face-centered cubic (fcc) design is compared in Table 1 for various "true models." All designs contain

$N = 16$ runs, and all Bayesian D-optimal designs were produced by SAS (Proc OPTEX). The metric used for design comparison is the scaled D criterion, $N[\mathrm{Det}(X'X)^{-1}]^{1/p}$, calculated for the true model in each case.

From Table 1 we see that in almost every case the Bayesian D-optimal designs outperform the fcc design. The performance of the two Bayesian D-optimal designs depends on the accuracy of the experimenter's prior knowledge about the relative significance of primary and potential terms, reflected through the choice of the parameter τ. For examploe, if it is believed that the quadratics are all within $\pm 2\sigma$ of zero (i.e., most likely insignificant) and therefore defines $\tau = 2/3$, the resulting design will be most efficient when the true model contains no quadratic terms. This design is not the best choice, however, if all quadratic terms truly belong in the model. In that case a larger value of τ, such as $\tau = 5$, would have been a better choice for controlling the design construction.

This weakness in the Bayesian D-optimal designs should not at all detract, however, from the work of DuMouchel and Jones. In fact, their greatest contribution was to provide a basis for the development of more efficient Bayesian design criteria, such as two-stage procedures, for the purpose of generating efficient designs under model (regressor) uncertainty. Consider the value of adopting the method of DuMouchel and Jones to produce a first-stage design with robustness to regressor uncertainty. Analysis of the first-stage data could then provide additional information about the relative importance of the $p + q$ regressors, enabling the remaining design points (second-stage design) to be chosen with greater efficiency. The second-stage design could then be generated from any optimality procedure that incorporates the improved model knowledge.

The two-stage approach described above was developed by Neff et al. (1997) for the purpose of developing numerous Bayesian two-stage design optimality procedures for the normal linear model under regressor uncertainty. Their work suggests that efficiency and robustness is gained from a two-stage design of size $N = 2(p + q + 2)$, with half of the design points

Table 1 Values of Determinant for Evaluation of FCC and Bayesian D-optimal Designs

Parameters contained in the true model	FCC	Bayesian D-opt ($\tau = 5$)	Bayesian D-opt ($= 2/3$)
$\beta_0, \beta_1, \beta_2, \beta_3, \beta_{12}, \beta_{13}, \beta_{23}$	1.65	1.42	1.20
$\beta_0, \beta_1, \beta_2, \beta_3, \beta_{12}, \beta_{13}, \beta_{23}, \beta_{33}$	1.85	1.68	1.56
$\beta_0, \beta_1, \beta_2, \beta_3, \beta_{12}, \beta_{13}, \beta_{23}, \beta_{11}, \beta_{22}, \beta_{33}$	2.33	2.18	2.47

allocated to each stage of the design. One such two-stage Bayesian approach is illustrated by a brief description of a Bayes *D-D* optimality procedure. Using this procedure, the first-stage design is chosen to be *D* optimal according to the method of DuMouchel and Jones. Consequently, the first-stage posterior distribution of $\beta = [\beta_{\text{pri}}, \beta_{\text{pot}}]'$ is normal, with $E(\beta_j/\mathbf{y}_1) = (\mathbf{X}_1'\mathbf{X}_1 + \mathbf{K}/\tau^2)^{-1}\mathbf{X}_1'\mathbf{y}_1$ and variance $\mathbf{V}_1 = \sigma^2(\mathbf{X}_1'\mathbf{X}_1 + \mathbf{K}/\tau^2)^{-1}$. Basing inferences on the first-stage posterior of β, the $p + q$ standardized estimates of the model parameters (coefficients) after the first stage are

$$\hat{\beta}_j^* = \frac{E(\beta_j|\mathbf{y}_1)}{\sigma\sqrt{c_{jj}}}, \qquad j = 1, 2, \ldots, p + q$$

where c_{jj} is the *j*th diagonal element of $(1/\sigma^2)\mathbf{V}_1$. Since the estimated effect of any regressor x_j is proportional to its standardized estimated coefficient, the relative importance of the various model terms can be estimated by the relative sizes of the $\hat{\beta}_j^*$'s (in absolute value). Normalizing these $\hat{\beta}_j^*$'s (in absolute value) produces a set of discrete scores or "weights of evidence" that quantify the relative importance of each model term. In other words, a new set of τ's, $\{\tau_1, \tau_2, \ldots, \tau_{p+q}\}$, is produced based on this updated prior information. Going into the second stage, beliefs about the relative importance of the $p + q$ model terms are expressed as $\beta|\sigma^2, \tau^2, \mathbf{y}_1(\mathbf{0}, \sigma^2\mathbf{T})$, where T is a $(p + q) \times (p + q)$ diagonal matrix with $\tau_1, \tau_2, \ldots, \tau_{p+q}$ appearing on the diagonals. Setting the prior mean to zero at this point is arbitrary, since it will have no impact on the second-stage design criterion. Still under the assumption of a normal linear model, the second-stage posterior distribution is also normal, with posterior covariance matrix $\mathbf{V}_2 = \sigma^2(\mathbf{X}_1'\mathbf{X}_1 + \mathbf{X}_2'\mathbf{X}_2 + \mathbf{T}^{-1})^{-1}$. Thus the second-stage conditionally *D*-optimal design is found by selecting the rows of $\mathbf{X}_2$ from a candidate list such that $|\mathbf{V}_2|$ is minimized. Due to the structure of $\mathbf{T}^{-1}$, the diagonals of $\mathbf{V}_2$ corresponding to less important regressors are already somewhat stabilized. Design points that provide information about the more important regressors and thus stabilize the corresponding diagonals will be chosen for the second-stage design. For a performance comparison of this procedure as well as other two-stage Bayesian design procedures relative to their single-stage competitors, the reader is referred to Neff et al. (1997).

3.4. Generalized Linear Models

The normal linear model is the model that has been most commonly used in response surface applications. The assumptions underlying this model are, of course, that the model errors are normally distributed with constant variance. In many quality improvement applications in industry, however, the quality characteristic or response most naturally follows a probability distribution other than the normal. Consider, for example, a quality improvement program at a plastics manufacturer focused on reducing the number of surface defects on injection-molded parts. The response in this case is the defect count per part, which most naturally follows a Poisson distribution, where the variance is not constant but is instead equal to the mean. Consider also applications in the field of reliability, in which the equipment's time to failure is the quality response under study. Again, the most natural error distribution is not the normal, but instead the exponential or gamma, both of which have nonconstant variance structures. These types of problems nicely parallel similar problems that exist in the biomedical field, particularly in the area of dose–response studies and survival analysis.

Regression models based on distributions such as the Poisson, gamma, exponential, and binomial fall into a family of distributions and models known as generalized linear models (GLM). See McCullough and Nelder (1989) for an excellent text on the subject. In addition the reader is referred to Myers and Montgomery (1997) for a tutorial on GLM. In fact, all distributions belonging to the exponential family are accommodated by GLM. These models have already been used a great deal in biomedical fields but are just now drawing interest in manufacturing areas. In the past, the approach has been to normalize the response through transformation, so that OLS model parameter estimates could be calculated. Hamada and Nelder (1997) show several examples in which the appropriate transformation either did not exist or produced unsatisfactory results compared to the appropriate GLM model. They also spoint out that with the progress that has been made in computing in this area, the GLM models are just as easily fit as the OLS model to the transformed data. A few example software packages with GLM capability are GLIM, SAS PROC GENMOD, S-plus, and ECHIP.

It is interesting that some work has been done that provides a connective tissue between generalized linear models and robust parameter design. This relationship between the two fields is extremely important, as it allows the response surface approach to Taguchi's parameter design to be generalized to clearly non-normal applications that were previously discussed in this section. Engel and Huele (1996) build a foundation for this important area, and there will certainly be other developments.

The difficulty comes in designing experiments for GLM models. Design optimality criteria become complex, and designs are not simple to construct even in the case of only two design variables. See, for example, Sitter and Torsney (1992) and Atkinson and Haines (1996). One most constantly be aware that even if an optimal design is found it requires parameter specifications. As a result, the use of robust or two-stage designs will likely, in the end, be the most practical approach.

3.5. Nonparametric and Semiparametric Response Surface Methods

Consider a response surface problem in which the quality characteristic (response) of interest is expected to behave in a highly nonlinear fashion as a function of a set of process variables. Although the model form is unknown, the model structure is of less importance than the ability to locate the process conditions that result in the optimum response value. The primary interest is in prediction of the response and understanding the general nature of the response surface. Additionally, in many of these kinds of problems the ranges in the design problems are wider than in traditional RSM in which local approximations are sought.

In the problem above, greater model flexibility is required than can be achieved with a low-order polynomial model. Nonparametric and semiparametric regression models can be combined with standard experimental design tools to provide a more flexible approach to the optimization of complex problems. Some of the nonparametric modeling methods that may be considered are thin-plate spline models, Gaussian stochastic process models, neural networks, generalized additive models (GAMs), and multiple adaptive regression splines (MARS). The reader is referred to Haaland et al. (1994) for a brief description of each model type. Vining and Bohn (1996) introduced a semiparametric as well as a nonparametric approach to mean and variance modeling. The semiparametric strategy involved the use of a nonparametric method to obtain variance estimates which then became inputs to modeling the response mean via weighted least squares. As an alternative approach they suggested utilizing a nonparametric method for modeling the response mean as well as the variance.

Haaland et al. (1996) point out that the experimental designs used for nonparametric response surface methods can include some of the traditional designs. For example, one may execute a series of fractional factorials followed by a central composite design, then develop a global model using a nonparametric method. An alternative to this design approach is to execute a single *space-filling design*, which covers the entire region of operability in one large experiment. This type of design is not based on any model form but instead contains points that are spread out uniformly (in some sense) over the experimental region. The intent is that no point in the experimental region will be very far from a design point. Space-filling designs have primarily been used in computer experiments but have also been applied in physical experiments in the pharmaceutical and biotechnology industries.

See Haaland et al. (1994) for references. Among the space-filling designs is a class of distance-based design criteria that focus on selection of a set of design points that have adequate coverage and spread over the experimental (or operability) region. Two software packages that will construct distance-based designs are SAS PROC OPTEX and Design-Expert.

3.6. Hard-to-Change or Hard-to-Control Design Variables

In the design and analysis of industrial experiments, one often encounters variables that are hard to change or hard to control. Consider the following example. A product engineer for a plastics manufacturer is conducting an experiment to determine the effect of extrusion conditions on various physical properties of the resulting plastic pellets. The three independent variables to be studied are screw design, screw speed, and extrusion rate. Minimal screw design changes can occur during the experiment, since each change requires costly line downtime. For this reason, screw design is referred to as a "hard-to-change" variable. Also, since screw designs vary between plant sites, the product engineer has no control over which screw design will ultimately be used at each site. For this reason, screw design is also labeled a "hard-to-control" variable.

This has been emphasized in recent years due to the important role of noise variables that are hard to control. Box and Jones (1992) investigated the use of split-plot designs as an alternative to Taguchi's crossed arrays for more efficiently studying noise and control variables. Lucas and Ju (1992) pointed out that often the designs for these situations are not completely randomized but are rather quite like a split plot and yet we analyze them incorrectly as CRDs.

Strictly speaking, the hard-to-control variables are whole-plot variables with levels that are randomly assigned to larger whole-plot experimental units (EUs). The appropriate levels of the easier to control variables are randomly assigned to smaller experimental units within each whole plot (thus making them subplot variables). As discussed by Letsinger et al. (1996), this birandomization structure leads to complications in analysis, since the error assumptions associated with the basic response surface model [i.e., all $\varepsilon_i \sim N(0, \sigma^2)$] are no longer valid. Let σ_δ^2 be the whole-plot error variance and σ_ε^2 the subplot error variance resulting from the first and seonc randomization, respectively. The model and error assumptions then become

$$\mathbf{y} = \mathbf{X}\beta + \delta + \varepsilon$$

where

$$\delta + \varepsilon \sim N(\mathbf{0}, \mathbf{V})$$

and

$$\mathbf{V} = \sigma_\delta^2 \mathbf{J} + \sigma_\varepsilon^2 \mathbf{I}$$

Assuming that there are j whole plots, then $\mathbf{J}$ is a block-diagonal matrix with nonzero blocks of the form $\mathbf{1}_{b_i \times 1} \times \mathbf{1}'_{b_i \times 1}$ and b_i is the number of observations in the ith whole plot, $i = 1, 2, ..., j$. Note that while observations belonging to different whole-plot EUs are independent, those b_i, observations within a given whole plot are correlated.

Practitioners may be tempted to ignore the birandomization error structure, analyzing the data as if they came from a completely randomized design (CRD). The analysis of a split-plot design as a CRD, however, can lead to erroneously concluding that whole-plot factors are significant when in fact they are not, while at the same time erroneously eliminating from the model significant subplot terms including whole-plot–subplot interactions. Unlike model estimation for the CRD, the error variances play a major role in the estimation of coefficients in the birandomization model. Under the assumption of normal errors, the maximum likelihood estimate (MLE) of the model is now obtained through the generalized least squares (GLS) estimation equations

$$(\hat{\beta}) = (\mathbf{X}'\mathbf{V}^{-1}\mathbf{X})^{-1}\mathbf{X}'\mathbf{V}^{-1}\mathbf{y}$$

and

$$\mathrm{Var}(\hat{\beta}) = (\mathbf{X}'\mathbf{V}^{-1}\mathbf{X})^{-1}$$

Note that both estimating equations depend on σ_δ^2 and σ_ε^2 through the matrix $\mathbf{V}$; therefore proper estimation of these error variances becomes a priority.

Appropriateness of various model and error estimation methods is dependent on the structure of the birandomization design (BRD). The general class of BRDs is divided into two subclasses: the crossed and the noncrossed. The distinguishing characteristic is that in the case of the crossed BRD, subplot conditions (i.e., factor level combinations) are identical across whole plots. This is the familiar split-plot design, which may result from restricted randomization of a 2^k, 3^k, or mixed-level factorial design. In the case of the noncrossed BRD, each whole plot may have a different number of subplot EUs as well as different factor combinations. Such a design could result from restricted randomization of a 2^{k-p} fractional factorial design or a second-order design such as the central composite design (CCD) or Box–Behnken design.

For the crossed BRD, Letsinger et al. (1996) show that GLS = OLS under certain model conditions, and therefore error variance knowledge is not essential for model estimation. Model editing, however, does depend on the availability of estimates of σ_δ^2 and σ_ε^2. One approach to estimating these variances makes use of whole-plot and subplot lack of fit. See Letsinger et al. (1986) for details.

In general, model estimation and editing are more complex for the noncrossed BRD. It is interesting to point out, however, that when the model is first-order, parameter estimation can be accomplished using the equivalency of GLS = OLS (as in the crossed case). Once again, model lack of fit can be used to develop estimators for the error variances, although the procedure is more complex than that for the crossed BRD. Both estimation and editing of a second-order model, however, depend on estimates of σ_δ^2 and σ_ε^2 through the matrix **V**. Three competing methods are mentioned here: OLS, iterated reweighted least squares (IRLS), and restricted maximum likelihood (REML).

One can argue that in some cases OLS is an acceptable method, even though it ignores the dependence among observations within each whole plot of the BRD. In fact, OLS provides an unbiased estimator of β. Also, for designs that provide little or no lack-of-fit information (for estimation of σ_ε^2 and σ_ε^2), the researcher may be better served by not trying to estimate **V** than by introducing more variability into the analysis. The IRLS method begins with an initial OLS estimate of β, then uses an iterative procedure for estimating σ_δ^2, σ_ε^2 and β until convergence is reached in $\hat{\beta}$. The REML method, first developed by Anderson and Bancroft (1952) and Russell and Bradley (1958), is similar to MLE except that it uses the likelihood of a transformation of the response, **y**. Refer to Searle et al. (1992) for a discussion on REML and its relationship to MLE. The PROC MIXED procedure in SAS (1992) can be adapted to calculate REML estimators. Letsinger et al. (1996) give details on the use of PROC MIXED for the analysis of a BRD.

The recent reminder that many RSM problems are accompanied by designs that are not completely randomized will hopefully produce new and useful tools for the practitioner. In that regard it is of great interest to note the similarity between the split-plot RSM problem (as far as analysis is concerned) and the approach taken with generalized estimating equations that find applications in the biostatistical and biomedical fields. The analysis is ver similar, though in the longitudinal data applications there generally is no designed experiment. Liang and Zeger (1986) and others extend this work to generalized linear models and indeed assume various correlation structures rather than the exchangeable correlation structure induced by the approach discussed above. The RSM practitioners can benefit greatly by borrowing from their colleagues in other fields.

4. CONCLUSION

Response surface methodology is growing. More statistical researchers are getting involved, dealing with a wider variety of complex problems. RSM will always play a large role in quality improvement. Much more development work is needed, however, to ensure that the methods are flexible enough to meet the challenges presented by other than the traditional fields of applications. In addition, strong communication is needed to solidify the growing interest of practitioners in the biological and biomedical fields.

REFERENCES

Abdelbasit KM, Plackett RL. (1983). Experimental design for binary data. J Am Stat Assoc 78:90–98.

Anderson RL, Bancroft TL. (1952) Statistical Theory in Research. New York: McGraw-Hill.

Atkinson AC, Haines LM. (1996). Designs for nonlinear and generalized linear models. In: Gosh S, Rao CR, eds. Handbook of Statistics, vol. 13. Amsterdam: Elsevier, pp. 437–475.

Box GEP, Jones S. (1992). Split-plot designs for robust product experimentation. J Appl Stat 19:3–26.

Chaloner K, Verdinelli I. (1995). Bayesian experimental design: A review. Stat Sci 10:273–304.

Dumouchel W, Jones B. (1994). A simple Bayesian modification of D-optimal designs to reduce dependence on an assumed model. Technometrics 36:37–47.

Engel J, Huele AF. (1996). A generalized linear modeling approach to robust design. Technometrics 38:365–373.

Giovannitti-Jensen A, Myers RH. (1989). Graphical assessment of the prediction capability of response surface designs. Technometrics 31:159–171.

Haaland PD, McMillan N., Nychka D, Welch W. (1994). Analysis of space-filling designs. Comput Sci Stat 26:111–120.

Haaland PD, Clarke RA, O'Connell MA, Nychka DW. (1996). Nonparametric response surface methods. Paper presented at 1996 ASA Meeting, Chicago, IL.

Hamada M, Nelder JA. (1997). Generalized linear models for quality improvement experiments. J Qual Technol 29:292–308.

Khattree R. (1996). Robust parameter design: A response surface approach. J Qual Technol 28:187–198.

Kiefer J. (1959). Optimum experimental designs (with discussion). J. Roy Stat Soc, Ser B 21:272–319.

Letsinger JD, Myers RH, Lentner M. (1996). Response surface methods for bi-randomization structures. J Qual Technol 28:381–397.

Letsinger WC. (1995). Optimal one and two-stage designs for the logistic regression model. Dissertation, Virgina Polytechnic Institute and State University.

Liang KY, Zeger SL. (1986). Longitudinal data analysis using generalized linear models. Biometrika 73(1):13–22.

Lin D, Tu W. (1995). Dual response surface optimization. J Qual Technol 27:34–39.

Lucas JM. (1994). How to achieve a robust process using response surface methodology. J Qual Technol 26:248–260.

Lucas JM, Ju HL. (1992). Split plotting and randomization in industrial experiments. *ASQC Quality Congress Transactions*, 27:34–39.

McCullough P, Nelder JA. (1989). Generalized Linear Models. 2nd ed. New York: Chapman and Hall.

Minkin S. (1987). Optimal designs for binary data. J Am Stat Assoc 82:1098–1103.

Montgomery DC. (1997). Design and Analysis of Experiments. 4th ed. New York: Wiley.

Myers RH, Montgomery DC. (1995). Response Surface Methodology: Process and Product Optimization Using Designed Experiments. New York: Wiley.

Myers RH, Montgomery DC. (1997). A tutorial on generalized linear models. J Qual Technol 29:274–291.

Myers RH, Khuri AI, Vining GG. (1992a). Response surface alternatives to the Taguchi robust parameter design approach. Am Stat 46:131–139.

Myers RH, Vining G, Giovannitti-Jensen A, Myers SL. (1992b). Variance dispersion properties of second-order response surface designs. J Qual Technol 24:1–11.

Myers RH, Kim Y, Griffiths KL. (1997). Response surface methods and the use of noise variables. J Qual Technol 29:429–440.

Myers WR, Myers RH, Carter, WH Jr, White KL Jr. (1996). Two stage designs for the logistic regression model in single agent bioassays. J Biopharm Stat, April issue.

Neff AR, Myers RH, Ye K. (1997). Bayesian two stage designs under model uncertainty. VPI & SU Tech Rep. 97–33.

Rozum MA. (1990). Effective design augmentation for prediction. PhD Thesis, Virginia Tech.

Rozum MA, Myers RH. (1991). Variance dispersion graphs for cuboidal regions. Paper presented at ASA Meeting, Atlanta, GA.

Russell TS, Bradley RA. (1958). One-way variances in the two-way classification. Biometrika 45:111–129.

SAS (1992). Tech Rep P-229. SAS/STAT Software, Release 6.07. Cary, NC.

Searle SR, Casella G, McCulloch CE. (1992). Variance Components. New York: Wiley.

Sitter RS. (1992). Robust designs for binary data. Biometrics 48:1145–1155.

Sitter RS, Torsney B. (1992). D Optimal designs for generalized linear models. In: Kitsos CP, Muller WG, eds. Advances in Model Oriented Data Analysis. Heidelberg: Physica–Verlag, pp 87–102.

Taguchi G, Wu Y. (1980). Introduction to Off-Line Quality Control. Central Japan Quality Control Association. (Available from American Supplier Institute, Dearborn, MI.)

Vining GG, Bohn L. (1996). Response surfaces for the mean and the process variance using a nonparametric approach. J Qual Technol 30:282–291.

Vining GG, Myers RH. (1990). Combining Taguchi and response surface philosophies: A dual response approach. J Qual Technol 22:38–45.

Index